# A COURSE IN
# ABSTRACT
# HARMONIC
# ANALYSIS
## Second Edition

# TEXTBOOKS in MATHEMATICS

## Series Editors: Al Boggess and Ken Rosen

# PUBLISHED TITLES CONTINUED

**TEXTBOOKS in MATHEMATICS**

# A COURSE IN
# ABSTRACT
# HARMONIC
# ANALYSIS

## Second Edition

## Gerald B. Folland

University of Washington
Seattle, Washington, USA

 CRC Press
Taylor & Francis Group
Boca Raton   London   New York

CRC Press is an imprint of the
Taylor & Francis Group  an **informa** business
A CHAPMAN & HALL BOOK

CRC Press
Taylor & Francis Group
6000 Broken Sound Parkway NW, Suite 300
Boca Raton, FL 33487-2742

© 2016 by Taylor & Francis Group, LLC
CRC Press is an imprint of Taylor & Francis Group, an Informa business

No claim to original U.S. Government works

Printed on acid-free paper
Version Date: 20150522

International Standard Book Number-13: 978-1-4987-2713-6 (Hardback)

**Library of Congress Cataloging-in-Publication Data**

Folland, G. B.
    A course in abstract harmonic analysis / Gerald B. Folland. -- Second edition.
        pages cm. --  (Textbooks in mathematics series ; 29)
    "A CRC title."
    Includes bibliographical references and index.
    ISBN 978-1-4987-2713-6 (alk. paper)
    1. Harmonic analysis. 2.  Mathematical analysis.  I. Title.

QA403.F58 2016
515'.2433--dc23                                                                      2015020328

**Visit the Taylor & Francis Web site at**
**http://www.taylorandfrancis.com**

**and the CRC Press Web site at**
**http://www.crcpress.com**

# Contents

# *Preface*

### Preface to the First Edition

The term "harmonic analysis" is a flexible one that has been used to denote a lot of different things. In this book I take it to mean those parts of analysis in which the action of a locally compact group plays an essential role: more specifically, the theory of unitary representations of locally compact groups, and the analysis of functions on such groups and their homogeneous spaces.

The purpose of this book is to give an exposition of the fundamental ideas and theorems of that portion of harmonic analysis that can be developed with minimal assumptions on the nature of the group with which one is working. This theory was mostly developed in the period from 1927 (the date of the Peter-Weyl theorem) through the 1960s. Since that time, research in harmonic analysis has proceeded in other directions, mostly on a more concrete level, so one may ask what is the excuse for a new book on the abstract theory at this time.

Well, in the first place, I submit that the material presented here is *beautiful.* I fell in love with it as a student, and this book is the fulfillment of a long-held promise to myself to return to it. In the second place, the abstract theory is still an indispensable foundation for the study of concrete cases; it shows what the general picture should look like and provides a number of results that are useful again and again. Moreover, the intervening years have produced few if any books with the scope of the present one. One can find expositions of various bits and pieces of this subject in a lot of places, and there are a few lengthy treatises in which one can perhaps learn more about certain aspects of it than one wants to know. But I have taken to heart the dictum propounded by R. F. Streater and A. S. Wightman in the preface of their book *PCT, Spin, Statistics, and All That*, that a book containing only Memorable Results is a Good Thing. The result, I hope, is a book that presents a rather large amount of important and interesting material in a concise and readable form.

The prerequisites for this book consist mostly of a familiarity with real analysis and elementary functional analysis. I use Folland [45] and Rudin [123] as standard references for this material; definitions and theorems in these books are used freely here, often without any specific reference. Rudin [123] also contains most of the material in Chapter 1, but the latter is included here because some of the concepts in it — especially projection-valued measures and the Gelfand transform — are an essential part of the fabric of ideas in later sections, and because I wished to include certain aspects of the spectral theorem that Rudin omits.

Chapters 2–6 are the core of the book. Chapter 2 develops the basic tools for doing analysis on groups and homogeneous spaces: invariant measures and the convolution product. Chapter 3 presents the rudiments of unitary representation theory, up through the Gelfand-Raikov existence theorem for irreducible unitary representations. In particular, it introduces the connection between representations and functions of positive type (or positive definite functions, as they are often called), an amazingly fruitful idea that also plays an important role in later chapters. Chapters 4 and 5 are devoted to analysis on Abelian and compact groups. Here the Fourier transform takes center stage, first as a straightforward generalization of the classical Fourier transform to locally compact Abelian groups, and then in the more representation-theoretic form that is appropriate to the non-Abelian case. Chapter 6 presents the theory of induced representations, including a complete proof of the Mackey imprimitivity theorem (something that is remarkably scarce in the expository literature) following the ideas of Blattner. In all these chapters, a number of specific examples are included to illustrate the general theory; they are interwoven with the rest of the text in Chapters 2–4 but are mostly collected in separate sections at the end in Chapters 5 and 6.

Chapter 7, on the theory of noncompact, non-Abelian groups, is of a somewhat different nature than the earlier chapters. To a considerable extent it is more like a survey article than a portion of a book, for many of the main results are stated without proof (but with references). To have given a complete treatment of the material in this chapter would have required the enlargement the book to an unwieldy size, involving a lengthy digression into the theory of von Neumann algebras and representations of C* algebras. (Indeed, many of the results are most naturally stated in this context, their application to groups coming via the group C* algebra.) The books of Dixmier [31], [32] already provide an excellent exposition of this theory, which I saw no reason to duplicate. Rather, I thought that many readers would appreciate a fairly detailed sketch of the Big Picture for noncompact, non-Abelian groups with the technical arguments omitted, especially since most of these results provide a background for the study of concrete cases rather than a set of working tools.

The Bibliography contains three kinds of items: original sources for the major results in the book, references for results stated without proof, and expository works to which readers can refer for more information on various topics. It makes no pretense of completeness. More extensive bibliographies can be found in Dixmier [32], Fell and Doran [40], [41], and Mackey [94], [98].

Chapters 2–5 are the embodiment of a course I gave at the University of Washington in the spring quarter of 1993. (The material of Chapter 1 was covered in a preceding course.) I wrote Chapters 6 and 7 while visiting the University of Colorado at Boulder for the fall semester of 1993, where I had the inestimable benefit of conversations with Arlan Ramsay and Larry Baggett. In addition, Baggett let me borrow some old handwritten notes by J. M. G. Fell, which were just what I needed to sort out many of the ideas in Chapter 6.

Many of the ideas in this book are an outgrowth of the study of the classical Fourier transform on the real line,

$$\mathcal{F}f(\xi) = \int_{-\infty}^{\infty} e^{-2\pi i x \xi} f(x)\, dx.$$

Indeed, $\mathbb{R}$ is a locally compact group; the functions $e^{2\pi i x \xi}$ out of which $\mathcal{F}$ is fashioned are its irreducible representations; and $\mathcal{F}$ gives the Gelfand transform on $L^1(\mathbb{R})$, the spectral resolution of the algebra of translation-invariant operators on $L^2(\mathbb{R})$, and the decomposition of the regular representation of $\mathbb{R}$ into its irreducible components. When I first thought of writing a book like this, I envisaged it as an essay on the group-theoretic aspects of the Fourier transform. The scope of the book as it finally turned out is a bit different, but the spirit of Fourier is still all-pervasive.

### Preface to the Second Edition

The main new features of this edition are as follows:

- In Chapter 1, I have added a short section (§1.6) on von Neumann algebras.

- In §4.5, I have included Mark Kac's simple proof of a restricted form of Wiener's theorem.

- I have rewritten part of §5.4 to explain the relation between $SU(2)$ and $SO(3)$ in terms of quaternions, an elegant method that also allows $SO(4)$ to be brought into the picture with little extra effort.

- In Chapter 6, I have added a discussion of the representations of the discrete Heisenberg group and its central quotients, which provides nice illustrations of both the Mackey machine for regular semi-direct products and the pathological phenomena that occur for nonregular ones. In consequence, the "Examples" section has now expanded into two sections (§§6.7–8).

- I have moved the background material on tensor products of Hilbert spaces and operators from §7.3 to a new appendix.

In addition, I have added a few items to the "Notes and References" sections and a few entries to the Bibliography, clarified some obscurities, and corrected a number of typographical and mathematical errors. Of course, there may be some errors still remaining; as they are brought to my attention, they will be listed on an errata sheet linked to my home page: www.math.washington.edu/~folland/Homepage/index.html.

## Some Matters of Notation and Terminology

The notation and terminology in this book agrees, for the most part, with that in Folland [45]. Here are a few specific items that are worthy of attention.

$\mathbb{T}$ denotes the multiplicative group of complex numbers of modulus one.

$\chi_E$ denotes the characteristic function or indicator function of the set $E$. If $\pi$ is a finite-dimensional unitary representation, $\chi_\pi$ denotes its character. These two uses of the letter $\chi$ will cause no confusion.

In a topological space, a **neighborhood** of a point $x$ or a set $E$ is a set whose interior contains $x$ or $E$. Thus, neighborhoods need not be open sets.

If $X$ is a locally compact Hausdorff space, $C(X)$, $C_0(X)$, and $C_c(X)$ denote the spaces of continuous (complex-valued) functions on $X$, continuous functions vanishing at infinity, and continuous functions of compact support, respectively. (Of course, these spaces coincide when $X$ is compact.) A **Radon measure** on $X$ is a Borel measure that is finite on compact sets, outer regular on all Borel sets, and inner regular on open sets. (Outer and inner regularity on a set mean that the set can be approximated in measure from the outside or inside by open or compact sets, respectively. $\sigma$-finite Radon measures are **regular**, that is, both outer and inner regular on all Borel sets.)

The uniform norm is denoted by $\| \; \|_{\sup}$. (In [45] it is denoted by $\| \; \|_u$, but I found that this led to an unsightly overuse of the letter $u$ in some situations.)

If $\mathcal{X}$ and $\mathcal{Y}$ are Banach spaces, the space of all bounded linear mappings from $\mathcal{X}$ to $\mathcal{Y}$ is denoted by $\mathcal{L}(\mathcal{X}, \mathcal{Y})$, and the space of all bounded linear mappings from $\mathcal{X}$ to itself is denoted by $\mathcal{L}(\mathcal{X})$.

In §§2.2–4, left and right Haar measures on a locally compact group

$G$ are denoted by $\lambda$ and $\rho$. However, in §2.5 and for the remainder of the book, $G$ is assumed to be equipped with a fixed left Haar measure, which is never given a name, and the symbols $\lambda$ and $\rho$ are freed for other purposes. The Haar measure of $E \subset G$ is denoted by $|E|$, the Lebesgue spaces of the Haar measure are denoted by $L^p(G)$ or simply $L^p$, and the Haar integral of $f \in L^1(G)$ is denoted by $\int f$ or $\int f(x)\,dx$.

# 1

# Banach Algebras and Spectral Theory

This chapter contains a brief exposition of that part of Banach algebra theory that will be needed in the rest of this book, including the spectral theorem for commutative C* algebras. Although these topics are not part of harmonic analysis as such, the Gelfand transform and the spectral theorem are embodiments of ideas that are also central to harmonic analysis: the conversion of operators into more transparent forms and the decomposition of operators into simpler pieces.

A **Banach algebra** is an algebra $\mathcal{A}$ over the field of complex numbers equipped with a norm with respect to which it is a Banach space and which satisfies $\|xy\| \leq \|x\| \, \|y\|$ for all $x, y$ in $\mathcal{A}$. $\mathcal{A}$ is called **unital** if it possesses a unit element or multiplicative identity, which we denote by $e$.

An **involution** on an algebra $\mathcal{A}$ is an anti-automorphism of $\mathcal{A}$ of order 2, that is, a map $x \mapsto x^*$ from $\mathcal{A}$ to $\mathcal{A}$ that satisfies

$$(1.1) \quad (x+y)^* = x^* + y^*, \quad (\lambda x)^* = \overline{\lambda} x^*, \quad (xy)^* = y^* x^*, \quad x^{**} = x$$

for all $x, y \in \mathcal{A}$ and $\lambda \in \mathbb{C}$. An algebra equipped with an involution is called a **\*-algebra**. A Banach \*-algebra that satisfies

$$(1.2) \qquad \|x^* x\| = \|x\|^2 \text{ for all } x$$

is called a **C\* algebra**.

We do not require an involution to satisfy $\|x^*\| = \|x\|$, although this holds for most of the examples we shall meet here. In particular it is true for C* algebras: the estimate $\|x\|^2 = \|x^* x\| \leq \|x^*\| \|x\|$ implies that $\|x\| \leq \|x^*\|$, and then $\|x^*\| \leq \|x^{**}\| = \|x\|$.

If $\mathcal{A}$ and $\mathcal{B}$ are Banach algebras, a (Banach algebra) **homomorphism** from $\mathcal{A}$ to $\mathcal{B}$ is a bounded linear map $\phi : \mathcal{A} \to \mathcal{B}$ such

1

that $\phi(xy) = \phi(x)\phi(y)$ for all $x, y \in \mathcal{A}$. If $\mathcal{A}$ and $\mathcal{B}$ are $*$-algebras, a $*$-**homomorphism** from $\mathcal{A}$ to $\mathcal{B}$ is a homomorphism $\phi$ such that $\phi(x^*) = \phi(x)^*$ for all $x \in \mathcal{A}$.

If $S$ is a subset of the Banach algebra $\mathcal{A}$, we say that $\mathcal{A}$ is **generated** by $S$ if the linear combinations of products of elements of $S$ are dense in $\mathcal{A}$.

We now describe four examples of Banach algebras. These examples barely begin to indicate how many different sorts of interesting Banach algebras there are, but they and their generalizations are the ones that will be important for us later.

*Example 1.* Let $X$ be a compact Hausdorff space. The space $C(X)$ of continuous complex-valued functions on $X$ is a unital Banach algebra with the usual pointwise algebra operations and the uniform norm. The map $f \mapsto \overline{f}$ is an involution that makes $C(X)$ into a C* algebra. Similarly, if $X$ is a noncompact, locally compact Hausdorff space, $C_0(X)$ is a nonunital C* algebra.

If $S$ is a set of functions in $C(X)$ (or $C_0(X)$) that separate points and have no common zeros, the Stone-Weierstrass theorem says that $C(X)$ (or $C_0(X)$) is generated by $S \cup \{\overline{f} : f \in S\}$.

*Example 2.* Let $\mathcal{H}$ be a Hilbert space. The set $\mathcal{L}(\mathcal{H})$ of all bounded linear operators on $\mathcal{H}$ is a unital Banach algebra, with the operator norm, and the map $T \mapsto T^*$ ($T^*$ being the adjoint of $T$) is an involution that makes $\mathcal{L}(\mathcal{H})$ into a C* algebra. Here is the verification of (1.2): On the one hand, we have $\|T^*T\| \leq \|T^*\|\,\|T\| = \|T\|^2$. On the other, for any unit vector $u \in \mathcal{H}$, $\|T^*T\| \geq \langle T^*Tu, u \rangle = \langle Tu, Tu \rangle = \|Tu\|^2$; taking the supremum over all such $u$ we get $\|T^*T\| \geq \|T\|^2$. Any subalgebra of $\mathcal{L}(\mathcal{H})$ that is closed in the operator norm and closed under taking adjoints is also a C* algebra.

*Example 3.* Let $l^1 = l^1(\mathbb{Z})$ be the space of all sequences $a = (a_n)_{-\infty}^{\infty}$ such that $\|a\| = \sum_{-\infty}^{\infty} |a_n| < \infty$. $l^1$ is a unital Banach algebra if we define multiplication to be convolution:

$$a * b = c, \text{ where } c_n = \sum_{-\infty}^{\infty} a_k b_{n-k}.$$

The unit element is $\delta$, defined by $\delta_0 = 1$ and $\delta_n = 0$ for $n \neq 0$. The standard involution on $l^1$ is defined by

$$(a^*)_n = \overline{a}_{-n}.$$

$l^1$ is not a C* algebra with this involution; we leave it as an exercise for the reader to find a counterexample to (1.2).

For $k \in \mathbb{Z}$, let $\delta^k \in l^1$ be defined by $(\delta^k)_n = 1$ if $n = k$, $(\delta^k)_n = 0$ otherwise. (In particular, $\delta^0 = \delta$.) It is easily verified that $\delta^j * \delta^k = \delta^{j+k}$. Hence $\delta^{-1}$ is the convolution inverse of $\delta^1$, and (by induction) $\delta^k = \delta^1 * \cdots * \delta^1$ and $\delta^{-k} = \delta^{-1} * \cdots * \delta^{-1}$ ($k$ factors) for $k \geq 1$. Moreover, for any $a = (a_n) \in l^1$ we clearly have $a = \sum_{-\infty}^{\infty} a_k \delta^k$. Thus $l^1$ is generated by $\delta^1$ and its inverse $\delta^{-1}$.

*Example 4.* The space $L^1(\mathbb{R})$ is a Banach algebra when multiplication is defined to be convolution,

$$(f * g)(x) = \int f(y)g(x - y)\, dy,$$

and as in Example 3 we can define an involution on it by $f^*(x) = \overline{f(-x)}$. $L^1(\mathbb{R})$ is not unital, nor is it a C* algebra.

For the remainder of this section we assume that $\mathcal{A}$ is a unital Banach algebra. In this case we can consider the elements of $\mathcal{A}$ possessing two-sided inverses, which we call **invertible** elements.

**1.3 Lemma.** *If $\|x\| < 1$ then $e - x$ is invertible, and*

$$(e - x)^{-1} = \sum_{0}^{\infty} x^n.$$

*Proof.* The usual proof that the geometric series $\sum_{0}^{\infty} t^n$ converges to $1/(1-t)$ for $|t| < 1$ works equally well in any unital Banach algebra. $\quad\square$

**1.4 Theorem.** *Let $\mathcal{A}$ be a unital Banach algebra.*
a. *If $|\lambda| > \|x\|$ then $\lambda e - x$ is invertible, and its inverse is $\sum_{0}^{\infty} \lambda^{-n-1} x^n$.*
b. *If $x$ is invertible and $\|y\| < \|x^{-1}\|^{-1}$ then $x - y$ is invertible, and its inverse is $x^{-1} \sum_{0}^{\infty} (yx^{-1})^n$.*
c. *If $x$ is invertible and $\|y\| \leq \frac{1}{2}\|x^{-1}\|^{-1}$ then $\|(x - y)^{-1} - x^{-1}\| \leq 2\|x^{-1}\|^2\|y\|$.*
d. *The set of invertible elements of $\mathcal{A}$ is open, and the map $x \mapsto x^{-1}$ is continuous on it.*

*Proof.* Since $\lambda e - x = \lambda(e - \lambda^{-1}x)$, (a) follows immediately from Lemma 1.3. So does (b), in view of the facts that $x - y = (e - yx^{-1})x$ and $\|yx^{-1}\| \leq \|y\| \|x^{-1}\| < 1$. (c) follows from (b), since

$$\|(x - y)^{-1} - x^{-1}\| = \left\| x^{-1} \sum_{1}^{\infty} (yx^{-1})^n \right\| \leq \|x^{-1}\| \sum_{1}^{\infty} (\|y\| \|x^{-1}\|)^n$$

$$\leq \|x^{-1}\|^2 \|y\| \sum_{0}^{\infty} 2^{-n} = 2\|x^{-1}\|^2\|y\|.$$

Finally, (d) is a direct consequence of (b) and (c). $\quad\square$

If $x \in \mathcal{A}$, the **spectrum** of $x$ is

$$\sigma(x) = \{\lambda \in \mathbb{C} : \lambda e - x \text{ is not invertible}\}.$$

$\sigma(x)$ is a closed subset of the disc $\{\lambda : |\lambda| \le \|x\|\}$ by Theorem 1.4(a,d). For $\lambda \notin \sigma(x)$, the **resolvent** of $x$ is the element

$$R(\lambda) = R_x(\lambda) = (\lambda e - x)^{-1}.$$

(We generally omit the subscript $x$ when no confusion will arise.) $R(\lambda)$ is continuous in $\lambda$ by Theorem 1.4(d). We shall now show that $R(\lambda)$ is an analytic $\mathcal{A}$-valued function on the open set $\mathbb{C} \setminus \sigma(x)$. By this we mean that the complex derivative $R'(\lambda)$ exists (and is continuous); this implies in particular that $\phi \circ R(\lambda)$ is an ordinary $\mathbb{C}$-valued analytic function of $\lambda$ for any bounded linear functional $\phi \in \mathcal{A}^*$.

**1.5 Lemma.** *$R(\lambda)$ is an analytic function of $\lambda \in \mathbb{C} \setminus \sigma(x)$.*

*Proof.* If $\lambda, \mu \notin \sigma(x)$, we have

$$\begin{aligned}
(\mu - \lambda)e &= (\mu e - x) - (\lambda e - x) \\
&= (\lambda e - x)R(\lambda)(\mu e - x) - (\lambda e - x)R(\mu)(\mu e - x) \\
&= (\lambda e - x)[R(\lambda) - R(\mu)](\mu e - x).
\end{aligned}$$

Multiplying both sides on the left by $R(\lambda)$ and on the right by $R(\mu)$, we see that

$$(\mu - \lambda)R(\lambda)R(\mu) = R(\lambda) - R(\mu),$$

and hence

$$\frac{R(\mu) - R(\lambda)}{\mu - \lambda} = -R(\lambda)R(\mu).$$

Letting $\mu \to \lambda$, we see that $R'(\lambda)$ exists and equals $-R(\lambda)^2$. $\qquad\square$

**1.6 Proposition.** *$\sigma(x)$ is nonempty for every $x \in \mathcal{A}$.*

*Proof.* If $\sigma(x)$ were empty, $R(\lambda)$ would be an entire function of $\lambda$. As $\lambda \to \infty$, $\|R(\lambda)\| = |\lambda|^{-1}\|(e - \lambda^{-1}x)^{-1}\| \to 0$ since $(e - \lambda^{-1}x)^{-1} \to e$. By Liouville's theorem (applied to $\phi \circ R(\lambda)$, for an arbitrary $\phi \in \mathcal{A}^*$), $R(\lambda)$ would be identically zero, which is absurd. $\qquad\square$

**1.7 Theorem** (The Gelfand-Mazur Theorem). *If $\mathcal{A}$ is a Banach algebra in which every nonzero element is invertible, then $\mathcal{A} \cong \mathbb{C}$.*

*Proof.* If $x \notin \mathbb{C}e$ then $\lambda e - x \ne 0$ for all $\lambda \in \mathbb{C}$ and hence $\lambda e - x$ is invertible for all $\lambda \in \mathbb{C}$. But then $\sigma(x) = \varnothing$, which is impossible. Hence $\mathcal{A} = \mathbb{C}e$. $\qquad\square$

If $x \in \mathcal{A}$, the **spectral radius** of $x$ is

$$\rho(x) = \sup\{|\lambda| : \lambda \in \sigma(x)\}.$$

We have $\rho(x) \leq \|x\|$ by Theorem 1.4(a). In fact, we can be more precise.

**1.8 Theorem.** $\rho(x) = \lim_{n \to \infty} \|x^n\|^{1/n}$.

*Proof.* We have $\lambda^n e - x^n = (\lambda e - x) \sum_0^{n-1} \lambda^j x^{n-1-j}$, from which it follows that if $\lambda^n e - x^n$ is invertible then so is $\lambda e - x$. In other words, if $\lambda \in \sigma(x)$ then $\lambda^n \in \sigma(x^n)$, so $|\lambda|^n \leq \|x^n\|$. It follows that $\rho(x) \leq \liminf \|x^n\|^{1/n}$.

On the other hand, if $\phi \in \mathcal{A}^*$, $\phi \circ R(\lambda)$ is analytic for $|\lambda| > \rho(x)$, and by Theorem 1.4(a) its Laurent series about infinity is $\sum_0^\infty \lambda^{-n-1} \phi(x^n)$. By standard complex variable theory, this series converges for $|\lambda| > \rho(x)$, so for any such $\lambda$ we have $|\lambda^{-n-1} \phi(x^n)| \leq C_\phi$ for all $n$. The uniform boundedness principle then implies the existence of a $C > 0$ such that $|\lambda|^{-n} \|x^n\| \leq C$ for all $n$, and hence $\|x^n\|^{1/n} \leq C^{1/n} |\lambda|$. Letting $n \to \infty$, we obtain $\limsup \|x^n\|^{1/n} \leq \rho(x)$. $\qquad \square$

We conclude with a couple of elementary observations about inverses and spectra in Banach $*$-algebras.

**1.9 Proposition.** *Let $\mathcal{A}$ be a unital Banach $*$-algebra.*

   *a.* $e = e^*$.
   *b. If $x$ is invertible, then so is $x^*$, and $(x^*)^{-1} = (x^{-1})^*$.*
   *c.* $\sigma(x^*) = \overline{\sigma(x)}$ *for any $x \in \mathcal{A}$.*

*Proof.* The relation $(xy)^* = y^* x^*$ shows that $e^*$ is another multiplicative identity and hence that $e^* = e$; it then also shows that $(x^*)^{-1} = (x^{-1})^*$. If $x \in \mathcal{A}$, $(\lambda e - x)^* = \overline{\lambda} e - x^*$ by (a), and so (c) follows from (b). $\qquad \square$

---

## 1.2   Gelfand Theory

In this section we study the spectrum (also called the *maximal ideal space* or *structure space*) of a commutative unital Banach algebra, a powerful tool that was first systematically exploited by Gelfand and his collaborators.

Let $\mathcal{A}$ be a commutative unital Banach algebra. By a **multiplicative functional** on $\mathcal{A}$ we shall mean a *nonzero* homomorphism from $\mathcal{A}$ to $\mathbb{C}$. The set of all multiplicative functionals on $\mathcal{A}$ is called the **spectrum** of $\mathcal{A}$; we denote it by $\sigma(\mathcal{A})$. (The relationship between this "spectrum" and the spectrum of an element defined in §1.1 will be explained in Proposition 1.15 below.)

**1.10 Proposition.** *Suppose* $h \in \sigma(\mathcal{A})$.
  a. $h(e) = 1$.
  b. *If* $x$ *is invertible in* $\mathcal{A}$ *then* $h(x) \neq 0$.
  c. $|h(x)| \leq \|x\|$ *for all* $x \in \mathcal{A}$.

*Proof.* (a): Pick $x \in \mathcal{A}$ with $h(x) \neq 0$; then $h(e)h(x) = h(ex) = h(x)$, so $h(e) = 1$. (b): If $x$ is invertible, $h(x^{-1})h(x) = h(x^{-1}x) = h(e) = 1$. (c): If $|\lambda| > \|x\|$ then $\lambda e - x$ is invertible by Theorem 1.4(a), so $\lambda - h(x) = h(\lambda e - x) \neq 0$ by (b). $\qquad\qquad\qquad\qquad\qquad\qquad\qquad\qquad\qquad\quad\square$

Proposition 1.10(c) says that $\sigma(\mathcal{A})$ is a subset of the closed unit ball $B$ of $\mathcal{A}^*$. We make $\sigma(\mathcal{A})$ into a topological space by imposing its weak* topology as a subset of $\mathcal{A}^*$, that is, the topology of pointwise convergence on $\mathcal{A}$. In view of (a), for an algebra homomorphism $h : \mathcal{A} \to \mathbb{C}$ the conditions $h \neq 0$ and $h(e) = 1$ are equivalent, so

$$\sigma(\mathcal{A}) = \big\{ h \in B : h(e) = 1 \text{ and } h(xy) = h(x)h(y) \text{ for all } x, y \in \mathcal{A} \big\}.$$

The conditions $h(e) = 1$ and $h(xy) = h(x)h(y)$ are clearly preserved under pointwise limits, so $\sigma(\mathcal{A})$ is a closed subset of $B$ in the weak* topology. By Alaoglu's theorem, then, $\sigma(\mathcal{A})$ is a compact Hausdorff space.

Multiplicative functionals are intimately connected with maximal ideals. We recall the terminology: if $\mathcal{A}$ is any algebra, a **left (right) ideal** of $\mathcal{A}$ is a subalgebra $\mathcal{I}$ of $\mathcal{A}$ such that $xy \in \mathcal{I}$ whenever $x \in \mathcal{A}$ and $y \in \mathcal{I}$ ($x \in \mathcal{I}$ and $y \in \mathcal{A}$). $\mathcal{I}$ is **proper** if $\mathcal{I} \neq \mathcal{A}$. If $\mathcal{A}$ is unital, $\mathcal{I}$ is proper if and only if $e \notin \mathcal{I}$, for if $e \in \mathcal{I}$ then $x = xe = ex \in \mathcal{I}$ for all $x \in \mathcal{A}$. If $\mathcal{A}$ is commutative, we can speak simply of ideals rather than left or right ideals; in this case, a **maximal ideal** is a proper ideal that is not contained in any larger proper ideal.

**1.11 Proposition.** *Let* $\mathcal{A}$ *be a commutative unital Banach algebra, and let* $\mathcal{I} \subset \mathcal{A}$ *be a proper ideal.*
  a. $\mathcal{I}$ *contains no invertible elements.*
  b. $\overline{\mathcal{I}}$ *(the closure of* $\mathcal{I}$*) is a proper ideal.*
  c. $\mathcal{I}$ *is contained in a maximal ideal.*
  d. *If* $\mathcal{I}$ *is maximal then* $\mathcal{I}$ *is closed.*

*Proof.* (a): If $x \in \mathcal{I}$ is invertible then $e = x^{-1}x \in \mathcal{I}$, so $\mathcal{I} = \mathcal{A}$. (b): If $\mathcal{I}$ is proper, it is contained in the set of nonivertible elements of $\mathcal{A}$, which is closed by Theorem 1.4(d); hence $e \notin \overline{\mathcal{I}}$, and it is easy to check that $\overline{\mathcal{I}}$ is an ideal. (c): This is a routine application of Zorn's lemma; the union of an increasing family of proper ideals is proper since it does not contain $e$. Finally, (d) follows from (b). $\qquad\qquad\qquad\qquad\qquad\qquad\quad\square$

**1.12 Theorem.** *Let* $\mathcal{A}$ *be a commutative unital Banach algebra. The map* $h \mapsto \ker(h)$ *is a one-to-one correspondence between* $\sigma(\mathcal{A})$ *and the set of maximal ideals in* $\mathcal{A}$.

*Proof.* If $h \in \sigma(\mathcal{A})$, $\ker(h)$ is an ideal that is proper since $h(e) = 1 \neq 0$ and is maximal since it has codimension 1. If $\ker(g) = \ker(h)$ then $g = h$, for if $x \in \mathcal{A}$ we have $x = h(x)e + y$ where $y \in \ker(h)$, so $g(x) = h(x)g(e) + g(y) = h(x)$. Thus $h \mapsto \ker(h)$ is an injection from $\sigma(\mathcal{A})$ to the set of maximal ideals.

On the other hand, suppose $\mathcal{M}$ is a maximal ideal, and let $\pi : \mathcal{A} \to \mathcal{A}/\mathcal{M}$ be the quotient mapping. $\mathcal{A}/\mathcal{M}$ inherits an algebra structure from $\mathcal{A}$, and it is a Banach space with the quotient norm

$$\|x + \mathcal{M}\| = \inf\{\|x + m\| : m \in \mathcal{M}\}.$$

(Here we need Proposition 1.11(d).) It is an easy exercise to check that $\mathcal{A}/\mathcal{M}$ is in fact a Banach algebra. It has no nontrivial ideals, for if $\mathcal{I} \subset \mathcal{A}/\mathcal{M}$ is an ideal then $\pi^{-1}(\mathcal{I})$ is an ideal in $\mathcal{A}$ such that $\mathcal{M} \subset \pi^{-1}(\mathcal{I}) \subset \mathcal{A}$; hence $\pi^{-1}(\mathcal{I}) = \mathcal{M}$ or $\mathcal{A}$ and $\mathcal{I} = \{0\}$ or $\mathcal{A}/\mathcal{M}$. But then every nonzero element of $\mathcal{A}/\mathcal{M}$ is invertible, for otherwise the ideal it generates would be nontrivial. By the Gelfand-Mazur theorem, $\mathcal{A}/\mathcal{M}$ is isomorphic to $\mathbb{C}$, and if we denote the isomorphism by $\phi$ then $\phi \circ \pi$ is a multiplicative functional on $\mathcal{A}$ whose kernel is $\mathcal{M}$. $\qquad\square$

If $x \in \mathcal{A}$, we define the function $\widehat{x}$ on $\sigma(\mathcal{A})$ by

$$\widehat{x}(h) = h(x).$$

$\widehat{x}$ is continuous on $\sigma(\mathcal{A})$ since the topology on $\sigma(\mathcal{A})$ is the topology of pointwise convergence on $\mathcal{A}$. The map $x \mapsto \widehat{x}$ from $\mathcal{A}$ to $C(\sigma(\mathcal{A}))$ is called the **Gelfand transform** on $\mathcal{A}$. We denote it by $\Gamma$ or $\Gamma_{\mathcal{A}}$ when necessary for clarity:
$$\Gamma x = \Gamma_{\mathcal{A}} x = \widehat{x}.$$

**1.13 Theorem.** *Suppose $\mathcal{A}$ is a commutative unital Banach algebra and $x \in \mathcal{A}$.*
  a. *The Gelfand transform is a homomorphism from $\mathcal{A}$ to $C(\sigma(\mathcal{A}))$, and $\widehat{e}$ is the constant function 1.*
  b. *$x$ is invertible if and only if $\widehat{x}$ never vanishes.*
  c. *$\operatorname{range}(\widehat{x}) = \sigma(x)$.*
  d. *$\|\widehat{x}\|_{\sup} = \rho(x) \leq \|x\|$.*

*Proof.* (a) is obvious — for example, $(xy)\widehat{\ }(h) = h(xy) = h(x)h(y) = \widehat{x}(h)\widehat{y}(h)$, and $\widehat{e} = 1$ by Proposition 1.10(a). For (b), we observe that $x$ is not invertible $\iff$ the ideal generated by $x$ is proper $\iff$ (by Proposition 1.11(c)) $x$ is contained in a maximal ideal $\iff$ (by Theorem 1.12) $h(x) = 0$ for some $h \in \sigma(\mathcal{A})$ $\iff$ $\widehat{x}$ has a zero. (c) follows from (b), for $\lambda \in \sigma(x) \iff \lambda e - x$ is not invertible $\iff \lambda - \widehat{x}(h) = 0$ for some $h \in \sigma(\mathcal{A})$. Finally, (d) follows immediately from (c). $\qquad\square$

If $\mathcal{A}$ is a *-algebra, one can ask whether the Gelfand transform takes the involution on $\mathcal{A}$ to the canonical involution (namely complex conjugation) on $C(\sigma(\mathcal{A}))$, that is, whether

$$\widehat{x^*} = \overline{\widehat{x}} \qquad (x \in \mathcal{A}).$$

This does not always happen (see the remarks following Corollary 1.18 for an example); when it does, $\mathcal{A}$ is called **symmetric**.

**1.14 Proposition.** *Suppose $\mathcal{A}$ is a commutative Banach *-algebra.*
 *a. $\mathcal{A}$ is symmetric if and only if $\widehat{x}$ is real-valued whenever $x = x^*$.*
 *b. If $\mathcal{A}$ is a $C^*$ algebra, $\mathcal{A}$ is symmetric.*
 *c. If $\mathcal{A}$ is symmetric, $\Gamma(\mathcal{A})$ is dense in $C(\sigma(\mathcal{A}))$.*

*Proof.* (a) If $\mathcal{A}$ is symmetric and $x = x^*$ then $\widehat{x} = \overline{\widehat{x}}$, so $\widehat{x}$ is real. To prove the converse, given $x \in \mathcal{A}$, let $u = (x + x^*)/2$ and $v = (x - x^*)/2i$. Then $u = u^*$ and $v = v^*$, so that $\widehat{u}$ and $\widehat{v}$ are real; also $x = u + iv$ and $x^* = u - iv$, so $\widehat{x^*} = \widehat{u} - i\widehat{v} = \overline{\widehat{x}}$.

(b) Suppose $\mathcal{A}$ is a C* algebra, $x = x^* \in \mathcal{A}$, and $h \in \sigma(\mathcal{A})$, and suppose $\widehat{x}(h) = h(x) = \alpha + i\beta$ with $\alpha, \beta$ real. For $t \in \mathbb{R}$, consider $z = x + ite$. We have $h(z) = \alpha + i(\beta + t)$ and $z^*z = x^2 + t^2 e$, so by Proposition 1.10(c),

$$\alpha^2 + (\beta + t)^2 = |h(z)|^2 \le \|z\|^2 = \|z^*z\| \le \|x^2\| + t^2.$$

Hence $\alpha^2 + \beta^2 + 2\beta t \le \|x^2\|$ for all $t \in \mathbb{R}$, which forces $\beta = 0$. Hence $\widehat{x}$ is real, so $\mathcal{A}$ is symmetric by (a).

(c) If $\mathcal{A}$ is symmetric, $\Gamma(\mathcal{A})$ is closed under complex conjugation. It contains the constants since $\widehat{e} = 1$, and it separates points on $\sigma(\mathcal{A})$ (trivially!). Hence $\Gamma(\mathcal{A})$ is dense in $C(\sigma(\mathcal{A}))$ by the Stone-Weierstrass therorem. $\square$

The motivation for calling $\sigma(\mathcal{A})$ the "spectrum" of $\mathcal{A}$ comes from the following result.

**1.15 Proposition.** *If $x_0 \in \mathcal{A}$, $\widehat{x}_0$ is a homeomorphism from $\sigma(\mathcal{A})$ to $\sigma(x_0)$ in each of the following cases:*
 *i. $\mathcal{A}$ is generated by $x_0$ and $e$, or*
 *ii. $x_0$ is invertible and $\mathcal{A}$ is generated by $x_0$ and $x_0^{-1}$, or*
 *iii. $\mathcal{A}$ is symmetric and $\mathcal{A}$ is generated by $x_0$, $x_0^*$, and $e$.*

*Proof.* $\widehat{x}_0$ maps $\sigma(\mathcal{A})$ onto $\sigma(x_0)$ by Theorem 1.13(c). Since $\sigma(\mathcal{A})$ and $\sigma(x_0)$ are both compact Hausdorff spaces, it suffices to prove that $\widehat{x}_0$ is injective. But in each of the three cases, any $h \in \sigma(\mathcal{A})$ is completely determined by its action on $x_0$ since $h(x_0^{-1}) = h(x_0)^{-1}$ in case (ii) and $h(x_0^*) = \overline{h(x_0)}$ in case (iii). Thus if $\widehat{x}_0(h_1) = \widehat{x}_0(h_2)$ then $h_1 = h_2$. $\square$

We now identify the spectrum and Gelfand transform for the two examples of commutative unital Banach algebras discussed in §1.1, namely $C(X)$ and $l^1$.

**1.16 Theorem.** *Let $X$ be a compact Hausdorff space. For each $x \in X$, define $h_x : C(X) \to \mathbb{C}$ by $h_x(f) = f(x)$. Then the map $x \mapsto h_x$ is a homeomorphism from $X$ to $\sigma(C(X))$. If we identify $x \in X$ with $h_x \in \sigma(C(X))$, the Gelfand transform on $C(X)$ becomes the identity map.*

*Proof.* It is clear that each $h_x$ is a mutliplicative functional on $C(X)$, and $h_x \neq h_y$ for $x \neq y$ since the continuous functions separate points on $X$. If $x_\alpha \to x$ then $f(x_\alpha) \to f(x)$ for each $f \in C(X)$, and this says that $h_{x_\alpha} \to h_x$ in the weak* topology. In short, $x \to h_x$ is a continuous injection of $X$ into $\sigma(C(X))$. Since these spaces are compact Hausdorff, it remains only to show that every multiplicative functional on $C(X)$ is of the form $h_x$ for some $x \in X$.

By Theorem 1.12, it is equivalent to show that every maximal ideal in $C(X)$ is of the form $\mathcal{M}_x = \{f : f(x) = 0\}$ for some $x \in X$, and this amounts to showing that every proper ideal $\mathcal{I} \subset C(X)$ is contained in some $\mathcal{M}_x$. Suppose to the contrary that for each $x \in X$ there exists $f_x \in \mathcal{I}$ such that $f_x(x) \neq 0$. The open sets $\{y : f_x(y) \neq 0\}$ then cover $X$, so by passing to a finite subcover we obtain $f_1, \ldots, f_n \in \mathcal{I}$ that have no common zero. Let $g = \sum_1^n |f_j|^2$. Then $g = \sum \overline{f_j} f_j \in \mathcal{I}$ and $g$ is invertible in $C(X)$ since $g > 0$ everywhere. By Proposition 1.11(a), this contradicts the assumption that $\mathcal{I}$ is proper. Thus $\mathcal{I} \subset \mathcal{M}_x$ for some $x$.

Finally, since $\widehat{f}(h_x) = h_x(f) = f(x)$, if we identify $h_x$ with $x$ we have $\widehat{f} = f$. $\qquad\qquad\qquad\qquad\qquad\qquad\qquad\qquad\qquad\qquad\qquad\qquad\qquad\qquad\square$

**1.17 Theorem.** *$\sigma(l^1)$ can be identified with the unit circle $\mathbb{T}$ in such a way that the Gelfand transform on $l^1$ becomes*

$$\widehat{a}(e^{i\theta}) = \sum_{-\infty}^{\infty} a_n e^{in\theta}.$$

*Proof.* Let $\delta^k$ and $\delta = \delta^0$ be as in the discussion of $l^1$ in §1.1. Then $l^1$ is generated by $\delta^1$ and its inverse $\delta^{-1}$, so by Proposition 1.15, $\sigma(l^1)$ is homeomorphic to $\sigma(\delta^1)$. We claim that $\sigma(\delta^1) = \mathbb{T}$.

Indeed, let us try to invert $\lambda\delta - \delta^1$ for $\lambda \in \mathbb{C}$. If $a \in l^1$ we have $[(\lambda\delta-\delta^1)*a]_n = \lambda a_n - a_{n-1}$, so $(\lambda\delta-\delta^1)*a = \delta$ if and only if $\lambda a_0 - a_{-1} = 1$ and $\lambda a_n = a_{n-1}$ for $n \neq 0$. Solving these equations recursively, we obtain

$$a_{-1} = \lambda a_0 - 1, \quad a_n = \lambda^{-n} a_0 \text{ for } n \geq 0, \quad a_{-n} = \lambda^{n-1} a_{-1} \text{ for } n \geq 1.$$

The condition $\sum |a_n| < \infty$ forces $a_0 = 0$ if $|\lambda| \leq 1$ and $a_{-1} = 0$ if $|\lambda| \geq 1$. Subject to these conditions there is a unique solution if $|\lambda| \neq 1$, namely

$a = -\sum_1^\infty \lambda^{n-1}\delta^{-n}$ if $|\lambda| < 1$ and $a = \sum_0^\infty \lambda^{-n-1}\delta^n$ if $|\lambda| > 1$, but there is no solution if $|\lambda| = 1$. Thus $\sigma(\delta^1) = \mathbb{T}$.

By Proposition 1.15, then, for each $e^{i\theta} \in \mathbb{T}$ there is a unique $h_\theta \in \sigma(l^1)$ such that $h_\theta(\delta^1) = e^{i\theta}$. But then $h_\theta$ is given by

$$h_\theta(a) = h_\theta\Big(\sum_{-\infty}^\infty a_n\delta^n\Big) = \sum_{-\infty}^\infty a_n h_\theta(\delta^1)^n = \sum_{-\infty}^\infty a_n e^{in\theta},$$

so if we identify $h_\theta$ with $e^{i\theta}$ we have

$$\widehat{a}(e^{i\theta}) = h_\theta(a) = \sum_{-\infty}^\infty a_n e^{in\theta}. \qquad \square$$

As an immediate corollary, we obtain a famous theorem of Wiener concerning absolutely convergent Fourier series.

**1.18 Corollary.** *If $f(e^{i\theta}) = \sum a_n e^{in\theta}$ with $\sum |a_n| < \infty$, and $f$ never vanishes, then $1/f(e^{i\theta}) = \sum b_n e^{in\theta}$ with $\sum |b_n| < \infty$.*

*Proof.* We are given $f = \widehat{a}$ with $a \in l^1$. If $f$ never vanishes, then $a$ is invertible in $l^1$ by Theorem 1.13(b). Let $b$ be its inverse; then $1/f = \widehat{b}$. $\qquad \square$

We observe that the algebras $C(X)$ and $l^1$ are both symmetric. For $C(X)$ this is immediate from Theorem 1.16, and for $l^1$ we have

$$\widehat{a^*}(e^{i\theta}) = \sum \overline{a}_{-n} e^{in\theta} = \sum \overline{a}_n e^{-in\theta} = \overline{\widehat{a}(e^{i\theta})}.$$

(This is why we chose the involution on $l^1$ as we did. If we define, for example, $(a^*)_n = \overline{a}_n$, $l^1$ becomes a nonsymmetric $*$-algebra.)

We now return to a general commutative unital Banach algebra $\mathcal{A}$. $\mathcal{A}$ is called **semisimple** if the Gelfand transform on $\mathcal{A}$ is injective, that is, if the intersection of all the maximal ideals of $\mathcal{A}$ is $\{0\}$. For example, Theorems 1.16 and 1.17 show that $C(X)$ and $l^1$ are semisimple. On the other hand, the algebra of $2 \times 2$ complex matrices of the form $\left(\begin{smallmatrix} a & b \\ 0 & a \end{smallmatrix}\right)$ is not: one easily verifies that it has precisely one nontrivial ideal, namely those matrices with $a = 0$.

A condition stronger than semisimplicity is for the Gelfand transform to be an isometry. It is easy to see when this happens:

**1.19 Proposition.** *Let $\mathcal{A}$ be a commutative unital Banach algebra.*
  *a. If $x \in \mathcal{A}$, $\|\widehat{x}\|_{\sup} = \|x\|$ if and only if $\|x^{2^k}\| = \|x\|^{2^k}$ for all $k \geq 1$.*
  *b. $\Gamma_{\mathcal{A}}$ is an isometry if and only if $\|x^2\| = \|x\|^2$ for all $x \in \mathcal{A}$.*

*Proof.* If $\|\widehat{x}\|_{\text{sup}} = \|x\|$ then

$$\|x^{2^k}\| \leq \|x\|^{2^k} = \|\widehat{x}\|_{\text{sup}}^{2^k} = \|\widehat{x}^{2^k}\|_{\text{sup}} \leq \|x^{2^k}\|,$$

so $\|x^{2^k}\| = \|x\|^{2^k}$. Conversely if $\|x^{2^k}\| = \|x\|^{2^k}$ for all $k$ then $\|\widehat{x}\|_{\text{sup}} = \lim \|x^{2^k}\|^{1/2^k} = \|x\|$ by Theorems 1.8 and 1.13(d). This proves (a), and (b) follows since if $\|x^2\| = \|x\|^2$ for all $x$ then $\|x^{2^k}\| = \|x\|^{2^k}$ for all $x$ and $k$ (by induction on $k$). $\qquad\square$

We now come to the most fundamental result of Gelfand theory.

**1.20 Theorem** (The Gelfand-Naimark Theorem). *If $\mathcal{A}$ is a commutative unital C\* algebra, $\Gamma_{\mathcal{A}}$ is an isometric \*-isomorphism from $\mathcal{A}$ to $C(\sigma(\mathcal{A}))$.*

*Proof.* If $x \in \mathcal{A}$, let $y = x^*x$. Then $y = y^*$, so

$$\|y^{2^k}\| = \|(y^{2^{k-1}})^* y^{2^{k-1}}\| = \|y^{2^{k-1}}\|^2.$$

It follows by induction that $\|y^{2^k}\| = \|y\|^{2^k}$, so $\|\widehat{y}\|_{\text{sup}} = \|y\|$ by Proposition 1.19(a). But then

$$\|x\|^2 = \|y\| = \|\widehat{y}\|_{\text{sup}} = \| |\widehat{x}|^2\|_{\text{sup}} = \|\widehat{x}\|_{\text{sup}}^2,$$

so $\Gamma_{\mathcal{A}}$ is an isometry. In particular, $\Gamma_{\mathcal{A}}$ is injective and has closed range. But by Proposition 1.14(b,c), $\Gamma_{\mathcal{A}}$ respects the involutions and has dense range. Combining these results, we are done. $\qquad\square$

We conclude this section with an application of the Gelfand theory to the study of spectra in general (noncommutative) C\* algebras.

Suppose $\mathcal{A}$ is a unital Banach algebra and $\mathcal{B} \subset \mathcal{A}$ is a closed subalgebra containing $e$. If $y \in \mathcal{B}$ and $y$ is invertible in $\mathcal{A}$, in general $y^{-1}$ will not lie in $\mathcal{B}$. Hence, if $x \in \mathcal{B}$, one must distinguish between the spectrum of $x$ with respect to $\mathcal{A}$ and the spectrum of $x$ with respect to $\mathcal{B}$; we denote these spectra by $\sigma_{\mathcal{A}}(x)$ and $\sigma_{\mathcal{B}}(x)$. If $\lambda e - x$ is invertible in $\mathcal{B}$ it is invertible in $\mathcal{A}$, so it is always true that $\sigma_{\mathcal{A}}(x) \subset \sigma_{\mathcal{B}}(x)$, and we are interested in knowing when equality holds. We give one result along these lines for general Banach algebras and then a stronger one for C\* algebras.

**1.21 Lemma.** *Suppose $\mathcal{A}$ is a unital Banach algebra and $x_0 \in \mathcal{A}$ is a boundary point of the set of invertible elements of $\mathcal{A}$. If $x_n$ is invertible for each $n$ and $x_n \to x_0$ as $n \to \infty$ then $\|x_n^{-1}\| \to \infty$.*

*Proof.* If $\|x_n^{-1}\| \nrightarrow \infty$, by passing to a subsequence we can assume that $\|x_n^{-1}\| \leq C < \infty$ for all $n$. For $n$ sufficiently large we have $\|x_n - x_0\| < C^{-1}$. But then $x_0 = x_n - (x_n - x_0)$ is invertible by Theorem 1.4(b), which is impossible since the set of invertible elements is open.      □

**1.22 Proposition.** *Suppose $\mathcal{A}$ is a unital Banach algebra and $\mathcal{B}$ is a closed subalgebra containing $e$. If $x \in \mathcal{B}$ and $\sigma_{\mathcal{B}}(x)$ is nowhere dense in $\mathbb{C}$, then $\sigma_{\mathcal{A}}(x) = \sigma_{\mathcal{B}}(x)$.*

*Proof.* If $\lambda_0 \in \sigma_{\mathcal{B}}(x)$, there is a sequence $\lambda_n \in \mathbb{C} \setminus \sigma_{\mathcal{B}}(x)$ that converges to $\lambda_0$. By Lemma 1.21, $\|(\lambda_n e - x)^{-1}\| \to \infty$. It follows that $\lambda_0 e - x$ is not invertible in $\mathcal{A}$ (otherwise $\|(\lambda_n e - x)^{-1}\| \to \|(\lambda_0 e - x)^{-1}\|$), so $\lambda_0 \in \sigma_{\mathcal{A}}(x)$.      □

**1.23 Proposition.** *Suppose $\mathcal{A}$ is a unital $C^*$ algebra and $\mathcal{B} \subset \mathcal{A}$ is a $C^*$ subalgebra containing $e$.*
  a. *If $x \in \mathcal{B}$ and $x$ is invertible in $\mathcal{A}$ then $x^{-1} \in \mathcal{B}$.*
  b. *If $x \in \mathcal{B}$ then $\sigma_{\mathcal{A}}(x) = \sigma_{\mathcal{B}}(x)$.*

*Proof.* Given $x \in \mathcal{B}$, let $y = x^*x$, and let $\mathcal{C}$ be the closed subalgebra of $\mathcal{A}$ generated by $y$ and $e$; thus $\mathcal{C} \subset \mathcal{B} \subset \mathcal{A}$. Since $x$ is invertible in $\mathcal{A}$, so is $y$, so $0 \notin \sigma_{\mathcal{A}}(y)$. On the other hand, $\mathcal{C}$ is a commutative $C^*$ algebra since $y = y^*$, and $\sigma_{\mathcal{C}}(y) \subset \mathbb{R}$ by Proposition 1.14(a,b) and Theorem 1.13(c), so $\sigma_{\mathcal{C}}(y) = \sigma_{\mathcal{A}}(y)$ by Proposition 1.22. Thus $0 \notin \sigma_{\mathcal{C}}(y)$, so $y$ is invertible in $\mathcal{C}$ and hence in $\mathcal{B}$. But then $x^{-1} = y^{-1}x^* \in \mathcal{B}$. This proves (a), and (b) follows by applying (a) to $\lambda e - x$.      □

**1.24 Proposition.** *Let $\mathcal{A}$ be a unital $C^*$ algebra.*
  a. *If $x \in \mathcal{A}$ and $xx^* = x^*x$ then $\rho(x) = \|x\|$.*
  b. *If $\mathcal{B}$ is a Banach $*$-algebra whose involution is an isometry and $\phi : \mathcal{B} \to \mathcal{A}$ is a $*$-homomorphism, then $\|\phi\| \leq 1$.*

*Proof.* (a) Let $\mathcal{C}$ be the closed subalgebra of $\mathcal{A}$ generated by $x$, $x^*$, and $e$. Then $\mathcal{C}$ is a commutative $C^*$ algebra, so by Theorem 1.13(d) and the Gelfand-Naimark theorem, $\rho(x) = \|\Gamma_{\mathcal{C}}x\|_{\sup} = \|x\|$.
  (b) $\phi$ is a bounded linear map, say $\|\phi\| = C < \infty$, so for any $y \in \mathcal{B}$,

$$\|\phi(y^*y)^n\| \leq C\|(y^*y)^n\| \leq C\|y^*y\|^n \leq C\|y\|^{2n}.$$

But then by (a) (applied to $x = \phi(y^*y)$, which satisfies $x^* = x$) and Theorem 1.8,

$$\|\phi(y)\|^2 = \|\phi(y^*y)\| = \lim \|\phi(y^*y)^n\|^{1/n} \leq \lim C^{1/n}\|y\|^2 = \|y\|^2.      □$$

## 1.3 Nonunital Banach Algebras

Let $\mathcal{A}$ be a nonunital Banach algebra. The results in the preceding sections that deal with inverses and spectra have no meaning in this situation, but a large part of the Gelfand theory still works.

The starting point is the fact that a nonunital algebra $\mathcal{A}$ can always be embedded in a unital algebra. Namely, let $\widetilde{\mathcal{A}}$ be the algebra whose underlying vector space is $\mathcal{A} \times \mathbb{C}$ and whose multiplication is given by

$$(x,a)(y,b) = (xy + ay + bx, \, ab).$$

It is easily verified that $\widetilde{\mathcal{A}}$ is an algebra with unit $(0,1)$, and that the norm

$$(1.25) \qquad\qquad \|(x,a)\| = \|x\| + |a|$$

makes it into a Banach algebra. Moreover, $\mathcal{A} \times \{0\}$ is a closed two-sided ideal in $\widetilde{\mathcal{A}}$ (a maximal one, since it has codimension one). We shall identify $\mathcal{A}$ with $\mathcal{A} \times \{0\}$ and thus think of $\mathcal{A}$ as a maximal ideal in $\widetilde{\mathcal{A}}$. The restriction of the norm (1.25) to $\mathcal{A}$ then coincides with the original norm on $\mathcal{A}$.

If $\mathcal{A}$ is a $*$-algebra, the involution on $\mathcal{A}$ extends uniquely to an involution on $\widetilde{\mathcal{A}}$:

$$(1.26) \qquad\qquad (x,a)^* = (x^*, \bar{a}).$$

In concrete instances, $\widetilde{\mathcal{A}}$ can often be realized in a more natural way. Here are two examples.

*Example 1.* Let $\mathcal{A} = L^1(\mathbb{R})$. The map $f \mapsto \mu_f$, where $d\mu_f(x) = f(x)\,dx$, embeds $L^1(\mathbb{R})$ into the space $M(\mathbb{R})$ of finite Borel measures on $\mathbb{R}$, which is also a Banach algebra with convolution defined by

$$\int f \, d(\mu * \nu) = \iint f(x+y) \, d\mu(x) \, d\nu(y).$$

$M(\mathbb{R})$ has a unit, namely the point mass at 0 or Dirac measure $\delta$. $\widetilde{L^1(\mathbb{R})}$ is isomorphic to the subalgebra of $M(\mathbb{R})$ spanned by $L^1(\mathbb{R})$ and $\delta$, and the norm (1.25) is the restriction to $\widetilde{L^1(\mathbb{R})}$ of the usual norm $\|\mu\| = |\mu|(\mathbb{R})$ on $M(\mathbb{R})$.

*Example 2.* If $\mathcal{A} = C_0(X)$ where $X$ is a noncompact, locally compact Hausdorff space, then $\widetilde{\mathcal{A}}$ is isomorphic to the algebra obtained by adjoining the constant functions to $\mathcal{A}$, or equivalently to the algebra $C(\widetilde{X})$ where $\widetilde{X}$ is the one-point compactification of $X$. However, in this case the norm (1.25) is not the uniform norm on $C(\widetilde{X})$.

Example 2 illustrates a general problem: if $\mathcal{A}$ is a C* algebra, $\widetilde{\mathcal{A}}$ is not a C* algebra with the norm (1.25). However, we can remedy this by choosing a different norm.

**1.27 Proposition.** *If $\mathcal{A}$ is a nonunital C* algebra, there is a unique norm on $\widetilde{\mathcal{A}}$ that makes $\widetilde{\mathcal{A}}$ into a C* algebra with involution (1.26). This norm agrees with the original norm on $\mathcal{A}$.*

*Proof.* Since $\mathcal{A}$ is an ideal in $\widetilde{\mathcal{A}}$, each $(x, a) \in \widetilde{\mathcal{A}}$ acts on $\mathcal{A}$ by left multiplication: $(x, a)(y, 0) = (xy + ay, 0)$. We define $\|(x, a)\|$ to be the norm of this bounded operator on $\mathcal{A}$:

$$(1.28) \qquad \|(x, a)\| = \sup\{\|xy + ay\| : y \in \mathcal{A}, \ \|y\| \le 1\}.$$

This clearly defines a seminorm on $\widetilde{\mathcal{A}}$ that satisfies

$$\|(x, a)(y, b)\| \le \|(x, a)\|\,\|(y, b)\|.$$

To see that it is a norm, suppose $(x, a)$ is a nonzero element of $\widetilde{\mathcal{A}}$ satisfying $\|(x, a)\| = 0$, so that $xy + ay = 0$ for all $y \in \mathcal{A}$. Clearly $x$ must be nonzero, and then $a$ must be nonzero since $xy \ne 0$ for $y = x^*$, so $z = -a^{-1}x$ is a left unit for $\mathcal{A}$. But then $z^*$ is a right unit for $\mathcal{A}$, so $z = zz^* = z^*$ is a unit for $\mathcal{A}$, contrary to assumption.

Since $\|xy\| \le \|x\|\|y\|$ with equality if $y = x^*$, we have $\|(x, 0)\| = \|x\|$, so the new and old norms agree on $\mathcal{A}$. In particular, since $\mathcal{A}$ is complete, it is closed in $\widetilde{\mathcal{A}}$, so the linear functional $(x, a) \mapsto a$ (which is essentially the quotient map from $\widetilde{\mathcal{A}}$ to $\widetilde{\mathcal{A}}/\mathcal{A}$) is continuous. It follows that $\widetilde{\mathcal{A}}$ is complete: if $\{(x_n, a_n)\}$ is a Cauchy sequence, so is $\{a_n\}$ and hence so is $\{x_n\}$; and then $\lim(x_n, a_n) = (\lim x_n, \lim a_n)$. Thus $\widetilde{\mathcal{A}}$ is a Banach algebra with the norm (1.28).

To see that $\widetilde{\mathcal{A}}$ is a C* algebra, suppose $(x, a) \ne (0, 0) \in \widetilde{\mathcal{A}}$. For any $\epsilon > 0$ there exists $y \in \mathcal{A}$ with $\|y\| = 1$ such that $\|xy + ay\| \ge (1 - \epsilon)\|(x, a)\|$. But then, since $\|z\| = \|(z, 0)\|$ for $z \in \mathcal{A}$ and $(xy + ay, 0) = (x, a)(y, 0)$,

$$(1 - \epsilon)^2\|(x, a)\|^2 \le \|xy + ay\|^2 = \|(xy + ay)^*(xy + ay)\|$$
$$= \|(xy + ay, 0)^*(xy + ay, 0)\| = \|(y^*, 0)(x, a)^*(x, a)(y, 0)\|$$
$$\le \|y\|^2\|(x, a)^*(x, a)\| = \|(x, a)^*(x, a)\|.$$

Since $\epsilon$ is arbitrary, $\|(x, a)\|^2 \le \|(x, a)^*(x, a)\|$. It follows that $\|(x, a)\|^2 \le \|(x, a)^*\|\,\|(x, a)\|$ and hence that $\|(x, a)\| \le \|(x, a)^*\|$ for all $x, a$. But then $\|(x, a)^*\| \le \|(x, a)^{**}\| = \|(x, a)\|$, and so $\|(x, a)\|^2 \le \|(x, a)^*(x, a)\| \le \|(x, a)\|^2$. Thus $\widetilde{\mathcal{A}}$ is a C* algebra.

The uniqueness of the norm follows from the fact that in a unital C* algebra, the norm of any element $\xi$ equals $\|\xi^*\xi\|^{1/2}$ and the norm of $\xi^*\xi$ is its spectral radius (Proposition 1.24(a)). $\qquad\square$

Now suppose $\mathcal{A}$ is a nonunital commutative Banach algebra. As in the unital case, we define the **spectrum** $\sigma(\mathcal{A})$ to be the set of multiplicative functionals on $\mathcal{A}$, i.e., the set of nonzero homomorphisms from $\mathcal{A}$ into $\mathbb{C}$. (The common term "maximal ideal space" for $\sigma(\mathcal{A})$ is not appropriate here, as Proposition 1.11 and Theorem 1.12 are no longer valid.)

Every $h \in \sigma(\mathcal{A})$ has an extension to a multiplicative functional $\tilde{h}$ on $\tilde{\mathcal{A}}$, namely

$$(1.29) \qquad \tilde{h}(x,a) = h(x) + a.$$

This extension is unique since $\tilde{h}(0,1)$ must be 1 by Proposition 1.10(a). Conversely, if $H \in \sigma(\tilde{\mathcal{A}})$, then either $H$ is the functional $H_0$ whose associated maximal ideal is $\mathcal{A}$, namely $H_0(x,a) = a$, or $H|\mathcal{A} \in \sigma(\mathcal{A})$. Moreover, $H_0$ is the extension $\tilde{0}$ of the zero functional given by (1.29).

In short, there is a natural one-to-one correspondence between $\sigma(\mathcal{A}) \cup \{0\}$ and $\sigma(\tilde{\mathcal{A}})$. In particular, by Proposition 1.10(c), every $h \in \sigma(\mathcal{A})$ satisfies

$$|h(x)| = |\tilde{h}(x,0)| \le \|(x,0)\| = \|x\|.$$

(This is true whether we use the norm (1.25) or the norm (1.28) in the case of a C* algebra.) Thus $\sigma(\mathcal{A}) \cup \{0\}$ is a weak* closed subset of the closed unit ball in $\mathcal{A}^*$, and as such is a compact Hausdorff space. If $\{0\}$ is an isolated point, then $\sigma(\mathcal{A})$ is also compact. If not — and this is the more common case — then $\sigma(\mathcal{A})$ is a locally compact Hausdorff space whose one-point compactification is $\sigma(\mathcal{A}) \cup \{0\}$, or equivalently $\sigma(\tilde{\mathcal{A}})$.

The Gelfand transform on $\mathcal{A}$ is defined as before, $\Gamma_{\mathcal{A}} x(h) = \hat{x}(h) = h(x)$, and it is related to the Gelfand transform on $\tilde{\mathcal{A}}$ by

$$\hat{x}(h) = h(x) = \tilde{h}(x,0) = (x,0)\widehat{\phantom{.}}(\tilde{h}).$$

In other words, if we identify $\sigma(\tilde{\mathcal{A}})$ with $\sigma(\mathcal{A}) \cup \{0\}$, $\hat{x} = \Gamma_{\mathcal{A}} x$ is just the restriction of $(x,0)\widehat{\phantom{.}} = \Gamma_{\tilde{\mathcal{A}}}(x,0)$ to $\sigma(\mathcal{A})$. Moreover, the value of $(x,0)\widehat{\phantom{.}}$ at the extra point $H_0 = \tilde{0}$ in $\sigma(\tilde{\mathcal{A}})$ is obviously 0. This means that when $\sigma(\mathcal{A})$ is noncompact, $\hat{x}$ vanishes at infinity for every $x \in \mathcal{A}$. Thus, if we agree that $C_0(\sigma(\mathcal{A})) = C(\sigma(\mathcal{A}))$ when $\sigma(\mathcal{A})$ is compact, we can summarize our results as follows.

**1.30 Theorem.** *Let $\mathcal{A}$ be a nonunital commutative Banach algebra. $\sigma(\mathcal{A})$ is a locally compact subset of the closed unit ball of $\mathcal{A}^*$ in the weak\* topology. If $\sigma(\mathcal{A})$ is noncompact, its weak\* closure in $\mathcal{A}^*$ is $\sigma(\mathcal{A}) \cup \{0\}$. The Gelfand transform on $\mathcal{A}$ is an algebra homomorphism from $\mathcal{A}$ to $C_0(\sigma(\mathcal{A}))$, and*

$$\|\hat{x}\|_{\sup} = \lim_{n \to \infty} \|x^n\|^{1/n} \qquad (x \in \mathcal{A}).$$

*Proof.* We have proved everything but the formula for $\|\hat{x}\|_{\sup}$. But the discussion above shows that $\|\hat{x}\|_{\sup} = \|(x,0)\hat{}\|_{\sup}$; $\|(x,0)\hat{}\|_{\sup}$ is the spectral radius of $(x,0)$ in $\tilde{\mathcal{A}}$ by Theorem 1.13(d); and the latter equals $\lim \|x^n\|^{1/n}$ by Theorem 1.8 since $\|(x,0)\| = \|x\|$. ☐

*Example 1 (continued).* Let $\mathcal{A} = L^1(\mathbb{R})$. The Fourier transform

$$\hat{f}(\xi) = \int e^{-2\pi i \xi x} f(x)\, dx$$

satisfies $(f * g)\hat{} = \hat{f}\hat{g}$, so for each $\xi \in \mathbb{R}$ the evaluation functional $h_\xi(f) = \hat{f}(\xi)$ belongs to $\sigma(L^1(\mathbb{R}))$. It is not hard to show that these are all the multiplicative functionals on $L^1(\mathbb{R})$; we shall give the proof in §4.1 (see Theorems 4.3 and 4.6(a)). Hence we can identify $\sigma(L^1(\mathbb{R}))$ with $\mathbb{R}$ by the map $h_\xi \mapsto \xi$, and when this is done, the Gelfand transform is the Fourier transform.

*Example 2 (continued).* Let $\mathcal{A} = C_0(X)$ where $X$ is a noncompact, locally compact Hausdorff space. Then, as we have observed, $\tilde{\mathcal{A}} \cong C(\tilde{X})$ where $\tilde{X}$ is the one-point compactification of $X$. Combining Theorem 1.16 with the discussion leading to Theorem 1.30, it is easy to see that $\sigma(C_0(X)) \cong X$ if we identify $x \in X$ with the evaluation functional $h_x(f) = f(x)$, and then the Gelfand transform is the identity map on $C_0(X)$.

Finally, we have a nonunital version of the Gelfand-Naimark theorem.

**1.31 Theorem.** *If $\mathcal{A}$ is a nonunital commutative $C^*$ algebra, $\Gamma_{\mathcal{A}}$ is an isometric $*$-isomorphism from $\mathcal{A}$ to $C_0(\sigma(\mathcal{A}))$.*

*Proof.* This is simply a matter of combining our previous results. We make $\tilde{\mathcal{A}}$ into a $C^*$ algebra according to Proposition 1.27. Then $\Gamma_{\tilde{\mathcal{A}}}$ is an isometric $*$-isomorphism from $\tilde{\mathcal{A}}$ to $C(\sigma(\tilde{\mathcal{A}}))$. $\mathcal{A}$ is a maximal ideal in $\tilde{\mathcal{A}}$ whose associated functional is $\tilde{0}$ defined by (1.29), and the corresponding maximal ideal in $C(\sigma(\tilde{\mathcal{A}}))$ is $\{f : f(\tilde{0}) = 0\}$. In view of the correspondence (1.29) between $\sigma(\mathcal{A})$ and $\sigma(\tilde{\mathcal{A}}) \setminus \{\tilde{0}\}$ and the relation $\hat{x}(h) = (x,0)\hat{}(\tilde{h})$, the result follows. ☐

## 1.4 The Spectral Theorem

In this section we use Gelfand theory to derive the spectral theorem for commutative $C^*$ algebras of operators on a Hilbert space.

The finite-dimensional spectral theorem, in its simplest form, says that if $T$ is a self-adjoint operator on a finite-dimensional Hilbert space $\mathcal{H}$, there is an orthonormal basis for $\mathcal{H}$ consisting of eigenvectors for $T$. In this form the theorem is false in infinite dimensions, where self-adjoint operators need not have any eigenvectors at all. (For example, consider $Tf(x) = xf(x)$ on $L^2(0,1)$.) However, there are ways of reformulating the theorem that do generalize.

*Formulation I.* Let $\Sigma$ be the spectrum of $T$, and for $\lambda \in \Sigma$ let $P_\lambda$ be the orthogonal projection onto the eigenspace for $\lambda$. Then

$$(1.32) \qquad P_\lambda P_\mu = 0 \text{ for } \lambda \neq \mu, \qquad I = \sum_{\lambda \in \Sigma} P_\lambda, \qquad T = \sum_{\lambda \in \Sigma} \lambda P_\lambda.$$

The first equation says that the ranges of the projections $P_\lambda$ are mutually orthogonal; the second says that they span $\mathcal{H}$, and the third gives the spectral decomposition of $T$. This can be generalized to infinite dimensions by replacing the sums in a suitable way by integrals.

*Formulation II.* Let $n = \dim \mathcal{H}$, and let us regard $\mathbb{C}^n$ as the set of complex-valued functions on $\{1, \ldots, n\}$. An orthonormal eigenbasis for $T$ determines a unitary map $U : \mathcal{H} \to \mathbb{C}^n$ and a function $\phi \in \mathbb{C}^n$ — namely, $\phi(j)$ is the eigenvalue for the $j$th eigenvector — such that $UTU^{-1}\psi = \phi\psi$ for $\psi \in \mathbb{C}^n$, where $\phi\psi$ is the pointwise product of the functions $\phi$ and $\psi$. This can be generalized to infinite dimensions by replacing $\mathbb{C}^n$ by $L^2(\Omega, \mu)$ for a suitable measure space $(\Omega, \mu)$.

A more sophisticated version of the finite-dimensional spectral theorem says that if $\mathcal{T}$ is a family of commuting self-adjoint operators on $\mathcal{H}$ ($\dim \mathcal{H} < \infty$), there is an orthonormal basis for $\mathcal{H}$ that is an eigenbasis for every $T \in \mathcal{T}$. Both the formulations above work in this more general situation. In the first one, $\Sigma$ is taken to be a list of the simultaneous eigenspaces for the operators in $\mathcal{T}$, the $P_\lambda$'s are the projections onto these spaces, and for each $T \in \mathcal{T}$ one has a function $\Phi_T$ on $\Sigma$ such that $T = \sum_{\lambda \in \Sigma} \Phi_T(\lambda) P_\lambda$. In the second, there is a unitary $U : \mathcal{H} \to \mathbb{C}^n$ and, for each $T \in \mathcal{T}$, a function $\phi_T \in \mathbb{C}^n$ such that $UTU^{-1}\psi = \phi_T\psi$. (The functions $\Phi_T$ and $\phi_T$ are of course closely related.)

If $\mathcal{T}$ is as above, the algebra of operators it generates is a commutative C* algebra, and the simultaneous eigenbasis for the members of $\mathcal{T}$ will also be an eigenbasis for every element of this algebra. Hence one might as well consider commutative C* algebras to begin with, and this is the context in which we shall develop the theorem in infinite dimensions. Here is the notation we shall be using:

   i.  $\mathcal{H}$ is a Hilbert space.

  ii.  $\mathcal{A}$ is a commutative C* subalgebra of $\mathcal{L}(\mathcal{H})$ containing $I$.

 iii.  $\Sigma = \sigma(\mathcal{A})$ is the spectrum of $\mathcal{A}$.

 iv.  For $T \in \mathcal{A}$, $\widehat{T} \in C(\Sigma)$ is the Gelfand transform of $T$.

v. For $f \in C(\Sigma)$, $T_f \in \mathcal{A}$ is the inverse Gelfand transform of $f$. By the Gelfand-Naimark theorem, this is well defined, and we have $\|T_f\| = \|f\|_{\sup}$.

vi. $B(\Sigma)$ is the space of bounded Borel measurable functions on $\Sigma$. Like $C(\Sigma)$, $B(\Sigma)$ is a commutative C* algebra under the pointwise algebra operations, complex conjugation, and the uniform norm.

vii. If $\{f_n\}$ is a sequence of complex-valued functions on a set $S$, we say that $f_n \to f$ **p.b.** (for "pointwise and boundedly"), or that $f$ is the **p.b. limit** of $f_n$, if $f_n(s) \to f(s)$ for every $s \in S$ and $\sup\{|f_n(s)| : s \in S, \ n \geq 1\} < \infty$.

The key to all our results is the following construction. If $u, v \in \mathcal{H}$, the map $f \mapsto \langle T_f u, v \rangle$ is a bounded linear functional on $C(\Sigma)$; in fact,

$$|\langle T_f u, v \rangle| \leq \|T_f\| \|u\| \|v\| = \|f\|_{\sup} \|u\| \|v\|.$$

Hence by the Riesz representation theorem, there is a unique regular complex Borel measure $\mu_{u,v}$ on $\Sigma$ such that

$$(1.33) \qquad \langle T_f u, v \rangle = \int f \, d\mu_{u,v} \quad (f \in C(\Sigma), \ u, v \in \mathcal{H}),$$

$$\|\mu_{u,v}\| \leq \|u\| \|v\|.$$

The map $(u, v) \mapsto \mu_{u,v}$ is a "measure-valued inner product" in the following sense.

**1.34 Proposition.** $(u, v) \mapsto \mu_{u,v}$ *is a sesquilinear map from* $\mathcal{H} \times \mathcal{H}$ *to* $M(\Sigma)$. *Moreover,* $\mu_{v,u} = \overline{\mu_{u,v}}$, *and* $\mu_{u,u}$ *is a positive measure for all* $u$.

*Proof.* Sesquilinearity is obvious. Since the Gelfand transform takes adjoints into complex conjugates, we have $T_f^* = T_{\overline{f}}$ for all $f \in C(\Sigma)$, so

$$\int f \, d\mu_{v,u} = \langle T_f v, u \rangle = \langle v, T_f^* u \rangle = \overline{\langle T_f^* u, v \rangle} = \overline{\int \overline{f} \, d\mu_{u,v}} = \int f \, d\overline{\mu_{u,v}}.$$

Hence $\mu_{v,u} = \overline{\mu_{u,v}}$. Finally, if $u \in \mathcal{H}$ and $f \geq 0 \in C(\Sigma)$, let $g$ be the positive square root of $f$. Then $g \in C(\Sigma)$ and $T_g^* T_g = T_g^2 = T_{g^2} = T_f$, so

$$\int f \, d\mu_{u,u} = \langle T_f u, u \rangle = \langle T_g^* T_g u, u \rangle = \|T_g u\|^2 \geq 0.$$

Hence $\mu_{u,u} \geq 0$. $\qquad\qquad\qquad\qquad\qquad\qquad\qquad\qquad\qquad\qquad\qquad\square$

The map $f \mapsto T_f$ gives a representation of the algebra $C(\Sigma)$ as bounded operators on $\mathcal{H}$. We now use (1.33) to extend this representation to the larger algebra $B(\Sigma)$. Namely, if $f \in B(\Sigma)$, we have

$$\left| \int f \, d\mu_{u,v} \right| \leq \|f\|_{\sup} \|\mu_{u,v}\| \leq \|f\|_{\sup} \|u\| \|v\|.$$

Hence there is a unique $T_f \in \mathcal{L}(\mathcal{H})$ such that

$$(1.35) \qquad \langle T_f u, v \rangle = \int f \, d\mu_{u,v} \quad (u, v \in \mathcal{H}), \qquad \|T_f\| \le \|f\|_{\sup}.$$

This definition of $T_f$ agrees with the previous one when $f \in C(\Sigma)$.

**1.36 Theorem.** *The map $f \mapsto T_f$ is a $*$-homomorphism from $B(\Sigma)$ to $\mathcal{L}(\mathcal{H})$. It has the following additional properties:*
   a. *If $S \in \mathcal{L}(\mathcal{H})$ commutes with every $T \in \mathcal{A}$, then $S$ commutes with $T_f$ for every $f \in B(\Sigma)$.*
   b. *If $f_n \in B(\Sigma)$ and $f_n \to f$ p.b., then $T_{f_n} \to T_f$ in the weak operator topology.*

*Proof.* Clearly $f \mapsto T_f$ is linear. By Proposition 1.34, if $f \in B(\Sigma)$,

$$\langle T_{\bar{f}} u, v \rangle = \int \bar{f} \, d\mu_{u,v} = \overline{\int f \, d\mu_{v,u}} = \overline{\langle T_f v, u \rangle} = \langle u, T_f v \rangle = \langle T_f^* u, v \rangle,$$

so $T_{\bar{f}} = T_f^*$. To see that $T_{fg} = T_f T_g$, we start with the fact that this relation is valid when $f, g \in C(\Sigma)$. In this case we have

$$\int fg \, d\mu_{u,v} = \langle T_f T_g u, v \rangle = \int f \, d\mu_{T_g u, v}.$$

This being true for all $f \in C(\Sigma)$, $d\mu_{T_g u, v} = g \, d\mu_{u,v}$ for $g \in C(\Sigma)$. Hence, for any $f \in B(\Sigma)$,

$$\int fg \, d\mu_{u,v} = \int f \, d\mu_{T_g u, v} = \langle T_f T_g u, v \rangle = \langle T_g u, T_f^* v \rangle = \int g \, d\mu_{u, T_f^* v}.$$

This being true for all $g \in C(\Sigma)$, $f \, d\mu_{u,v} = d\mu_{u, T_f^* v}$. But then, for any $g \in B(\Sigma)$,

$$\langle T_f T_g u, v \rangle = \langle T_g u, T_f^* v \rangle = \int g \, d\mu_{u, T_f^* v} = \int fg \, d\mu_{u,v} = \langle T_{fg} u, v \rangle,$$

so $T_f T_g = T_{fg}$.

If $S$ commutes with $T_f$ for every $f \in C(\Sigma)$, we have

$$\int f \, d\mu_{u, S^* v} = \langle T_f u, S^* v \rangle = \langle S T_f u, v \rangle = \langle T_f S u, v \rangle = \int f \, d\mu_{Su, v},$$

so $\mu_{u, S^* v} = \mu_{Su, v}$. Hence, for any $f \in B(\Sigma)$,

$$\langle T_f S u, v \rangle = \int f \, d\mu_{Su, v} = \int f \, d\mu_{u, S^* v} = \langle T_f u, S^* v \rangle = \langle S T_f u, v \rangle.$$

This proves (a). Finally, if $f_n \to f$ p.b., then $\int f_n \, d\mu_{u,v} \to \int f \, d\mu_{u,v}$ by the dominated convergence theorem. In other words, $\langle T_{f_n} u, v \rangle \to \langle T_f u, v \rangle$ for all $u, v$, so (b) is proved. $\qquad \square$

We shall obtain a stronger form of assertion (b) in Proposition 1.48 below.

If $E \subset \Sigma$ is a Borel set, let $\chi_E$ be the characteristic function of $E$ and

$$(1.37) \qquad\qquad P(E) = T_{\chi_E}.$$

**1.38 Theorem.** *The correspondence $E \mapsto P(E)$ defined by (1.37) has the following properties.*
  a. *Each $P(E)$ is an orthogonal projection.*
  b. *$P(\varnothing) = 0$ and $P(\Sigma) = I$.*
  c. *$P(E \cap F) = P(E)P(F)$.*
  d. *If $E_1, E_2, \ldots$ are disjoint then $P(\bigcup E_j) = \sum P(E_j)$, where the sum converges in the strong operator topology.*

*Proof.* Since $\chi_E^2 = \chi_E = \overline{\chi}_E$, we have $P(E)^2 = P(E) = P(E)^*$. The first equation says that $P(E)$ is a projection, and the second one implies that its range is orthogonal to its nullspace. This proves (a); (b) is obvious, and (c) follows from the fact that $\chi_{E \cap F} = \chi_E \chi_F$. (d) is true when the sequence $E_1, E_2, \ldots$ is finite since $\chi_{\bigcup_1^n E_j} = \sum_1^n \chi_{E_j}$. For the infinite case, let us write $F_n = \bigcup_1^n E_j$ and $F = \bigcup_1^\infty E_j$. Then $\chi_{F_n} \to \chi_F$ p.b., so by Theorem 1.36(b),

$$\sum_1^n P(E_j) = P(F_n) \to P(F) = P\left(\bigcup_1^\infty E_j\right) \text{ weakly.}$$

But also $F$ is the disjoint union of $F_n$ and $F \setminus F_n$, so $P(F) = P(F_n) + P(F \setminus F_n)$. Hence, for any $u \in \mathcal{H}$, since $P(F \setminus F_n)$ is an orthogonal projection we have

$$\|[P(F) - P(F_n)]u\|^2 = \|P(F \setminus F_n)u\|^2 = \langle P(F \setminus F_n)u, P(F \setminus F_n)u \rangle$$
$$= \langle P(F \setminus F_n)u, u \rangle = \langle [P(F) - P(F_n)]u, u \rangle \to 0.$$

Thus the series actually converges strongly, and we are done.  $\square$

**1.39 Corollary.** *If $E$ and $F$ are disjoint, the ranges of $P(E)$ and $P(F)$ are mutually orthogonal.*

*Proof.* For any $u$ and $v$, $\langle P(E)u, P(F)v \rangle = \langle P(F)P(E)u, v \rangle = \langle P(E \cap F)u, v \rangle = \langle P(\varnothing)u, v \rangle = 0.$  $\square$

The situation described in Theorem 1.38 can be formulated in an abstract setting. Namely, suppose $\Omega$ is a set equipped with a $\sigma$-algebra $\mathcal{M}$, and $\mathcal{H}$ is a Hilbert space. An $\mathcal{H}$-**projection-valued measure** (or just a **projection-valued measure** if $\mathcal{H}$ is understood) on $(\Omega, \mathcal{M})$ is a map $P : \mathcal{M} \to \mathcal{L}(\mathcal{H})$ that satisfies properties (a–d) of Theorem 1.38 (with $\Sigma$ replaced by $\Omega$). If $P$ is an $\mathcal{H}$-projection-valued measure on $(\Omega, \mathcal{M})$ and $u, v \in \mathcal{H}$, the map

$$(1.40) \qquad P_{u,v}(E) = \langle P(E)u, v \rangle$$

is an ordinary complex measure. The correspondence $(u, v) \mapsto P_{u,v}$ is a "measure-valued inner product" as in Proposition 1.34, and

$$(1.41) \qquad \|P_{v,v}\| = P_{v,v}(\Omega) = \|v\|^2.$$

If $\Omega$ is a locally compact Hausdorff space and $\mathcal{M}$ is the $\sigma$-algebra of Borel sets, a projection-valued measure $P$ on $(\Omega, \mathcal{M})$ is called **regular** if each of the measures $P_{u,v}$ is regular.

Let $B(\Omega) = B(\Omega, \mathcal{M})$ be the space of bounded $\mathcal{M}$-measurable functions on $\Omega$. If $f \in B(\Omega)$, one can define the integral of $f$ with respect to a projection-valued measure $P$ as follows. If $v \in \mathcal{H}$, by (1.41) we have

$$\left| \int f \, dP_{v,v} \right| \le \|f\|_{\sup} \|P_{v,v}\| = \|f\|_{\sup} \|v\|^2.$$

Thus, by polarization (see Appendix 1), if $u, v \in \mathcal{H}$ and $\|u\| = \|v\| = 1$,

$$\left| \int f \, dP_{u,v} \right| \le \tfrac{1}{4} \|f\|_{\sup} \big[ \|u + v\|^2 + \|u - v\|^2 + \|u + iv\|^2 + \|u - iv\|^2 \big]$$

$$\le 4\|f\|_{\sup}.$$

(As we shall see shortly, the factor of 4 on the right is superfluous.) By homogeneity, it then follows that

$$\left| \int f \, dP_{u,v} \right| \le 4\|f\|_{\sup} \|u\| \|v\| \qquad (u, v \in \mathcal{H}).$$

Hence there is a bounded operator $T$ on $\mathcal{H}$ such that $\langle Tu, v \rangle = \int f \, dP_{u,v}$ for all $u, v \in \mathcal{H}$. We denote $T$ by $\int f \, dP$:

$$(1.42) \qquad \left\langle \left( \int f \, dP \right) u, v \right\rangle = \int f \, dP_{u,v}.$$

If $f$ is a simple function, say $f = \sum_1^n c_j \chi_{E_j}$, then

$$\int f \, dP_{u,v} = \sum c_j P_{u,v}(E_j) = \sum c_j \langle P(E_j)u, v \rangle = \left\langle \sum c_j P(E_j)u, v \right\rangle,$$

so that $\int f \, dP = \sum c_j P(E_j)$ as one would expect. Moreover, since every $f \in B(\Omega)$ is a uniform limit of a sequence $\{f_n\}$ of simple functions, and since $\| \int f \, dP - \int f_n \, dP \| \le 4\|f - f_n\|_{\sup}$, we can obtain $\int f \, dP$ as a limit (in the norm topology) of "Riemann sums" just as we do for ordinary integrals.

**1.43 Theorem.** *If $P$ is an $\mathcal{H}$-projection-valued measure on $(\Omega, \mathcal{M})$, the map $f \mapsto \int f\, dP$ is a $*$-homomorphism from $B(\Omega)$ to $\mathcal{L}(\mathcal{H})$.*

*Proof.* The map $f \mapsto \int f\, dP$ is clearly linear, and we have seen that $\| \int f\, dP \| \le 4\|f\|_{\sup}$. If $f = \sum_1^n c_j \chi_{E_j}$ and $g = \sum_1^m d_k \chi_{F_k}$, then $fg = \sum_{j,k} c_j d_k \chi_{E_j \cap F_k}$, so

$$\int fg\, dP = \sum c_j d_k P(E_j \cap F_k) = \sum c_j d_k P(E_j) P(F_k) = \int f\, dP \int g\, dP.$$

By passing to uniform limits, we see that $\int fg\, dP = (\int f\, dP)(\int g\, dP)$ for all $f, g \in B(\Omega)$. Similarly, $\int \bar{f}\, dP = (\int f\, dP)^*$.    $\square$

It now follows from Proposition 1.24(b) that $\| \int f\, dP \| \le \|f\|_{\sup}$. It is also easy to see this directly: if $T = \int f\, dP$ then $T^*T = \int |f|^2\, dP$, so by (1.41),

$$\|Tv\|^2 = \langle T^*Tv, v \rangle = \int |f|^2\, dP_{v,v} \le \|f\|_{\sup}^2 \|v\|^2.$$

Let us now return to the case where $(\Omega, \mathcal{M})$ is the spectrum $\Sigma$ of the algebra $\mathcal{A}$ equipped with its Borel $\sigma$-algebra and $P$ is given by (1.37). Then

$$P_{u,v}(E) = \langle P(E)u, v \rangle = \int \chi_E\, d\mu_{u,v} = \mu_{u,v}(E)$$

for any $E$, so $P_{u,v} = \mu_{u,v}$. In particular, $P$ is regular. Moreover, it follows from (1.35) and (1.42) that $\int f\, dP = T_f$ for all $f \in B(\Sigma)$. We have now arrived at the promised generalization of (1.32):

**1.44 Theorem** (Spectral Theorem I). *Let $\mathcal{A}$ be a commutative $C^*$ subalgebra of $\mathcal{L}(\mathcal{H})$ containing $I$, and let $\Sigma$ be its spectrum. There is a unique regular projection-valued measure $P$ on $\Sigma$ such that $T = \int \widehat{T}\, dP$ for all $T \in \mathcal{A}$, and $T_f = \int f\, dP$ for all $f \in B(\Sigma)$. Moreover, if $S \in \mathcal{L}(\mathcal{H})$, the following are equivalent:*

*i. $S$ commutes with every $T \in \mathcal{A}$.*
*ii. $S$ commutes with $P(E)$ for every Borel set $E \subset \Sigma$.*
*iii. $S$ commutes with $\int f\, dP$ for every $f \in B(\Sigma)$.*

*Proof.* We have proved everything except the uniqueness and the final assertion. Uniqueness holds because of the uniqueness in the Riesz representation: the operators in $\mathcal{A}$ determine the measures $\mu_{u,v} = P_{u,v}$ through (1.33), and the measures $P_{u,v}$ determine $P$ through (1.40). As for the final assertion, (iii) clearly implies (i) and (ii), and we proved that (i) implies (iii) in Theorem 1.36. (ii) implies (iii) by Theorem 1.43 since every $f \in B(\Sigma)$ is a uniform limit of simple functions, and the norm limit of operators that commute with $S$ also commutes with $S$.    $\square$

We now give the generalization of the second reformulation of the finite-dimensional spectral theorem. Here, too, the crucial ingredient is the measures $\mu_{u,v}$ defined by (1.33). But first, a couple of lemmas.

**1.45 Lemma.** *Suppose $\mathcal{A} \subset \mathcal{L}(\mathcal{H})$ is a $C^*$ algebra, and $\mathcal{X} \subset \mathcal{H}$ is a closed subspace such that $T(\mathcal{X}) \subset \mathcal{X}$ for all $T \in \mathcal{A}$.*

 a. *We have $T(\mathcal{X}^\perp) \subset \mathcal{X}^\perp$ for all $T \in \mathcal{A}$.*
 b. *Let $\mathcal{A}^\mathcal{X} = \{T|\mathcal{X} : T \in \mathcal{A}\}$, $\Sigma = \sigma(\mathcal{A})$, and $\Sigma^\mathcal{X} = \sigma(\mathcal{A}^\mathcal{X})$. Then $\Sigma^\mathcal{X}$ can be naturally identified with a compact subset of $\Sigma$, and when this is done, the Gelfand transforms on $\mathcal{A}$ and $\mathcal{A}^\mathcal{X}$ are related by $(T|\mathcal{X})\hat{} = \widehat{T}|\Sigma^\mathcal{X}$.*
 c. *We extend any Borel measure $\mu$ on $\Sigma^\mathcal{X}$ (considered as a subset of $\Sigma$) to a measure $\widetilde{\mu}$ on $\Sigma$ by setting $\widetilde{\mu}(E) = \mu(E \cap \Sigma^\mathcal{X})$ for all Borel $E \subset \Sigma$. If $v \in \mathcal{X}$ and $\mu_{v,v}$ and $\mu_{v,v}^\mathcal{X}$ are the measures on $\Sigma$ and $\Sigma^\mathcal{X}$ associated to $\mathcal{A}$ and $\mathcal{A}^\mathcal{X}$ by (1.33), then $\mu_{v,v} = (\mu_{v,v}^\mathcal{X})\widetilde{}$.*

*Proof.* (a) If $u \in \mathcal{X}^\perp$, $v \in \mathcal{X}$, and $T \in \mathcal{A}$, then $\langle Tu, v \rangle = \langle u, T^*v \rangle = 0$ since $\mathcal{A}$ is closed under adjoints; hence $Tu \in \mathcal{X}^\perp$.

(b) Any multiplicative functional $\lambda$ on $\mathcal{A}^\mathcal{X}$ defines a multiplicative functional $\widetilde{\lambda}$ on $\mathcal{A}$ by $\widetilde{\lambda}(T) = \lambda(T|\mathcal{X})$, and the quantities on the left and right sides of this equality are, respectively, $\widehat{T}(\widetilde{\lambda})$ and $(T|\mathcal{X})\hat{}(\lambda)$.

(c) The restriction map $f \mapsto f|\Sigma^\mathcal{X}$ from $C(\Sigma)$ to $C(\Sigma^\mathcal{X})$ is a surjection by the Tietze extension theorem, so it induces an injection on the dual spaces, $M(\Sigma^\mathcal{X}) \to M(\Sigma)$. It is an easy exercise to check that this is the map $\mu \mapsto \widetilde{\mu}$. The desired result follows in view of (b): if $f \in C(\Sigma)$ and $T_f$ is its inverse Gelfand transform in $\mathcal{A}$,

$$\int_\Sigma f \, d\mu_{v,v} = \langle T_f v, v \rangle = \langle (T|\mathcal{X})v, v \rangle = \int_{\Sigma^\mathcal{X}} f \, d\mu_{v,v}^\mathcal{X} = \int_\Sigma f \, d(\mu_{v,v}^\mathcal{X})\widetilde{}.$$

$\square$

**1.46 Lemma.** *Suppose $(\Omega, \mu)$ is a semi-finite[1] measure space and $\phi \in L^\infty(\mu)$. If $T \in \mathcal{L}(L^2(\mu))$ is defined by $Tf = \phi f$, then $\|T\| = \|\phi\|_\infty$.*

*Proof.* Since $|\phi f| \leq \|\phi\|_\infty |f|$ a.e., it is clear that $\|\phi f\|_2 \leq \|\phi\|_\infty \|f\|_2$, so $\|T\| \leq \|\phi\|_\infty$. On the other hand, given $\epsilon > 0$, let $E = \{\omega : |\phi(\omega)| > \|\phi\|_\infty - \epsilon\}$. Then $\mu(E) > 0$, so there exists $F \subset E$ with $0 < \mu(F) < \infty$. But then $\chi_F \in L^2(\mu)$ and

$$\int |\phi \chi_F|^2 \, d\mu \geq (\|\phi\|_\infty - \epsilon)^2 \mu(F) = (\|\phi\|_\infty - \epsilon)^2 \|\chi_F\|_2^2,$$

so $\|T\| \geq \|\phi\|_\infty - \epsilon$. $\square$

---

[1]A measure space is **semi-finite** if every set of infinite measure contains a subset of positive finite measure.

**1.47 Theorem** (Spectral Theorem II). *Let $\mathcal{A}$ be a commutative $C^*$ subalgebra of $\mathcal{L}(\mathcal{H})$ containing $I$. There is a semi-finite measure space $(\Omega, \mathcal{M}, \mu)$, a unitary map $U : \mathcal{H} \to L^2(\mu)$, and an isometric $*$-homomorphism $T \mapsto \phi_T$ from $\mathcal{A}$ into $L^\infty(\mu)$ such that $UTU^{-1}\psi = \phi_T\psi$ for all $\psi \in L^2(\mu)$ and $T \in \mathcal{A}$. $\Omega$ can be taken as the disjoint union of copies of the spectrum $\Sigma$ of $\mathcal{A}$ in such a way that $\mu$ is finite on each copy and $\phi_T = \widehat{T}$ on each copy.*

*Proof.* First suppose there exists $v \in \mathcal{H}$ such that $\mathcal{A}v = \{Tv : T \in \mathcal{A}\}$ is dense in $\mathcal{H}$, and let $\mu = \mu_{v,v}$ as in (1.33). Then, for any $T \in \mathcal{A}$,

$$\|Tv\|^2 = \langle T^*Tv, v\rangle = \int |\widehat{T}|^2 \, d\mu.$$

In particular, if $Tv = Sv$ then $\widehat{T} - \widehat{S} = 0$ $\mu$-a.e., so $Tv \mapsto \widehat{T}$ is a well-defined linear isometry from $\mathcal{A}v$ into $L^2(\mu)$, and it extends uniquely to a linear isometry $U : \mathcal{H} \to L^2(\mu)$. The range of $U$ is necessarily closed, and it includes $C(\Sigma)$, which is dense in $L^2(\mu)$ since $\mu$ is regular, so $U$ is unitary. If $\psi \in C(\Sigma)$ and $T \in \mathcal{A}$ then

$$UTU^{-1}\psi = UTT_\psi v = (TT_\psi)^\frown = \widehat{T}\psi,$$

and it follows that $UTU^{-1}\psi = \widehat{T}\psi$ for all $\psi \in L^2(\mu)$.

For the general case, let $\{v_i\}_{i \in I}$ be a maximal collection of nonzero vectors in $\mathcal{H}$ such that the subspaces $\mathcal{H}_i = \overline{\mathcal{A}v_i}$ are mutually orthogonal; such a set exists by Zorn's lemma. Then each $\mathcal{H}_i$ is invariant under every $T \in \mathcal{A}$; hence so is $\bigoplus_{i \in I} \mathcal{H}_i$. It follows from Lemma 1.45(a) that $\bigoplus_{i \in I} \mathcal{H}_i = \mathcal{H}$, for otherwise one could add any $v \perp \bigoplus \mathcal{H}_i$ to the set $\{v_i\}$. For each $i \in I$, let $\Sigma_i$ be a copy of $\Sigma$, let $\mu_i$ be the measure $\mu_{v_i,v_i}$ on $\Sigma_i$ and let $\Omega$ be the disjoint union of the $\Sigma_i$; moreover, let $\mathcal{M}$ be the $\sigma$-algebra of sets $E \subset \Omega$ such that $E \cap \Sigma_i$ is Borel in $\Sigma_i$ for every $i$, and define $\mu$ on $\mathcal{M}$ by $\mu(E) = \sum_{i \in I} \mu_i(E \cap \Sigma_i)$. Since each $\mu_i$ is finite, $\mu$ is semi-finite, and $L^2(\mu) \cong \bigoplus_{i \in I} L^2(\mu_i)$. For each $i$, the algebra $\mathcal{A}_i = \{T|\mathcal{H}_i : T \in \mathcal{A}\}$ satisfies the conditions of the preceding paragraph, and the construction there together with the identifications of spectra, Gelfand transforms, and measures in Lemma 1.45(b,c) yield a unitary map $U_i : \mathcal{H}_i \to L^2(\mu_i)$ such that $U_i T U_i^{-1}$ is multiplication by $\widehat{T}$. Let $U = \bigoplus U_i : \mathcal{H} \to L^2(\mu)$; then, if $T \in \mathcal{A}$, $UTU^{-1}$ is multiplication by $\phi_T$ where $\phi_T = \widehat{T}$ on each $\Sigma_i$. Since the Gelfand transform is a $*$-homomorphism, and since $\|T\| = \|UTU^{-1}\| = \|\phi_T\|_\infty$ by Lemma 1.46, we are done. $\qquad\square$

We remark that if $\mathcal{H}$ is separable, there can only be countably many summands $\mathcal{H}_i$, so the measure $\mu$ is $\sigma$-finite. In fact, $\mu$ can be taken to be finite: simply multiply the $v_i$ by small scalars so that $\sum_{i \in I} \|v_i\|^2 < \infty$.

In comparison to Spectral Theorem I, Spectral Theorem II has the disadvantage that the measure space $(\Omega, \mu)$ and the unitary map $U$, unlike the projection-valued measure $P$, are not canonically determined by $\mathcal{A}$. However, it is frequently more useful, because important properties of operators are often quite transparent when the operators are multiplication by bounded functions on an $L^2$ space. As an example, we obtain a significant improvement on Theorem 1.36(b).

**1.48 Proposition.** *If $\{f_n\} \subset B(\Sigma)$ and $f_n \to f$ p.b., then $T_{f_n} \to T_f$ in the strong operator topology.*

*Proof.* With notation as in the proof of Spectral Theorem II, it is an easy exercise (left to the reader) to check that for any $f \in B(\Sigma)$, $UT_f U^{-1}\psi = \phi_f \psi$ where $\phi_f = f$ on each $\Sigma_i$. If $f_n \to f$ p.b., clearly $\phi_{f_n} \to \phi_f$ p.b. But then it is obvious from the dominated convergence theorem that $\|\phi_{f_n}\psi - \phi_f \psi\|_2 \to 0$ for any $\psi \in L^2(\mu)$. Taking $\psi = Uv$, this means that $\|T_{f_n} v - T_f v\| \to 0$ for every $v \in \mathcal{H}$. □

Let us now reduce all this general theory to the case of a single self-adjoint operator, or more generally a single normal operator. (Recall that $T \in \mathcal{L}(\mathcal{H})$ is **normal** if $TT^* = T^*T$.) If $T$ is normal, let $\mathcal{A}_T$ be the C* algebra generated by $T$, $T^*$, and $I$. Then $\mathcal{A}_T$ is commutative, and by Proposition 1.15 we can identify $\sigma(\mathcal{A}_T)$ with $\sigma(T) \subset \mathbb{C}$ in such a way that the Gelfand transform of $T$ is the function $\iota(\lambda) = \lambda$ on $\sigma(T)$. Once we have done this, Spectral Theorem I gives a projection-valued measure $P_T$ on $\sigma(T)$ such that

$$T = \int \lambda \, dP_T(\lambda).$$

If $p(\lambda) = \sum_{j,k=1}^{n} c_{jk} \lambda^j \overline{\lambda}^k$ is a polynomial in $\lambda$ and $\overline{\lambda}$ (or equivalently a polynomial in $\operatorname{Re} \lambda$ and $\operatorname{Im} \lambda$), since the correspondence $f \mapsto \int f \, dP_T$ is a *-homomorphism we have $\int p \, dP_T = \sum c_{jk} T^j T^{*k}$. In other words, $\int p \, dP$ is obtained by formally substituting $T$ and $T^*$ for $\lambda$ and $\overline{\lambda}$ in $p$. It is therefore natural to define the operator $f(T)$ for any $f \in B(\sigma(T))$ by

$$(1.49) \qquad f(T) = \int f \, dP_T,$$

and we obtain in this way a Borel functional calculus for the operator $T$. Its main properties are summarized in Theorem 1.51 below.

In view of Spectral Theorem II, it is important to see what these constructions yield when $\mathcal{H} = L^2(\Omega, \mu)$ and $T$ is multiplication by $\phi \in L^\infty(\Omega, \mu)$. In the first place, the spectrum of $T$ is the "essential range" of $\phi$, namely, the set of $\lambda \in \mathbb{C}$ for which $\{\omega : |\phi(\omega) - \lambda| < \epsilon\}$ has positive measure for every $\epsilon > 0$, or equivalently the set of $\lambda$ for which $(\phi - \lambda)^{-1} \notin L^\infty$. From this it is easy to see that $\phi(\omega) \in \sigma(T)$ for a.e. $\omega$, so by modifying $\phi$ on a nullset we can always assume that range$(\phi) \subset \sigma(T)$. Clearly $T^n$ is multiplication by $\phi^n$ and $T^*$ is multiplication by $\overline{\phi}$, so $p(T)$ is multiplication by $p \circ \phi$ for any polynomial $p$ in $\lambda$ and $\overline{\lambda}$. One is therefore led to guess that $f(T)$ will be multiplication by $f \circ \phi$ for every $f \in B(\sigma(T))$; in particular, the projections $P_T(E)$ will be multiplication by $\chi_{\phi^{-1}(E)}$. It is not too hard to prove this directly from the definitions, but we shall give a somewhat slicker proof in the following theorem.

**1.50 Lemma.** *Let $K$ be a compact subset of $\mathbb{R}^n$, and let $\mathcal{B}$ be the smallest algebra of functions on $K$ that contains all polynomials and is closed under p.b. limits. Then $\mathcal{B} = B(K)$, the algebra of bounded Borel functions on $K$.*

*Proof.* $\mathcal{B}$ contains $C(K)$ by the Stone-Weierstrass theorem, and then by taking p.b. limits one can easily see that it contains the characteristic function of every open set. Let $\mathcal{M} = \{E \subset K : \chi_E \in \mathcal{B}\}$. Then $\mathcal{M}$ is closed under complements since $\chi_{K \setminus E} = 1 - \chi_E$; it is closed under finite intersections since $\chi_{E \cap F} = \chi_E \chi_F$; and it is then closed under countable intersections since $\chi_{\bigcap_1^\infty E_j}$ is the p.b. limit of $\chi_{\bigcap_1^n E_j}$. Hence $\mathcal{M}$ is a $\sigma$-algebra, and it contains the open sets, so it contains all Borel sets. But then $\mathcal{B}$ contains all simple Borel functions and hence (by taking p.b. limits once again) all bounded Borel functions. Since $B(K)$ is an algebra that is closed under p.b. limits, we are done. $\qquad\square$

**1.51 Theorem.** *Suppose $T \in \mathcal{L}(\mathcal{H})$ is normal. There is a unique $*$-homomorphism $f \mapsto f(T)$ from $B(\sigma(T))$ to $\mathcal{L}(\mathcal{H})$ such that (1) $f(T) = T$ when $f(\lambda) = \lambda$, and (2) $f_n(T) \to f(T)$ in the strong operator topology whenever $f_n \to f$ p.b. The correspondence $f \mapsto f(T)$ has the following additional properties:*

    *a. If $\mathcal{A}$ is any commutative $C^*$ algebra containing $T$, $\widehat{T}$ is the Gelfand transform of $T$ with respect to $\mathcal{A}$, and $P_\mathcal{A}$ is the associated projection-valued measure on $\sigma(\mathcal{A})$, then $f(T) = \int f \circ \widehat{T} \, dP_\mathcal{A}$.*

    *b. If $\mathcal{H} = L^2(\mu)$ and $T$ is multiplication by $\phi \in L^\infty(\mu)$ (with range$(\phi) \subset \sigma(T)$), then $f(T)$ is multiplication by $f \circ \phi$ for every $f \in B(\sigma(T))$.*

    *c. If $S \in \mathcal{L}(\mathcal{H})$ commutes with $T$ and $T^*$, then $S$ commutes with $f(T)$ for every $f \in B(\sigma(T))$.*

*Proof.* To prove existence, of a $*$-homomorphism $f \mapsto f(T)$ satisfying (1) and (2), we define $f(T)$ by (1.49); the desired properties follow from Theorem 1.36 (or (1.43)) and Proposition 1.48. To prove uniqueness, suppose $f \mapsto \widetilde{f}(T)$ is another such correspondence, and let $\mathcal{B} = \{f : f(T) = \widetilde{f}(T)\}$. Then $\mathcal{B}$ contains all polynomials in $\lambda$ and $\overline{\lambda}$, or equivalently all polynomials in Re $\lambda$ and Im $\lambda$, since the correspondences are *-homomorphisms. Also, $\mathcal{B}$ is clearly closed under sums, products, and p.b. limits. Hence $\mathcal{B} = B(\sigma(T))$ by Lemma 1.50.

Properties (a) and (b) follow from the uniqueness, since the maps $f \mapsto \int f \circ \widehat{T} \, dP_{\mathcal{A}}$ and $f \mapsto$ (mult. by $f \circ \phi$) have all the asserted properties. (This follows from Spectral Theorem I and Proposition 1.48 in the case (a), and by simple direct arguments in the case (b).) Finally, (c) is true by Theorem 1.36, since if $S$ commutes with $T$ and $T^*$, it clearly commutes with every operator in the C* algebra they generate. $\qquad\square$

Let us observe that the finite-dimensional spectral theorem is an easy corollary of Spectral Theorem I. Indeed, if $\mathcal{A}$ is a commutative C* algebra of operators on $\mathcal{H}$ and dim $\mathcal{H} < \infty$, then $\sigma(\mathcal{A})$ is a set of cardinality dim $\mathcal{A} < \infty$, say $\sigma(\mathcal{A}) = \{\sigma_1, \ldots, \sigma_n\}$. The ranges of the projections $P(\{\sigma_j\})$, $1 \le j \le n$, are mutually orthogonal subspaces of $\mathcal{H}$ whose direct sum is $\mathcal{H}$, and they are eigenspaces for every $T \in \mathcal{A}$, viz., $TP(\{\sigma_j\}) = \widehat{T}(\sigma_j)P(\{\sigma_j\})$ for all $j$.

We conclude by showing that for compact operators, the spectral theorem becomes a direct generalization of the finite-dimensional case.

**1.52 Theorem.** *If $T$ is a compact normal operator on $\mathcal{H}$, there is an orthonormal basis for $\mathcal{H}$ consisting of eigenvectors for $T$.*

*Proof.* Let $E_0 = \{0\}$ and $E_n = \{\lambda \in \sigma(T) : n^{-1} \le |\lambda| < (n-1)^{-1}\}$ for $n \ge 1$, and let $\mathcal{H}_n$ be the range of the projection $P(E_n) = \chi_{E_n}(T)$. The spaces $\mathcal{H}_n$ are mutually orthogonal and invariant under $T$, and $\mathcal{H} = \bigoplus_0^\infty \mathcal{H}_n$. $\mathcal{H}_0$ is already an eigenspace for $T$ (with eigenvalue 0), so it suffices to show that each $\mathcal{H}_n$ $(n \ge 1)$ has an orthonormal basis consisting of eigenvectors for $T$. Now, $T|\mathcal{H}_n$ is an invertible operator on $\mathcal{H}_n$. (Use Spectral Theorem II: $T$ is unitarily equivalent to multiplication by a function $\phi$ on a space $L^2(\Omega, \mu)$. The subspace of $L^2(\Omega, \mu)$ corresponding to $\mathcal{H}_n$ is $L^2(\Omega_n, \mu)$ where $\Omega_n = \{\omega : \phi(\omega) \in E_n\}$, and $1/\phi$ is bounded by $n$ on this set.) But $T|\mathcal{H}_n$ is also compact, so the identity operator on $\mathcal{H}_n$ is compact and hence dim $\mathcal{H}_n < \infty$. The proof is therefore concluded by applying the finite-dimensional spectral theorem to $T|\mathcal{H}_n$. $\qquad\square$

## 1.5    Spectral Theory of *-Representations

The object of this section is to derive a version of the spectral theorem that applies to *-homomorphisms from a Banach *-algebra $\mathcal{A}$ into the algebra of bounded operators on a Hilbert space. For this purpose, we shall wish to apply the spectral theorem to C* subalgebras of $\mathcal{L}(\mathcal{H})$ that do not contain $I$, and there is one minor pitfall to be avoided. Namely, suppose $\mathcal{H}_1$ and $\mathcal{H}_2$ are Hilbert spaces and $\mathcal{A}$ is a C* subalgebra of $\mathcal{L}(\mathcal{H}_1)$ containing $I_1$, the identity operator on $\mathcal{H}_1$. For $T \in \mathcal{A}$, define $T' \in \mathcal{L}(\mathcal{H}_1 \oplus \mathcal{H}_2)$ by $T'(x_1, x_2) = (Tx_1, 0)$. Then $\mathcal{A}' = \{T' : T \in \mathcal{A}\}$ is a C* subalgebra of $\mathcal{L}(\mathcal{H}_1 \oplus \mathcal{H}_2)$ that has a unit (namely $I_1'$, the orthogonal projection onto $\mathcal{H}_1$) but does not contain the identity operator on $\mathcal{H}_1 \oplus \mathcal{H}_2$!

To avoid this situation, we define a *-subalgebra $\mathcal{A}$ of $\mathcal{L}(\mathcal{H})$ to be **nondegenerate** if there is no nonzero $v \in \mathcal{H}$ such that $Tv = 0$ for all $T \in \mathcal{A}$. In this case, if $\mathcal{A}$ contains a unit $E$, $E$ is an orthogonal projection ($E^2 = E^* = E$) and every $v$ in the nullspace of $E$ satisfies $Tv = TEv = 0$ for all $T \in \mathcal{A}$, whence $v = 0$; it follows that $E = I$. In other words, if $\mathcal{A}$ does not contain $I$ then $\mathcal{A}$ is nonunital.

Suppose now that $\mathcal{A}$ is a nondegenerate commutative C* subalgebra of $\mathcal{L}(\mathcal{H})$ that does not contain $I$. The unital augmentation $\widetilde{\mathcal{A}}$ of $\mathcal{A}$, given abstractly by Proposition 1.27, is realized concretely as $\widetilde{\mathcal{A}} = \mathcal{A} \oplus \mathbb{C}I$. Then, as explained in §1.3, we can identify $\sigma(\widetilde{\mathcal{A}})$ with $\sigma(\mathcal{A}) \cup \{0\}$, and we now do so. The spectral theorem associates to $\widetilde{\mathcal{A}}$ a regular projection-valued measure $P$ on $\sigma(\widetilde{\mathcal{A}})$. If $T \in \mathcal{A}$ we have $\widehat{T}(0) = 0$, so $TP(\{0\}) = 0$, and hence every $v$ in the range of $P(\{0\})$ satisfies $Tv = 0$ for all $T \in \mathcal{A}$. Since $\mathcal{A}$ is nondegenerate, it follows that $P(\{0\}) = 0$, so we can think of $P$ as a projection-valued measure on $\sigma(\mathcal{A})$. As such it is clearly still regular, so we have the following extension of Spectral Theorem I:

**1.53 Theorem.** *Spectral Theorem I remains valid if the assumption that $I \in \mathcal{A}$ is replaced by the assumption that $\mathcal{A}$ is nondegenerate.*

Now let $\mathcal{A}$ be an abstract Banach *-algebra. A *-**representation** of $\mathcal{A}$ on a Hilbert space $\mathcal{H}$ is a *-homomorphism $\phi$ from $\mathcal{A}$ to $\mathcal{L}(\mathcal{H})$. In this case the norm-closure $\mathcal{B}$ of $\phi(\mathcal{A})$ in $\mathcal{L}(\mathcal{H})$ is a C* subalgebra of $\mathcal{L}(\mathcal{H})$, and we say that $\phi$ is **nondegenerate** if $\mathcal{B}$ is nondegenerate, i.e., if there is no nonzero $v \in \mathcal{H}$ such that $\phi(x)v = 0$ for all $x \in \mathcal{A}$.

Suppose now that $\mathcal{A}$ is commutative, and that $\phi$ is a nondegenerate *-representation of $\mathcal{A}$ on $\mathcal{H}$. We first consider the case where $\mathcal{A}$ has a unit $e$. Then $\phi(e)$ is a unit for $\mathcal{B} = \overline{\phi(\mathcal{A})}$, so the nondegeneracy implies that $\phi(e) = I$. $\phi$ induces a continuous map $\phi^* : \sigma(\mathcal{B}) \to \sigma(\mathcal{A})$, namely $\phi^* h = h \circ \phi$. If $\phi^* h_1 = \phi^* h_2$ then $h_1$ and $h_2$ agree on $\phi(\mathcal{A})$ and hence everywhere, so $\phi^*$ is an injection. Since $\sigma(\mathcal{B})$ is a compact Hausdorff space, $\phi^*$ is a homeomorphism onto its range, which is a compact subset of $\sigma(\mathcal{A})$. Moreover, the spectral theorem associates to $\mathcal{B}$ a unique regular projection-valued measure $P_0$ on $\sigma(\mathcal{B})$ such that $T = \int \widehat{T} \, dP_0$ for all $T \in \mathcal{B}$. The map $\phi^*$ can be used to pull $P_0$ back to a projection-valued measure $P$ on $\sigma(\mathcal{A})$, namely $P(E) = P_0(\phi^{*-1}(E))$, and in view of the properties of $\phi^*$ stated above it is easy to check that $P$ is regular. Moreover, the Gelfand transforms on $\mathcal{A}$ and $\mathcal{B}$ are related by $(\phi(x))^\frown(h) = \widehat{x}(\phi^*h)$, so we have $\phi(x) = \int \widehat{x}(\phi^*h)\,dP_0(h) = \int \widehat{x}\,dP$ for all $x \in \mathcal{A}$.

Much the same thing works if $\mathcal{A}$ is nonunital. It may still happen that $I \in \mathcal{B}$, in which case the preceding discussion goes through with no change. If not, $\phi$ can be extended to a nondegenerate $*$-representation of the unital augmentation $\widetilde{\mathcal{A}}$ of $\mathcal{A}$ in the obvious way, $\phi(x,a) = \phi(x) + aI$, and the norm closure of $\phi(\widetilde{\mathcal{A}})$ is $\widetilde{\mathcal{B}} = \mathcal{B} \oplus \mathbb{C}I$. As above, we have the map $\phi^*h = h \circ \phi$ from $\sigma(\mathcal{B})$ to $\sigma(\mathcal{A})$, which extends continuously to a map from $\sigma(\widetilde{\mathcal{B}}) \cong \sigma(\mathcal{B}) \cup \{0\}$ to $\sigma(\widetilde{\mathcal{A}}) \cong \sigma(\mathcal{A}) \cup \{0\}$ by setting $\phi^*(0) = 0$. $\phi^*$ is a homeomorphism from $\sigma(\widetilde{\mathcal{B}})$ onto a compact subset of $\sigma(\widetilde{\mathcal{A}})$, and hence is a homeomorphism from $\sigma(\mathcal{B})$ onto a closed subset of $\sigma(\mathcal{A})$. By Theorem 1.53 there is a unique regular projection-valued measure $P_0$ on $\sigma(\mathcal{B})$ such that $T = \int \widehat{T}\,dP_0$ for all $T \in \mathcal{B}$, and just as above, this induces a regular projection-valued measure $P$ on $\sigma(\mathcal{A})$ such that $\phi(x) = \int \widehat{x}\,dP$ for all $x \in \mathcal{A}$. In short, we have:

**1.54 Theorem.** *Let $\mathcal{A}$ be a commutative Banach $*$-algebra, and let $\phi$ be a nondegenerate $*$-representation of $\mathcal{A}$ on $\mathcal{H}$. There is a unique regular projection-valued measure $P$ on $\sigma(\mathcal{A})$ such that $\phi(x) = \int \widehat{x}\,dP$ for all $x \in \mathcal{A}$. If $T \in \mathcal{L}(\mathcal{H})$, $T$ commutes with $\phi(x)$ for every $x \in \mathcal{A}$ if and only if $T$ commutes with $P(E)$ for every Borel $E \subset \sigma(\mathcal{A})$.*

In particular, suppose that $\mathcal{A} = C_0(S)$ where $S$ is a locally compact Hausdorff space. Then $\sigma(\mathcal{A})$ is naturally homeomorphic to $S$ in such a way that the Gelfand transform on $\mathcal{A}$ becomes the identity map. We then have:

**1.55 Corollary.** *Let $S$ be a locally compact Hausdorff space, and let $\phi$ be a nondegenerate $*$-representation of $C_0(S)$ on $\mathcal{H}$. There is a unique regular projection-valued measure $P$ on $S$ such that $\phi(f) = \int f\,dP$ for all $f \in C_0(S)$. If $T \in \mathcal{L}(\mathcal{H})$, $T$ commutes with $\phi(f)$ for every $f \in C_0(S)$ if and only if $T$ commutes with $P(E)$ for every Borel $E \subset S$.*

## 1.6  Von Neumann Algebras

A **von Neumann algebra** is a C\* algebra of operators on a Hilbert space that contains the identity and is closed in the *weak* operator topology. The theory of von Neumann algebras is a large subject; here we present only a few basic results that will be of use later on and discuss the connection with the spectral functional calculus.

One obvious way to obtain a von Neumann algebra is to start with any $*$-subalgebra of $\mathcal{L}(\mathcal{H})$ containing $I$ and take its weak closure. Another is as follows. If $\mathcal{S}$ is any subset of $\mathcal{L}(\mathcal{H})$, its **commutant** or **centralizer** is the set $\mathcal{S}'$ defined by

$$\mathcal{S}' = \{T \in \mathcal{L}(\mathcal{H}) : ST = TS \text{ for all } S \in \mathcal{S}\}.$$

It is simple to check that $\mathcal{S}'$ is a unital algebra, and it is weakly closed because if $T_\alpha \to T$ weakly then $ST_\alpha \to ST$ and $T_\alpha S \to TS$ weakly for any $S$. If, in addition, $\mathcal{S}$ is closed under taking adjoints, so is $\mathcal{S}'$, and hence $\mathcal{S}'$ is a von Neumann algebra. We may then go a step further and consider the **bicommutant** $\mathcal{S}'' = (\mathcal{S}')'$; this is a von Neumann algebra that contains the original set $\mathcal{S}$.

The following theorem is the most fundamental result about von Neumann algebras. To prepare for it, we need to make some remarks about direct sums that will also be useful elsewhere.

Let $\mathcal{H}$ be a Hilbert space, and let $J$ be an index set. For notational convenience we shall take $J$ to be either $\{1, 2, \ldots, N\}$ $(N > 1)$ or $\{1, 2, 3, \ldots\}$, but this is not really necessary. For each $i \in J$ let $\mathcal{H}_i$ be a copy of $\mathcal{H}$, and let $\mathcal{H}^J = \bigoplus_{i \in J} \mathcal{H}_i$. Then every $T \in \mathcal{L}(\mathcal{H}^J)$ can be represented as a matrix $[T_{ij}]$ of operators in $\mathcal{L}(\mathcal{H})$. That is, for $x \in \mathcal{H}$, $T_{ij}x$ is the $i$-th component of $T(0, \ldots, 0, x, 0, \ldots)$ where $x$ appears in the $j$th slot, so that for $\mathbf{x} = (x_1, x_2, \ldots) \in \mathcal{H}^J$, $(T\mathbf{x})_i = \sum_j T_{ij}x_j$.

Next, for $A \in \mathcal{L}(\mathcal{H})$, let $A^J$ be the direct sum of $J$ copies of $A$ acting on $\mathcal{H}^J$:

$$A^J(x_1, x_2, \ldots) = (Ax_1, Ax_2, \ldots).$$

Suppose $\mathcal{A}$ is a subset of $\mathcal{L}(\mathcal{H})$; we set $\mathcal{A}^J = \{A^J : A \in \mathcal{A}\}$. It is an easy exercise to check that an operator $T = [T_{ij}]$ on $\mathcal{H}^J$ belongs to $(\mathcal{A}^J)'$ if and only if every $T_{ij}$ belongs to $\mathcal{A}'$. It follows that if $S \in \mathcal{A}''$, then $S^J \in (\mathcal{A}^J)''$.

**1.56 Theorem** (The von Neumann Density Theorem). *Let $\mathcal{A}$ be a nondegenerate $*$-subalgebra of $\mathcal{L}(\mathcal{H})$. Then:*

a. *$\mathcal{A}$ is dense in $\mathcal{A}''$ in the strong operator topology.*
b. *The following conditions are equivalent:*
   i. *$\mathcal{A} = \mathcal{A}''$.*
   ii. *$\mathcal{A}$ is a von Neumann algebra.*
   iii. *$\mathcal{A}$ is closed in the strong operator topology.*

*Proof.* For (a), suppose $S \in \mathcal{A}''$. We must show that for any $\epsilon > 0$ and any $x_1, \ldots, x_N \in \mathcal{H}$ there exists $A \in \mathcal{A}$ such that $\sum \|Sx_i - Ax_i\|^2 < \epsilon^2$.

First, consider the case $N = 1$. Given $x \in \mathcal{H}$, let $\mathcal{X}$ be the closure of $\{Ax : A \in \mathcal{A}\}$, a closed subspace of $\mathcal{H}$. Since $\mathcal{X}$ is invariant under $\mathcal{A}$, it is easy to check that the orthogonal projection $P$ onto $\mathcal{X}$ lies in $\mathcal{A}'$. This implies, first, that for all $A \in \mathcal{A}$, $A(I - P)x = (I - P)Ax = 0$; since $\mathcal{A}$ is nondegenerate, we have $(I - P)x = 0$, that is, $x = Px$. But it also implies that $SP = PS$, so $Sx = SPx = PSx \in \mathcal{X}$. In other words, for any $\epsilon > 0$ there is an $A \in \mathcal{A}$ such that $\|Sx - Ax\| < \epsilon$.

Now for the case $N > 1$, we simply apply the result for $N = 1$ to the algebra $\mathcal{A}^J$ with $J = \{1, \ldots, N\}$. As we observed above, if $S \in \mathcal{A}''$ then $S^J \in (\mathcal{A}^J)''$, so given $\mathbf{x} = (x_1, \ldots, x_N) \in \mathcal{H}^J$ there is an $A^J \in \mathcal{A}^J$ with $\|S^J \mathbf{x} - A^J \mathbf{x}\| < \epsilon$; in other words, $\sum \|Sx_i - Ax_i\|^2 < \epsilon^2$, as required.

Part (b) is an immediate consequence. The implications (i) $\Rightarrow$ (ii) and (ii) $\Rightarrow$ (iii) are trivial, and (iii) $\Rightarrow$ (i) by part (a). $\qquad\square$

Commutative von Neumann algebras are intimately connected with the spectral functional calculus developed earlier in this chapter: In the notation of §1.4, the final assertion of Spectral Theorem I (1.44) says that if $\mathcal{A}$ is a commutative C* subalgebra of $\mathcal{L}(\mathcal{H})$ with spectrum $\Sigma$, the operators $T_f = \int f \, dP$, $f \in B(\Sigma)$, all belong to $\mathcal{A}''$, which by Theorem 1.56 is the von Neumann algebra generated by $\mathcal{A}$.

This has the consequence that von Neumann algebras are rich in orthogonal projections. Indeed, any von Neumann algebra contains all the spectral projections of each of its self-adjoint elements. This contrasts strongly with the situation for C* algebras, which may contain no nontrivial projections at all. This is the case, for example, for any commutative C* algebra whose spectrum is connected, as follows easily from the Gelfand-Naimark theorem.

When $\mathcal{H}$ is separable, the connection between the spectral functional calculus for a C* algebra $\mathcal{A} \subset \mathcal{L}(\mathcal{H})$ and the von Neumann algebra $\mathcal{A}''$ goes both ways:

**1.57 Theorem.** *Suppose $\mathcal{H}$ is a separable Hilbert space, $\mathcal{A}$ is a commutative C* subalgebra of $\mathcal{L}(\mathcal{H})$ with spectrum $\Sigma$, and $P$ is the associated projection-valued measure on $\Sigma$. Then the von Neumann algebra generated by $\mathcal{A}$ consists precisely of the operators $T_f = \int f \, dP$ as $f$ ranges over the space $B(\Sigma)$ of bounded Borel functions on $\Sigma$.*

The proof that every $S \in \mathcal{A}''$ is of the form $T_f$ is easy when $\mathcal{A}$ has a cyclic vector $v$. In this case, by Spectral Theorem II (1.47) (and its proof), we may assume that $\mathcal{H} = L^2(\Sigma, \mu)$ where $\mu = \mu_{v,v}$ is a finite Borel measure on $\Sigma$, so that for $f \in B(\Sigma)$, $T_f$ is simply the operator of multiplication by $f$. Since $\mu$ is finite, the constant function 1 is in $L^2(\mu)$; let $\phi = S1 \in L^2(\mu)$. Since $S \in \mathcal{A}''$ and $\mathcal{A} \subset \mathcal{A}'$, for any $f \in B(\Sigma)$ we have $Sf = S(T_f 1) = T_f S1 = f\phi$. Since $S$ is bounded on $L^2(\mu)$ we must have $\phi \in L^\infty(\mu)$, so we can take $\phi \in B(\Sigma)$; and since $B(\Sigma)$ is dense in $L^2(\mu)$, it follows that $S = T_\phi$.

In the general case, $\mathcal{H}$ is the direct sum of cyclic subspaces $\mathcal{H}_i$, and for each $i$ we can assume that $\mathcal{H}_i = L^2(\mu_i)$; the preceding argument yields a function $\phi_i$ on $\Sigma$ so that $S|\mathcal{H}_i$ is multiplication by $\phi_i$. What must be shown is that one can take the $\phi_i$'s all to be the same function $\phi$ so that $S = T_\phi$. Since each one is determined only a.e. with respect to $\mu_i$, and the $\mu_i$'s may have different sets of measure zero, this is not entirely straightforward, and indeed not always true when $\mathcal{H}$ is inseparable so that uncountably many cyclic subspaces are needed. We refer the reader to Dixmier [31, §I.7] or Pedersen [111, Corollary 2.8.8]; also Riesz and Sz.-Nagy [120, §129] for the case where $\mathcal{A}$ is generated by one self-adjoint operator.

In the setting of Theorem 1.57, it follows that the von Neumann algebra $\mathcal{A}''$ is isometrically $*$-isomorphic to the algebra $L^\infty(\Sigma, P)$. (Since the $L^\infty$ space of a Borel measure $\mu$ depends on $\mu$ only to the extent that $\mu$ determines which sets have measure zero, it doesn't matter that $P$ is projection-valued rather than scalar-valued.) This much carries over to the inseparable case: every commutative von Neumann algebra is isometrically $*$-isomorphic to an $L^\infty$ space. See Dixmier [31, §I.7].

Incidentally, suppose the algebra $\mathcal{A}$ in Theorem 1.57 is a von Neumann algebra to begin with. Then $\mathcal{A} = \mathcal{A}''$, so on the one hand, $\mathcal{A} \cong C(\Sigma)$, and on the other, $\mathcal{A} \cong L^\infty(\Sigma, P)$, so every bounded Borel function on $\Sigma$ — and in particular, the characteristic function of every Borel set — agrees $P$-a.e. with a continuous function! Since the support of $P$ is all of $\Sigma$, it follows that the topology on $\Sigma$ is extremely strong and, if not actually discrete, quite bizarre. The reader might enjoy exploring the pathologies of the case $\mathcal{A} = L^\infty([0,1])$, acting on $L^2([0,1])$ (with Lebesgue measure) in the obvious way.

At the other extreme from the commutative von Neumann algebras (which satisfy $\mathcal{A} \subset \mathcal{A}'$) are the von Neumann algebras whose center is trivial, that is, which satisfy $\mathcal{A} \cap \mathcal{A}' = \mathbb{C}I$. Such algebras are called **factors**. Note that since the condition $\mathcal{A} \cap \mathcal{A}' = \mathbb{C}I$ is symmetric in $\mathcal{A}$ and $\mathcal{A}'$, $\mathcal{A}$ is a factor if and only if $\mathcal{A}'$ is a factor. The most obvious example of a factor in $\mathcal{L}(\mathcal{H})$ is $\mathcal{L}(\mathcal{H})$ itself. (The verification that $\mathcal{L}(\mathcal{H})' = \mathbb{C}I$ is an easy exercise.)

More generally, with the notation introduced in the remarks preceding Theorem 1.56, for any index set $J$, $\mathcal{L}(\mathcal{H})^J$ is a factor on the direct sum $\mathcal{H}^J$ of copies of $\mathcal{H}$. Indeed, as we observed before Theorem 1.56, an operator $T = [T_{ij}]$ on $\mathcal{H}^J$ belongs to $(\mathcal{L}(\mathcal{H})^J)'$ if and only if $T_{ij} \in \mathcal{L}(\mathcal{H})'$ — that is, $T_{ij} = t_{ij}I$ for some $t_{ij} \in \mathbb{C}$ — for all $i$ and $j$. On the other hand, $T \in \mathcal{L}(\mathcal{H})^J$ if and only if $T_{ij} = \delta_{ij}T_0$ for some $T_0 \in \mathcal{L}(\mathcal{H})$. These two conditions are simultaneously satisfied only when $T$ is a scalar multiple of the identity on $\mathcal{H}^J$.

This situation can be neatly described in the language of tensor products. As discussed in Appendix 3, the direct sum $\mathcal{H}^J$ can be naturally identified with $\mathcal{H} \otimes l^2(J)$ (Proposition A.18). The algebra $\mathcal{L}(\mathcal{H})^J$ then becomes $\mathcal{L}(\mathcal{H}) \otimes I = \{T \otimes I : T \in \mathcal{L}(\mathcal{H})\}$, and its commutant becomes $I \otimes \mathcal{L}(l^2(J)) = \{I \otimes S : S \in \mathcal{L}(l^2(J))\}$ since linear operators on $l^2(J)$ can be identified with matrices $[t_{ij}]$ with scalar entries. This result is presented more formally in Theorem 7.11.

Factors are classified as type I, II, or III depending on the structure of their lattice of orthogonal projections. To describe this situation in detail would take us too far outside the scope of this book; we shall encounter no factors of type III in the sequel, and factors of type II make only a fleeting appearance in Chapter 7. However, we shall say a few things about factors of type I.

There are several equivalent definitions of factors of type I, among which the following one will be most convenient for us: a factor $\mathcal{A} \subset \mathcal{L}(\mathcal{H})$ is of **type I** if it is isometrically $*$-isomorphic to $\mathcal{L}(\mathcal{H}_1)$ for some (possibly different) Hilbert space $\mathcal{H}_1$. For example, the algebra $\mathcal{L}(\mathcal{H})^J$ is a factor of type I on $\mathcal{H}^J$, for the map $A \to A^J$ is an isomorphism from $\mathcal{L}(\mathcal{H})$ to $\mathcal{L}(\mathcal{H})^J$. In fact, these are essentially the only examples; we state the result in terms of tensor products.

**1.58 Theorem.** *If $\mathcal{A}$ is a factor of type I on $\mathcal{H}$, isomorphic to $\mathcal{L}(\mathcal{H}_1)$, there is another Hilbert space $\mathcal{H}_2$ and a unitary map $U : \mathcal{H} \to \mathcal{H}_1 \otimes \mathcal{H}_2$ such that $U\mathcal{A}U^{-1} = \mathcal{L}(\mathcal{H}_1) \otimes I$ and $U\mathcal{A}'U^{-1} = I \otimes \mathcal{L}(\mathcal{H}_2)$.*

A proof may be found in Dixmier [31, §I.8.2, Corollary 3].

**1.59 Corollary.** *A factor $\mathcal{A}$ is of type I if and only if $\mathcal{A}'$ is of type I.*

## 1.7  Notes and References

For more extensive treatments of Banach algebras and spectral theory, we refer the reader to Rudin [123], Reed and Simon [116], Dixmier [32], Rickart [118], Loomis [84], Pedersen [111], and Dunford and Schwartz [34]. The latter book is a good source for historical references.

Banach algebras are sometimes called normed rings in the older literature, and C* algebras are sometimes called B* algebras. (The latter two names originally referred to distinct concepts, which were eventually proved to be essentially identical. The distinction between them has not been found to be worth preserving, and the name "C*" has won out. See Rickart [118, p. 248].)

The Gelfand-Naimark theorem is the definitive structure theorem for commutative C* algebras. There is also a structure theorem for general C* algebras:

**1.60 Theorem.** *Every $C^*$ algebra is isometrically $*$-isomorphic to a $C^*$ subalgebra of the algebra of bounded operators on some Hilbert space.*

The proof of this theorem — the so-called Gelfand-Naimark-Segal construction — can be found in Rudin [123] or Pedersen [111]. It consists of showing that every $C^*$ algebra $\mathcal{A}$ has a large supply of positive linear functionals (a bounded linear functional $\phi$ on a Banach $*$-algebra is **positive** if $\phi(x^*x) \geq 0$ for all $x$), and then using the positive functionals to construct $*$-representations of $\mathcal{A}$ in much the same way that we shall use functions of positive type to construct representations of a group in §3.3.

Halmos [58] defines a Borel projection-valued measure $P$ on a locally compact Hausdorff space $\Omega$ to be regular if, for any Borel set $E \subset \Omega$, $P(E)$ is the projection onto the closed linear span of the ranges of $P(K)$ as $K$ varies over compact subsets of $E$. It is a fairly simple exercise to see that this condition holds if and only if the measures $P_{v,v}$ are all inner regular, i.e., $P_{v,v}(E) = \sup_{K \subset \subset E} P_{v,v}(K)$. But inner and outer regularity are equivalent for finite measures, and the complex measures $P_{u,v}$ are obtained from the $P_{v,v}$ by polarization, so Halmos's definition of regularity for projection-valued measures is equivalent to ours.

Despite the non-uniqueness of $\Omega$, $\mu$, and $U$ in Spectral Theorem II, there is a canonical form for these objects, given by the theory of spectral multiplicity. See Halmos [58] or Nelson [109]; the latter treatment is particularly elegant.

In Theorem 1.51(c), the hypothesis that $S$ commutes with $T$ and $T^*$ can be replaced by the apparently weaker hypothesis that $S$ merely commutes with $T$. This is the Fuglede commutativity theorem; two quite different but equally entertaining proofs can be found in Rudin [123] and Halmos [59].

The classic reference for von Neumann algebras is Dixmier [31]. The theory has expanded considerably since it was written (the original French version was published in 1969), but it is still a very good source for the topics it contains. More recent books such as Pedersen [111], Takesaki [134], and Blackadar [12] include accounts of newer developments.

# 2

## Locally Compact Groups

This chapter contains the basic theory of the objects on which harmonic analysis is performed, namely the locally compact topological groups and their homogeneous spaces. The fundamental features, without which little else is possible, are the existence and uniqueness of a translation-invariant measure $\lambda$ on any locally compact group and the endowment of $L^1(\lambda)$ with the structure of a Banach $*$-algebra.

## 2.1  Topological Groups

A **topological group** is a group $G$ equipped with a topology with respect to which the group operations are continuous; that is, $(x, y) \mapsto xy$ is continuous from $G \times G$ to $G$ and $x \to x^{-1}$ is continuous from $G$ to $G$.

If $G$ is a topological group, we shall denote the unit element of $G$ by 1. If $A \subset G$ and $x \in G$, we define

$$Ax = \{yx : y \in A\}, \qquad xA = \{xy : y \in A\}, \qquad A^{-1} = \{y^{-1} : y \in A\},$$

and if also $B \subset G$, we define

$$AB = \{xy : x \in A, y \in B\}.$$

(Note: we shall refrain from writing $A^2$ for $AA$, as $A^2$ might equally denote $\{x^2 : x \in A\}$, which is in general a proper subset of $AA$.) We say that $A$ is **symmetric** if $A = A^{-1}$. It is a useful observation that $A \cap B = \varnothing$ if and only if $1 \notin A^{-1}B$.

The following proposition lists several basic properties of topological groups that we shall use, often without comment, in the sequel.

**2.1 Proposition.** *Let $G$ be a topological group.*
 a. *The topology of $G$ is invariant under translations and inversion; that is, if $U$ is open then so are $xU$, $Ux$, and $U^{-1}$ for any $x \in G$. Moreover, if $U$ is open then so are $AU$ and $UA$ for any $A \subset G$.*

b. *For every neighborhood $U$ of 1 there is a symmetric neighborhood $V$ of 1 such that $VV \subset U$.*
c. *If $H$ is a subgroup of $G$, so is $\overline{H}$.*
d. *Every open subgroup of $G$ is closed.*
e. *If $A$ and $B$ are compact sets in $G$, so is $AB$.*

*Proof.* (a) The first assertion is equivalent to the separate continuity of the map $(x, y) \to xy$ and the continuity of the map $x \to x^{-1}$. The second one follows since $AU = \bigcup_{x \in A} xU$ and $UA = \bigcup_{x \in A} Ux$.

(b) Continuity of $(x, y) \to xy$ at 1 means that for every neighborhood $U$ of 1 there are neighborhoods $W_1, W_2$ of 1 with $W_1 W_2 \subset U$. The desired set $V$ can be taken to be $W_1 \cap W_2 \cap W_1^{-1} \cap W_2^{-1}$.

(c) If $x, y \in \overline{H}$ there are nets $\{x_\alpha\}$, $\{y_\beta\}$ in $H$ converging to $x, y$. Then $x_\alpha y_\beta \to xy$ and $x_\alpha^{-1} \to x^{-1}$, so $xy$ and $x^{-1}$ are in $\overline{H}$.

(d) If $H$ is open, so are all its cosets $xH$; its complement $G \setminus H$ is the union of all these cosets except $H$ itself; hence $G \setminus H$ is open and $H$ is closed.

(e) $AB$ is the image of the compact set $A \times B$ under the continuous map $(x, y) \to xy$, hence is compact. $\qquad\square$

Suppose $H$ is a subgroup of the topological group $G$. Let $G/H$ be the space of left cosets of $H$, and let $q : G \to G/H$ be the canonical quotient map. We impose the quotient topology on $G/H$; that is, $U \subset G/H$ is open if and only if $q^{-1}(U)$ is open in $G$. $q$ maps open sets in $G$ to open sets in $G/H$, for if $V$ is open in $G$ then $q^{-1}(q(V)) = VH$ is also open by Proposition 2.1(a); hence $q(V)$ is open.

**2.2 Proposition.** *Suppose $H$ is a subgroup of the topological group $G$.*
a. *If $H$ is closed, $G/H$ is Hausdorff.*
b. *If $G$ is locally compact, so is $G/H$.*
c. *If $H$ is normal, $G/H$ is a topological group.*

*Proof.* (a) Suppose $\overline{x} = q(x)$, $\overline{y} = q(y)$ are distinct points of $G/H$. If $H$ is closed, $xHy^{-1}$ is a closed set that does not contain 1, so by Proposition 2.1(b) there is a symmetric neighborhood $U$ of 1 with $UU \cap xHy^{-1} = \emptyset$. Since $U = U^{-1}$ and $H = HH$, $1 \notin UxH(Uy)^{-1} = (UxH)(UyH)^{-1}$, so $(UxH) \cap (UyH) = \emptyset$. Thus $q(Ux)$ and $q(Uy)$ are disjoint neighborhoods of $\overline{x}$ and $\overline{y}$.

(b) If $U$ is a compact neighborhood of 1 in $G$, $q(Ux)$ is a compact neighborhood of $q(x)$ in $G/H$.

(c) If $x, y \in G$ and $U$ is a neighborhood of $q(xy)$ in $G/H$, the continuity of multiplication in $G$ at $(x, y)$ implies that there are neighborhoods

$V, W$ of $x, y$ such that $VW \subset q^{-1}(U)$. Then $q(V)$ and $q(W)$ are neighborhoods of $q(x)$ and $q(y)$ such that $q(V)q(W) \subset U$, so multiplication is continuous on $G/H$. Similarly, inversion is continuous. $\qquad \square$

**2.3 Corollary.** *If $G$ is $T_1$ then $G$ is Hausdorff. If $G$ is not $T_1$ then $\overline{\{1\}}$ is a closed normal subgroup, and $G/\overline{\{1\}}$ is a Hausdorff topological group.*

*Proof.* The first assertion follows by taking $H = \{1\}$ in Proposition 2.2(a). $\overline{\{1\}}$ is a subgroup by Proposition 2.1(c); it is clearly the smallest closed subgroup of $G$. It is therefore normal, for otherwise one would obtain a smaller closed subgroup by intersecting it with one of its conjugates. The second assertion therefore follows from Proposition 2.2(a,c) by taking $H = \overline{\{1\}}$. $\qquad \square$

In view of Corollary 2.3, it is essentially no restriction to assume that a topological group is Hausdorff (simply work with $G/\overline{\{1\}}$ instead of $G$), and we do so henceforth. In particular, by a **locally compact group** we shall mean a topological group whose topology is locally compact *and Hausdorff.*

**2.4 Proposition.** *Every locally compact group $G$ has a subgroup $G_0$ that is open, closed, and $\sigma$-compact.*

*Proof.* Let $U$ be a symmetric compact neighborhood of 1. Let $U_n = UU \cdots U$ ($n$ factors), and let $G_0 = \bigcup_1^\infty U_n$. Then $G_0$ is a group (namely the group generated by $U$); it is open since $U_{n+1}$ is a neighborhood of $U_n$ for all $n$, and hence closed by Proposition 2.1(d). Moreover, each $U_n$ is compact by Proposition 2.1(e); hence $G_0$ is $\sigma$-compact. $\qquad \square$

With the notation of Proposition 2.4, $G$ is the disjoint union of the cosets of $G_0$, each of which is closed and open in $G$ and homeomorphic to $G_0$. Hence, from a topological point of view, $G$ is just a disjoint union of copies of a $\sigma$-compact space. In particular, if $G$ is connected then $G$ is $\sigma$-compact.

If $f$ is a function on the topological group $G$ and $y \in G$, we define the left and right translates of $f$ through $y$ by

$$(2.5) \qquad L_y f(x) = f(y^{-1}x), \qquad R_y f(x) = f(xy).$$

The reason for using $y^{-1}$ in $L_y$ and $y$ in $R_y$ is to make the maps $y \to L_y$ and $y \to R_y$ group homomorphisms:

$$L_{yz} = L_y L_z, \qquad R_{yz} = R_y R_z.$$

We shall say that $f$ is **left** (resp. **right**) **uniformly continuous** if $\|L_y f - f\|_{\sup} \to 0$ (resp. $\|R_y f - f\|_{\sup} \to 0$) as $y \to 1$.

*Warning*: Some authors reverse the role of the words "left" and "right" in these definitions. The justification is that according to our definition, *right* uniform continuity of $f$ has to do with the behavior of $f(xy) - f(x)$ when $y$ is in a neighborhood $U$ of the identity; but the latter condition means that $xy$ is in the neighborhood $xU$ of $x$ obtained by *left* translation of $U$. Our convention agrees with Weil [146] but not with Bourbaki [18], Hewitt and Ross [66], [67], or Reiter [117]. Fortunately, this issue will arise only in a couple of places.

**2.6 Proposition.** *If $f \in C_c(G)$ then $f$ is left and right uniformly continuous.*

*Proof.* We give the proof for $R_y f$; the argument for $L_y f$ is similar. Given $f \in C_c(G)$ and $\epsilon > 0$, let $K = \operatorname{supp} f$. For every $x \in K$ there is a neighborhood $U_x$ of 1 such that $|f(xy) - f(x)| < \frac{1}{2}\epsilon$ for $y \in U_x$, and there is a symmetric neighborhood $V_x$ of 1 such that $V_x V_x \subset U_x$. The sets $xV_x$ ($x \in K$) cover $K$, so there exist $x_1, \ldots x_n \in K$ such that $K \subset \bigcup_1^n x_j V_{x_j}$. Let $V = \bigcap_1^n V_{x_j}$; we claim that $\|R_y f - f\|_{\sup} < \epsilon$ for $y \in V$.

If $x \in K$ then there is some $j$ for which $x_j^{-1} x \in V_{x_j}$, so that $xy = x_j(x_j^{-1} x) y \in x_j U_{x_j}$. But then

$$|f(xy) - f(x)| \le |f(xy) - f(x_j)| + |f(x_j) - f(x)| < \tfrac{1}{2}\epsilon + \tfrac{1}{2}\epsilon = \epsilon.$$

Similarly, if $xy \in K$ then $|f(xy) - f(x)| < \epsilon$. But if $x$ and $xy$ are not in $K$ then $f(x) = f(xy) = 0$, so we are done. $\square$

The locally compact groups that arise most frequently in practice are the (finite-dimensional) Lie groups. Examples include the additive group $\mathbb{R}^n$ and all closed subgroups of the group $GL(n, \mathbb{R})$ of invertible linear transformations of $\mathbb{R}^n$. The group $\mathbb{T}$ of complex numbers of modulus one,

$$\mathbb{T} = \{ z \in \mathbb{C} : |z| = 1 \},$$

will play a particularly important role for us. It is isomorphic to $\mathbb{R}/\mathbb{Z}$ or $\mathbb{R}/2\pi\mathbb{Z}$, and we shall often not distinguish between $\mathbb{T}$ and $\mathbb{R}/2\pi\mathbb{Z}$.

In addition, there are a few classes of non-Lie groups that are of considerable importance. One such class is the class of compact groups obtained by taking infinite products of compact Lie groups or finite groups. For example, the product of a countable number of copies of the 2-element group $Z_2$ turns up in several contexts. In probability theory, it is the sample space for the simplest sort of stochastic process, an infinite sequence of coin tosses. In Fourier analysis, it is the basis for the theory of Walsh functions on the unit interval, as we shall see later.

The other important class of examples is the local fields and the matrix groups associated to them. Here we shall work out the basic example, the field $\mathbb{Q}_p$ of $p$-adic numbers.

Fix a prime $p$. By the unique factorization theorem, any nonzero rational number $r$ can be written uniquely as $r = p^m q$ where $m \in \mathbb{Z}$ and $q$ is a rational number whose numerator and denominator are not divisible by $p$. We define the $p$-**adic norm** of $r$, denoted by $|r|_p$, to be $p^{-m}$, and we set $|0|_p = 0$. The $p$-adic norm clearly satisfies

$$(2.7) \qquad |r_1 + r_2|_p \leq \max(|r_1|_p, |r_2|_p), \qquad |r_1 r_2|_p = |r_1|_p |r_2|_p.$$

It follows that the $p$-adic distance function $d_p(r_1, r_2) = |r_1 - r_2|_p$ is a metric on $\mathbb{Q}$ with respect to which the arithmetic operations are continuous. These operations therefore extend to the completion of $\mathbb{Q}$ with respect to the metric $d_p$, yielding a field that is called the field of $p$-**adic numbers** and is denoted by $\mathbb{Q}_p$. $\mathbb{Q}_p$ can be described more concretely as follows.

**2.8 Proposition.** *If $m \in \mathbb{Z}$ and $c_j \in \{0, 1, \ldots, p-1\}$ for $j \geq m$, the series $\sum_m^\infty c_j p^j$ converges in $\mathbb{Q}_p$. Moreover, every $p$-adic number is the sum of such a series.*

*Proof.* The first assertion is easy: the partial sums of the series $\sum_m^\infty c_j p^j$ are Cauchy, because $|\sum_M^N c_j p^j|_p \leq p^{-M} \to 0$ as $M \to \infty$. To prove the second one, we observe that the series $\sum_m^\infty c_j p^j$ looks just like the decimal expansion of a positive real number in base $p$ except that there are only finitely many digits to the *right* of the decimal point and perhaps infinitely many to the *left*. (If the sum is finite, it *is* the base-$p$ decimal expansion of a rational number.) The usual algorithms one learns in grade school for adding, multiplying, and dividing positive real numbers in decimal form work equally well here, with the obvious modifications. In the present situation, however, subtraction is also possible: if $c_m \neq 0$,

$$- \sum_m^\infty c_j p^j = (p - c_m) p^m + \sum_{m+1}^\infty (p - 1 - c_j) p^j.$$

It follows that the set of series $\sum_m^\infty c_j p^j$ forms a field. To see that it is all of $\mathbb{Q}_p$, we need only show that it is complete.

Suppose then that $x_n = \sum c_{jn} p^j$ and $\{x_n\}$ is Cauchy. Then for each $k$ there exists $N$ such that $|x_{n_1} - x_{n_2}|_p < p^{-k}$ if $n_1, n_2 > N$. But since $|\sum c_j p^j|_p = p^{-m}$, where $m$ is the smallest integer for which $c_m \neq 0$, this can only happen if $c_{jn_1} = c_{jn_2}$ for $j \leq k$. Thus the sequences $\{c_{jn}\}$ are eventually constant in $n$ for each $j$. In other words, $c_j = \lim_{n \to \infty} c_{jn}$ exists for every $j$, and then $\lim x_n = \sum c_j p^j$. $\qquad \square$

For $r \geq 0$ and $x \in \mathbb{Q}_p$, consider the closed ball

$$\overline{B}(r,x) = \{y \in \mathbb{Q}_p : |x - y|_p \leq r\}.$$

Since the norm $|\cdot|_p$ takes on only the values $p^k$, $k \in \mathbb{Z}$, and 0, for any $r > 0$ there is some $\epsilon > 0$ for which the condition $|x - y|_p \leq r$ is equivalent to $|x - y|_p < r + \epsilon$. Hence the balls $\overline{B}(r,x)$, $r > 0$, are both open and closed. (It follows that $\mathbb{Q}_p$ is totally disconnected but has no isolated points: it is a Cantor set.) Moreover, by (2.7), if $|x - y|_p \leq r$ and $|y - z|_p \leq r$ then $|x - z|_p \leq r$, so any point of a ball is a "center" of that ball, and if two balls intersect then one is contained in the other.

By (2.7) again, $\overline{B}(r,0)$ is an additive subgroup of $\mathbb{Q}_p$ for any $r \geq 0$, and for $r \leq 1$ it is a subring. $\overline{B}(1,0)$ is called the ring of $p$-**adic integers** and is denoted by $\mathbb{Z}_p$; it is the completion of the ordinary integers in the $p$-adic metric. $p\mathbb{Z}_p = \overline{B}(p^{-1},0)$ is a subgroup of $\mathbb{Z}_p$, and the quotient is the cyclic group $Z_p$ of order $p$. In particular, $\mathbb{Z}_p$ is the disjoint union of $p$ cosets of $p\mathbb{Z}_p$, i.e., of $p$ balls of radius $p^{-1}$. It follows easily that if $m > n \in \mathbb{Z}$, every ball $\overline{B}(p^m,x)$ is the disjoint union of $p^{m-n}$ balls of radius $p^n$. In particular, these balls are totally bounded, so since they are closed in the complete space $\mathbb{Q}_p$, they are compact. We have therefore proved that $\mathbb{Q}_p$ is a locally compact group under addition, and $\mathbb{Q}_p \setminus \{0\}$ is a locally compact group under multiplication.

------

## 2.2   Haar Measure

Let $G$ be a locally compact group. We recall that $C_c(G)$ is the space of compactly supported continuous functions on $G$, and we set

$$C_c^+(G) = \{f \in C_c(G) : f \geq 0 \text{ and } f \not\equiv 0\}.$$

Since the positive and negative parts of a real continuous function are continuous, the linear span of $C_c^+(G)$ is $C_c(G)$.

A **left** (resp. **right**) **Haar measure** on $G$ is a nonzero Radon measure $\mu$ on $G$ that satisfies $\mu(xE) = \mu(E)$ (resp. $\mu(Ex) = \mu(E)$) for every Borel set $E \subset G$ and every $x \in G$.

**2.9 Proposition.** *Let $\mu$ be a Radon measure on the locally compact group $G$, and let $\widetilde{\mu}(E) = \mu(E^{-1})$.*
*a. $\mu$ is a left Haar measure if and only if $\widetilde{\mu}$ is a right Haar measure.*
*b. $\mu$ is a left Haar measure if and only if $\int L_y f \, d\mu = \int f \, d\mu$ for every $f \in C_c^+(G)$ and every $y \in G$.*

*Proof.* (a) is obvious. As for (b), for any Radon measure $\mu$ one has $\int L_y f \, d\mu = \int f \, d\mu_y$ for all $f$ where $\mu_y(E) = \mu(yE)$, as one sees by approximating $f$ by simple functions. Thus if $\mu$ is a Haar measure then $\int L_y f \, d\mu = \int f \, d\mu$, and if the latter condition holds for all $f \in C_c^+(G)$ and hence all $f \in C_c(G)$, one has $\mu = \mu_y$ by the uniqueness in the Riesz representation theorem. □

In view of Proposition 2.9(a), it is of little importance whether one chooses to study left or right Haar measure. The more common choice, to which we shall adhere, is the left one. The first order of business is to establish the existence and uniqueness of Haar measure. The following existence theorem is of fundamental theoretical importance; however, one must admit that for most specific groups that arise in practice one can construct Haar measure in a simpler and more explicit fashion. We shall discuss a number of classes of examples later in this section.

**2.10 Theorem.** *Every locally compact group $G$ possesses a left Haar measure $\lambda$.*

*Proof.* Taking our cue from Proposition 2.9(b), we shall construct $\lambda$ as a linear functional on $C_c(G)$. The idea is as follows. Imagine a function $\phi \in C_c^+(G)$ that is bounded by 1, equals 1 on a small open set, and is supported in a very slightly larger open set $U$. If $f \in C_c^+(G)$ is sufficiently slowly varying so that it is essentially constant on the left translates of $U$, $f$ can be well approximated by a linear combination of left translates of $\phi$: $f \approx \sum c_j L_{x_j} \phi$. If $\lambda$ were a left Haar measure on $G$, we would then have $\int f \, d\lambda \approx (\sum c_j) \int \phi \, d\lambda$. This approximation will get better and better as the support of $\phi$ shrinks to a point, and if we introduce a normalization to cancel out the factor of $\int \phi \, d\lambda$ on the right we will obtain $\int f \, d\lambda$ as a limit of the sums $\sum c_j$. (Think of the case $G = \mathbb{R}$: $\phi$ is essentially the characteristic function of a small interval, $f \approx \sum c_j L_{x_j} \phi$ is essentially an approximation of $f$ by step functions, and $(\sum c_j) \int \phi \, d\lambda$ is essentially a Riemann sum for $\int f \, d\lambda$.)

We now make this precise. If $f, \phi \in C_c^+(G)$, we define $(f : \phi)$ to be the infimum of all finite sums $\sum_1^n c_j$ such that $f \leq \sum_1^n c_j L_{x_j} \phi$ for some $x_1, \ldots, x_n \in G$. (This makes sense because the support of $f$ can be covered by some finite number $N$ of left translates of the set where $\phi > \frac{1}{2} \|\phi\|_{\sup}$, and it follows that $(f : \phi) \leq 2N \|f\|_{\sup} / \|\phi\|_{\sup}$.) The quantity $(f : \phi)$ has the following properties:

(2.11)          $(f : \phi) = (L_y f : \phi)$ for any $y \in G$.

(2.12)          $(f_1 + f_2 : \phi) \leq (f_1 : \phi) + (f_2 : \phi)$.

(2.13)          $(cf : \phi) = c(f : \phi)$ for any $c > 0$.

(2.14)          $(f_1 : \phi) \leq (f_2 : \phi)$ whenever $f_1 \leq f_2$.

(2.15)          $(f : \phi) \geq \|f\|_{\sup}/\|\phi\|_{\sup}$.

(2.16)          $(f : \phi) \leq (f : \psi)(\psi : \phi)$ for any $\psi \in C_c^+(G)$.

All of these are obvious with a moment's thought except perhaps the last one, which follows from the observation that if $f \leq \sum_i c_i L_{x_i} \psi$ and $\psi \leq \sum_j b_j L_{y_j} \phi$ then $f \leq \sum_{i,j} c_i b_j L_{x_i y_j} \phi$.

We now make a normalization by fixing an $f_0 \in C_c^+(G)$ and defining

$$I_\phi(f) = \frac{(f : \phi)}{(f_0 : \phi)} \qquad (f, \phi \in C_c^+(G)).$$

By (2.11)–(2.14), for each $\phi$ the functional $I_\phi$ is left-invariant, subadditive, homogeneous of degree 1, and monotone. Moreover, by (2.16), it satisfies

(2.17)          $(f_0 : f)^{-1} \leq I_\phi(f) \leq (f : f_0)$.

If only $I_\phi$ were additive rather than subadditive, it would be the restriction to $C_c^+(G)$ of a left-invariant positive linear functional on $C_c(G)$ and we would be in business. But this is not far from being the case:

**2.18 Lemma.** *If $f_1, f_2 \in C_c^+(G)$ and $\epsilon > 0$, there is a neighborhood $V$ of $1$ in $G$ such that $I_\phi(f_1) + I_\phi(f_2) \leq I_\phi(f_1 + f_2) + \epsilon$ whenever* $\operatorname{supp}(\phi) \subset V$.

*Proof.* Fix $g \in C_c^+(G)$ such that $g = 1$ on $\operatorname{supp}(f_1 + f_2)$ and let $\delta$ be a positive number (to be determined later). Let $h = f_1 + f_2 + \delta g$ and $h_i = f_i/h$ $(i = 1, 2)$, with the understanding that $h_i = 0$ wherever $f_i = 0$. Then $h_i \in C_c^+(G)$, so by Proposition 2.6 there is a neighborhood $V$ of $1$ in $G$ such that $|h_i(x) - h_i(y)| < \delta$ for $i = 1, 2$ and $y^{-1}x \in V$. Suppose $\phi \in C_c^+(G)$ and $\operatorname{supp}(\phi) \subset V$. If $h \leq \sum c_j L_{x_j} \phi$ then

$$f_i(x) = h(x)h_i(x) \leq \sum c_j \phi(x_j^{-1} x) h_i(x) \leq \sum c_j \phi(x_j^{-1} x)[h_i(x_j) + \delta],$$

because $|h_i(x) - h_i(x_j)| < \delta$ whenever $x_j^{-1} x \in \operatorname{supp}(\phi)$. Since $h_1 + h_2 \leq 1$, this gives

$$(f_1 : \phi) + (f_2 : \phi) \leq \sum c_j [h_1(x_j) + \delta] + \sum c_j [h_2(x_j) + \delta] \leq \sum c_j [1 + 2\delta].$$

Taking the infimum of all such sums $\sum c_j$, by (2.12) and (2.13) we obtain

$$I_\phi(f_1) + I_\phi(f_2) \le (1 + 2\delta)I_\phi(h) \le (1 + 2\delta)[I_\phi(f_1 + f_2) + \delta I_\phi(g)].$$

By (2.17), then, we can reach the desired conclusion by taking $\delta$ small enough so that

$$2\delta(f_1 + f_2 : f_0) + \delta(1 + 2\delta)(g : f_0) < \epsilon. \qquad \square$$

Now we can complete the proof of Theorem 2.10. For each $f \in C_c^+(G)$ let $X_f$ denote the interval $[(f_0 : f)^{-1}, (f : f_0)]$, and let $X$ be the Cartesian product of all the $X_f$'s. $X$ is a compact Hausdorff space consisting of all functions from $C_c^+(G)$ into $(0, \infty)$ whose value at $f$ lies in $X_f$; hence, by (2.17), $I_\phi \in X$ for all $\phi \in C_c^+(G)$. For each neighborhood $V$ of 1 in $G$, let $K(V)$ be the closure in $X$ of $\{I_\phi : \text{supp}(\phi) \subset V\}$. The sets $K(V)$ have the finite intersection property, since $\bigcap_1^n K(V_j) \supset K(\bigcap_1^n V_j)$, so compactness implies that there is a point $I \in X$ that lies in every $K(V)$. This means that every neighborhood of $I$ in $X$ contains $I_\phi$'s with $\text{supp}(\phi)$ arbitrarily small. In other words, for any neighborhood $V$ of 1, any $\epsilon > 0$, and any $f_1, \ldots, f_n \in C_c^+(G)$ there exists $\phi \in C_c^+(G)$ with $\text{supp}(\phi) \subset V$ and $|I(f_j) - I_\phi(f_j)| < \epsilon$ for all $j$. Hence, by (2.11)–(2.13) and Lemma 2.18, $I$ commutes with left translations, addition, and multiplication by positive scalars.

Any $f \in C_c(G)$ can be written as $f = g - h$ with $g, h \in C_c^+(G)$. If also $f = g' - h'$, we have $g + h' = h + g'$, hence $I(g) + I(h') = I(h) + I(g')$. It follows that the formula $I(f) = I(g) - I(h)$ gives a well-defined extension of $I$ to $C_c(G)$, and this extension is a nonzero positive linear functional on $C_c(G)$. The desired measure $\lambda$ is then the one associated to $I$ by the Riesz representation theorem. $\qquad \square$

Before proceeding to the uniqueness theorem, we point out a couple of important properties of all Haar measures that are true by construction for the ones produced in the preceding proof.

**2.19 Proposition.** *If $\lambda$ is a left Haar measure on $G$, then $\lambda(U) > 0$ for every nonempty open set $U$, and $\int f \, d\lambda > 0$ for every $f \in C_c^+(G)$.*

*Proof.* Suppose $U$ is open and nonempty, and $\lambda(U) = 0$. Then $\lambda(xU) = 0$ for all $x$, and since any compact set $K$ can be covered by finitely many translates of $U$, we have $\lambda(K) = 0$ for every compact set $K$. But then $\lambda(G) = 0$ by inner regularity, contradicting the condition $\lambda \neq 0$. Next, given $f \in C_c^+(G)$, let $U = \{x : f(x) > \frac{1}{2}\|f\|_{\sup}\}$. Then $\int f \, d\lambda > \frac{1}{2}\|f\|_{\sup}\lambda(U) > 0$. $\qquad \square$

**2.20 Theorem.** *If $\lambda$ and $\mu$ are left Haar measures on $G$, there exists $c \in (0, \infty)$ such that $\mu = c\lambda$.*

*Proof.* In view of Proposition 2.19, the assertion that $\mu = c\lambda$ is equivalent to the assertion that the ratio $\int f \, d\lambda / \int f \, d\mu$ is the same for all $f \in C_c^+(G)$. Suppose then that $f, g \in C_c^+(G)$. Fix a symmetric compact neighborhood $V_0$ of 1 and set

$$A = (\operatorname{supp} f)V_0 \cup V_0(\operatorname{supp} f), \qquad B = (\operatorname{supp} g)V_0 \cup V_0(\operatorname{supp} g).$$

Then $A$ and $B$ are compact, and for $y \in V_0$, $f(xy) - f(yx)$ and $g(xy) - g(yx)$ are supported in $A$ and $B$ respectively, as functions of $x$.

Given $\epsilon > 0$, by Proposition 2.6 there is a symmetric neighborhood $V \subset V_0$ of 1 such that $|f(xy) - f(yx)| < \epsilon$ and $|g(xy) - g(yx)| < \epsilon$ for all $x$ when $y \in V$. Pick $h \in C_c^+(G)$ with $h(x) = h(x^{-1})$ and $\operatorname{supp}(h) \subset V$. Then

$$\int h \, d\mu \int f \, d\lambda = \iint h(y)f(x) \, d\lambda(x) \, d\mu(y)$$

$$= \iint h(y)f(yx) \, d\lambda(x) \, d\mu(y),$$

and since $h(x) = h(x^{-1})$,

$$\int h \, d\lambda \int f \, d\mu = \iint h(x)f(y) \, d\lambda(x) \, d\mu(y)$$

$$= \iint h(y^{-1}x)f(y) \, d\lambda(x) \, d\mu(y)$$

$$= \iint h(x^{-1}y)f(y) \, d\mu(y) \, d\lambda(x)$$

$$= \iint h(y)f(xy) \, d\mu(y) \, d\lambda(x)$$

$$= \iint h(y)f(xy) \, d\lambda(x) \, d\mu(y).$$

(Fubini's theorem is applicable since all the integrals are effectively over sets that are compact and hence of finite measure.) Therefore,

$$\left| \int h \, d\lambda \int f \, d\mu - \int h \, d\mu \int f \, d\lambda \right| = \left| \iint h(y)[f(xy) - f(yx)] \, d\lambda(x) \, d\mu(y) \right|$$

$$\leq \epsilon \lambda(A) \int h \, d\mu.$$

In the same way,

$$\left| \int h \, d\lambda \int g \, d\mu - \int h \, d\mu \int g \, d\lambda \right| \leq \epsilon \lambda(B) \int h \, d\mu.$$

Dividing these inequalities by $\int h\,d\mu \int f\,d\mu$ and $\int h\,d\mu \int g\,d\mu$, respectively, and adding them, we obtain

$$\left| \frac{\int f\,d\lambda}{\int f\,d\mu} - \frac{\int g\,d\lambda}{\int g\,d\mu} \right| \le \epsilon \left[ \frac{\lambda(A)}{\int f\,d\mu} + \frac{\lambda(B)}{\int g\,d\mu} \right].$$

Since $\epsilon$ is arbitrary, the ratio of the integrals with respect to $\lambda$ and $\mu$ is the same for $f$ and $g$, which is what we needed to show. $\qquad\square$

The remainder of this section is devoted to some more explicit constructions of Haar measure for various important classes of groups.

First, if $G$ is a Lie group, Haar measure can be obtained by a simple differential-geometric construction. Namely, start by picking an inner product on the tangent space at 1; transport it to the tangent space at every other point by left translation, obtaining a left-invariant Riemannian metric; and then take the associated Riemannian volume element. Alternatively, start by picking a nonzero $n$-covector at 1 (where $n = \dim G$); transport it to every other point by left translation, obtaining a left-invariant differential $n$-form $\omega$; then Haar measure is defined by the functional $f \to \int f\omega$. Either way, one sees that Haar measure on a Lie group is given by a smooth density times Lebesgue measure in any local coordinates, and the construction via differential forms leads to a calculation of this density as a Jacobian determinant. Here is a simple result along these lines that covers many cases of interest.

**2.21 Proposition.** *Suppose the underlying manifold of $G$ is an open subset of $\mathbb{R}^N$ and left translations are given by affine maps: $xy = A(x)y + b(x)$, where $A(x)$ is a linear transformation of $\mathbb{R}^N$ and $b(x) \in \mathbb{R}^N$. Then $|\det A(x)|^{-1}\,dx$ is a left Haar measure on $G$, where $dx$ denotes Lebesgue measure on $\mathbb{R}^N$. (Similarly for right translations and right Haar measure.)*

The proof is an exercise in using the change-of-variable formula for multiple integrals, which we leave to the reader. The following examples, whose verification we also leave to the reader, are applications of this proposition.

1. $dx/|x|$ is a Haar measure on the multiplicative group $\mathbb{R} \setminus \{0\}$.
2. $dx\,dy/(x^2+y^2)$ is a Haar measure on the multiplicative group $\mathbb{C}\setminus\{0\}$, with coordinates $z = x + iy$.
3. Lebesgue measure $\prod_{i<j} d\alpha_{ij}$ is a left and right Haar measure on the group of $n \times n$ real matrices $(\alpha_{ij})$ such that $\alpha_{ij} = 0$ for $i > j$ and $\alpha_{ii} = 1$ for $1 \le i \le n$.
4. $|\det T|^{-n}\,dT$ is a left and right Haar measure on the group $GL(n, \mathbb{R})$ of invertible linear transformations of $\mathbb{R}^n$, where $dT$ is Lebesgue measure on the vector space of all real $n \times n$ matrices. (To see that

the determinant of the map $S \to TS$ on matrix space is $(\det T)^n$, observe that it maps the matrix with columns $\mathbf{x}_1, \ldots, \mathbf{x}_n$ to the matrix with columns $T\mathbf{x}_1, \ldots, T\mathbf{x}_n$, so it is the direct sum of $n$ copies of $T$ acting on $\mathbb{R}^n$.)

5. The **ax+b group** $G$ is the group of all affine transformations $x \to ax + b$ of $\mathbb{R}$ with $a > 0$ and $b \in \mathbb{R}$. On $G$, $da\,db/a^2$ is a left Haar measure and $da\,db/a$ is a right Haar measure.

If $G_1, \ldots, G_n$ are locally compact groups with left Haar measures $\lambda_1, \ldots, \lambda_n$, then left Haar measure on $G = \prod_1^n G_j$ is obviously the Radon product of $\lambda_1, \ldots, \lambda_n$, that is, the Radon measure on $G$ associated to the linear functional

$$f \to \int \cdots \int f(x_1, \ldots, x_n)\, d\lambda_1(x_1) \ldots d\lambda_n(x_n).$$

When the $G_j$'s are $\sigma$-compact, this is an extension of the ordinary product $\lambda_1 \times \cdots \times \lambda_n$, and when the $G_j$'s are second countable, it is equal to this product. (See Folland [45, §7.4], and also the next section. The technicalities implicit here are that if the $G_j$'s are not second countable, the Borel $\sigma$-algebras on the factors do not generate the whole Borel $\sigma$-algebra on the product; and if the $G_j$'s are not $\sigma$-compact, outer regularity is incompatible with the usual definition of product.)

In the case of compact groups one can also allow infinitely many factors, provided that one normalizes the Haar measures to have total mass 1. Indeed, suppose $\{G_\alpha\}_{\alpha \in A}$ is a family of compact groups, and let $\lambda_\alpha$ be the Haar measure on $G_\alpha$ such that $\lambda_\alpha(G_\alpha) = 1$. (As we shall shortly see, left and right Haar measures are the same on compact groups.) Let $G = \prod_{\alpha \in A} G_\alpha$, and let $C_F(G)$ be the space of continuous functions on $G$ that depend on only finitely many coordinates. If $f \in C_F(G)$ depends only on $x_{\alpha_1}, \ldots, x_{\alpha_n}$, we set

$$I(f) = \int \cdots \int f(x_{\alpha_1}, \ldots, x_{\alpha_n})\, d\lambda_{\alpha_1}(x_{\alpha_1}) \cdots d\lambda_{\alpha_n}(x_{\alpha_n}).$$

This is well-defined: it does not depend on the order of the coordinates, and it is unaffected if we add in some extra coordinates on which $f$ does not depend. $I$ is thus a left-invariant positive linear functional on $C_F(G)$ such that $|I(f)| \le \|f\|_{\sup}$ and $I(1) = 1$. It follows easily from the Stone-Weierstrass theorem that $C_F(G)$ is dense in $C(G)$, so $I$ extends uniquely to a left-invariant positive linear functional on $C(G)$, and the associated Radon measure on $G$ is its Haar measure.

An example of particular interest is the group $(Z_2)^\omega$, the product of an infinite sequence of copies of the integers mod 2, where Haar measure on each factor assigns measure $\frac{1}{2}$ to each of the two points 0 and 1.

The elements of $(Z_2)^\omega$ are sequences $(a_1, a_2, \ldots)$ where each $a_j$ is 0 or 1. Consider the map $\Phi : (Z_2)^\omega \to [0, 1]$ that assigns to such a sequence the real number $\sum a_j 2^{-j}$ whose base-2 decimal expansion is $0.a_1 a_2 \ldots$. This map is not a group homomorphism. However, it is almost bijective — if $x \in [0, 1]$, $\Phi^{-1}(\{x\})$ is one point unless $x = j2^{-k}$ with $0 < j < 2^k$, in which case it is two points. Moreover, $\Phi$ *is measurable and maps Haar measure* $\mu$ *on* $G$ *to Lebesgue measure* $\lambda$ *on* $[0, 1]$. One can see this by the following steps:

i. If $I = [j2^{-k}, (j+1)2^{-k}]$ where $0 \leq j < 2^k$, $\Phi^{-1}(I)$ is a set of the form $\prod_1^\infty E_i$, where $E_i = Z_2$ for $i > k$ and $E_i$ is $\{0\}$ or $\{1\}$ for $i \leq k$. Clearly $\lambda(I) = \mu(\phi^{-1}(I)) = 2^{-k}$.

ii. The finite disjoint unions of half-open intervals $[j2^{-k}, (j+1)2^{-k})$ form an algebra $\mathcal{A}_1$ that generates the Borel $\sigma$-algebra on $[0, 1)$, and the finite disjoint unions of sets $E$ as in (i) form an algebra $\mathcal{A}_2$ that generates the Borel $\sigma$-algebra on $(Z_2)^\omega$. If $A \in \mathcal{A}_1$ then the union of $\Phi^{-1}(A)$ and some finite set is in $\mathcal{A}_2$, and $\lambda(A) = \mu(\Phi^{-1}(A))$.

iii. It now follows that $\Phi^{-1}(B)$ is a Borel set for any Borel $B \subset [0, 1]$ and that $\lambda(B) = \mu(\Phi^{-1}(B))$. (Cf. Folland [45, Theorem 1.14].)

Finally, we show how to construct Haar measure $\lambda$ on the $p$-adic numbers. With notation as in §2.1, we fix the normalization of $\lambda$ by declaring that $\lambda(\mathbb{Z}_p) = 1$. Then the measure of any ball $\overline{B}(1, x)$ of radius 1 must be 1. If $m > 0$, a ball of radius $p^m$ is a disjoint union of $p^m$ balls of radius 1, and a ball of radius 1 is the disjoint union of $p^m$ balls of radius $p^{-m}$. It follows that $\lambda(\overline{B}(p^k, x)) = p^k$ for any $k \in \mathbb{Z}$, $x \in \mathbb{Q}_p$. Any open set is a countable disjoint union of such balls, so its measure is the sum of their measures. Finally, the measure of any Borel set $E$ is determined by outer regularity, so we end up with a formula like that for Lebesgue measure on the line:

$$\lambda(E) = \inf \left\{ \sum_{j=1}^\infty p^{m_j} : E \subset \bigcup_1^\infty \overline{B}(p^{m_j}, x_j) \right\}.$$

---

## 2.3   Interlude: Some Technicalities

In order not to artificially restrict the generality of our investigations, we have not assumed that our locally compact groups $G$ are $\sigma$-compact. In the non-$\sigma$-compact case, Haar measure is not $\sigma$-finite, a fact that results in certain technical complications in the measure theory. The purpose of this section is to point these out and to explain why they are not serious.

The reader who is content to add a standing assumption that all groups under consideration are $\sigma$-compact can omit this material.

Let $G$ be a non-$\sigma$-compact locally compact group, with left Haar measure $\lambda$. By Proposition 2.4, $G$ has a subgroup $G_0$ that is open, closed, and $\sigma$-compact. Let $Y$ be a subset of $G$ that contains exactly one element of each left coset of $G_0$, so that $G$ is the disjoint union of the sets $yG_0$, $y \in Y$. A moment's thought shows that the restriction of $\lambda$ to the Borel subsets of $G_0$ is a left Haar measure on $G_0$. Moreover, this restriction determines $\lambda$ completely, but not in quite the obvious way. In the first place, it determines $\lambda$ on the Borel subsets of each coset $yG_0$, since $\lambda(yE) = \lambda(E)$. One might then think that for any Borel $E \subset G$ one would have $\lambda(E) = \sum_{y \in Y} \lambda(E \cap yG_0)$. But in fact, what happens is the following.

**2.22 Proposition.** *Suppose $E \subset G$ is a Borel set. If $E \subset \bigcup_1^\infty y_j G_0$ for some countable set $\{y_j\} \subset Y$, then $\lambda(E) = \sum_1^\infty \lambda(E \cap y_j G_0)$. If $E \cap yG_0 \neq \varnothing$ for uncountably many $y$, then $\lambda(E) = \infty$.*

*Proof.* The first assertion is true simply by countable additivity. To prove the second one, by outer regularity it suffices to assume that $E$ is open. In this case, by Proposition 2.19, $\lambda(E \cap yG_0) > 0$ whenever $E \cap yG_0 \neq \varnothing$. If this happens for uncountably many $y$, for some $\epsilon > 0$ there are uncountably many $y$ for which $\lambda(E \cap yG_0) > \epsilon$, and it follows that $\lambda(E) = \infty$. $\qquad\square$

Here is a useful example to keep in mind: let $G = \mathbb{R} \times \mathbb{R}_d$, where $\mathbb{R}_d$ is $\mathbb{R}$ with the discrete topology. Here we can take $G_0 = \mathbb{R} \times \{0\}$ and $Y = \{0\} \times \mathbb{R}_d$. To obtain Haar measure $\lambda$ on $G$, simply take Lebesgue measure on each horizontal line $\mathbb{R} \times \{y\}$ and add them together according to Proposition 2.22. In particular, observe that $Y$ is closed and that $\lambda(Y) = \infty$ although the intersection of $Y$ with any coset of $G_0$, or with any compact set, has measure 0. (This is also true for $Y$ in the general situation above, as long as $G$ is not discrete.) This shows that $\lambda$ is not inner regular on $Y$. It also shows that $\lambda$ is not quite the product of the Haar measures on $\mathbb{R}$ and $\mathbb{R}_d$. Indeed, the latter are Lebesgue measure $\mu$ and counting measure $\nu$, and one has $(\mu \times \nu)(Y) = \mu(\{0\})\nu(\mathbb{R}) = 0 \cdot \infty = 0$ — unless one modifies the usual convention that $0 \cdot \infty = 0$ to say that $\mu(E)\nu(F) = \infty$ when $E \neq \varnothing$ and $F$ is non-$\sigma$-finite.

There are three fundamental theorems of measure theory that break down for general non-$\sigma$-finite measures: Fubini's theorem, the Radon-Nikodym theorem, and the duality of $L^1$ and $L^\infty$. We shall need to deal with each of these, and the rest of this section is devoted to explaining how.

We shall need Fubini's theorem to reverse the order of integration in double integrals $\int_G \int_G f(x,y)\,d\lambda(x)\,d\lambda(y)$. *There is no problem in doing this as long as $f$ vanishes outside a $\sigma$-compact set $E \subset G \times G$.* Indeed, in this case the projections $E_1$ and $E_2$ of $E$ onto the first and second factors are also $\sigma$-compact, and $E \subset E_1 \times E_2$. We can therefore replace $G \times G$ by the $\sigma$-finite space $E_1 \times E_2$, to which the Fubini theorem for Radon products (Folland [45, Theorem 7.27]) is applicable. Moreover, this hypothesis on $f$ is almost always satisfied when $f$ is built up out of functions on $G$ that belong to $L^p(\lambda)$ for some $p < \infty$, for such functions vanish outside some $\sigma$-compact set $\bigcup_1^\infty y_j G_0$ by Proposition 2.22. For example, in dealing with convolutions one is concerned with functions of the form $f(x,y) = g(x)h(x^{-1}y)$. If $g$ vanishes outside $A$ and $h$ vanishes outside $B$ then $f$ vanishes outside $A \times AB$, and $AB$ is $\sigma$-compact whenever $A$ and $B$ are.

We shall therefore use Fubini's theorem without comment in the sequel whenever the conditions of the preceding paragraph are satisfied, leaving their verification to the mistrustful reader. (We have already encountered this situation once, in the proof of Theorem 2.20.)

As for the Radon-Nikodym theorem: Radon measures — or rather their Carathéodory extensions, which are complete and saturated — enjoy a property called "decomposability," which implies a version of the Radon-Nikodym theorem that is adequate for virtually all purposes. (See Hewitt-Ross [66, Theorem 12.17], or Folland [45, Exercises 1.22 and 3.15]. For Haar measures, decomposability is an easy consequence of Propositions 2.4 and 2.22.) However, all we shall really need to consider are the following more restricted situations.

The first one is that of two Radon measures $\mu$ and $\nu$ on a locally compact Hausdorff space $X$ such that $\nu \ll \mu$ and $\nu$ is $\sigma$-finite. By inner regularity there is a $\sigma$-compact set $E$ such that $\nu(X \setminus E) = 0$. But the restriction of $\mu$ to subsets of $E$ is $\sigma$-finite, so there is a measurable $f : E \to [0, \infty)$ such that $d\nu = f\,d\mu$ on $E$. If we set $f = 0$ on $X \setminus E$, we then have $d\nu = f\,d\mu$ everywhere.

The second situation is that of two Radon measures $\mu$ and $\nu$ on a locally compact Hausdorff space $X$ that are **equivalent**, that is, mutually absolutely continuous. In the cases we are interested in, we will not need to invoke the Radon-Nikodym theorem, but will be able to produce a continuous $f : X \to (0, \infty)$ such that $\int \phi\,d\nu = \int \phi f\,d\mu$ for all $\phi \in C_c(X)$, and the following proposition completes the picture.

**2.23 Proposition.** *Suppose $\mu$ and $\nu$ are Radon measures on $X$, and there is a continuous $f : X \to (0, \infty)$ such that $\int \phi\,d\nu = \int \phi f\,d\mu$ for all $\phi \in C_c(X)$. Then $\nu(E) = \int_E f\,d\mu$ for all Borel sets $E \subset X$.*

*Proof.* Let $\tilde{\nu}(E) = \int_E f\,d\mu$; then $\tilde{\nu}$ is a Borel measure on $X$. To show that $\widetilde{\nu} = \nu$, it will suffice to show that $\widetilde{\nu}$ is outer regular and that $\widetilde{\nu}(U) = \nu(U)$ for all open sets $U$.

Suppose $E$ is a Borel set with $\widetilde{\nu}(E) < \infty$, and $\epsilon > 0$. For $j \in \mathbb{Z}$, let $V_j = \{x : 2^{j-2} < f(x) < 2^j\}$. The open sets $V_j$ cover $X$, so we can write $E = \bigcup_{-\infty}^{\infty} E_j$ where $E_j \subset V_j$. Since $\mu$ is outer regular and $\mu(E_j) < 2^{2-j} \int_{E_j} f \, d\mu = 2^{2-j} \widetilde{\nu}(E_j) < \infty$, for each $j$ we can find an open $U_j \subset V_j$ such that $U_j \supset E_j$ and $\mu(U_j \setminus E_j) < \epsilon 2^{-2|j|}$. Then $\widetilde{\nu}(U_j \setminus E_j) < 2^j \mu(U_j \setminus E_j) < \epsilon 2^{-|j|}$, so $U = \bigcup_{-\infty}^{\infty} U_j$ is an open set containing $E$ such that $\widetilde{\nu}(U \setminus E) < 3\epsilon$. Hence $\widetilde{\nu}$ is outer regular.

Next, if $U$ is open, let $\Phi = \{\phi \in C_c(X) : 0 \leq \phi \leq 1, \text{ supp}(\phi) \subset U\}$. Then $\nu(U) = \sup_\Phi \int \phi \, d\nu = \sup_\Phi \int \phi f \, d\mu$. But $\sup_\Phi \phi f = \chi_U f$, so by the monotone convergence theorem for nets of lower semi-continuous functions (Folland [45, Prop. 7.12]), $\sup_\Phi \int \phi f \, d\mu = \int \chi_U f \, d\mu = \widetilde{\nu}(U)$. $\square$

If $\mu$ and $\nu$ satisfy the conditions of Proposition 2.23, we shall say that they are **strongly equivalent**. Lest the reader should suspect that our arguments here are more complicated than necessary, we point out that Proposition 2.23 is no longer true if we allow $f$ to assume the value 0. For example, if $\lambda$ is Haar measure on the group $\mathbb{R} \times \mathbb{R}_d$ discussed above, and $f(x,y) = |x|$, then the measure $\nu(E) = \int_E f \, d\lambda$ is *not* a Radon measure. Indeed, if $Y = \{0\} \times \mathbb{R}_d$ then $\nu(Y) = 0$ but $\nu(U) = \infty$ for any open $U \supset Y$, by the argument in Proposition 2.22.

Finally, we consider the duality of $L^1(\mu)$ and $L^\infty(\mu)$. When $\mu$ is not $\sigma$-finite it is generally false that $L^\infty(\mu) = L^1(\mu)^*$ with the usual definition of $L^\infty$, but when $\mu$ is a Radon measure on a locally compact Hausdorff space $X$ the result can be salvaged by modifying the definition of $L^\infty$. Here's how. A set $E \subset X$ is **locally Borel** if $E \cap F$ is Borel whenever $F$ is Borel and $\mu(F) < \infty$. A locally Borel set $E$ is **locally null** if $\mu(E \cap F) = 0$ whenever $F$ is Borel and $\mu(F) < \infty$. An assertion about points in $X$ is true **locally almost everywhere** if it is true except on a locally null set. A function $f : X \to \mathbb{C}$ is **locally measurable** if $f^{-1}(A)$ is locally Borel for every Borel set $A \subset \mathbb{C}$. We now (re-)define $L^\infty(\mu)$ to be the set of all locally measurable functions that are bounded except on a locally null set, modulo functions that are zero locally a.e. $L^\infty(\mu)$ is a Banach space with norm

$$\|f\|_\infty = \inf\{c : |f(x)| \leq c \text{ locally a.e.}\}.$$

We then have $L^\infty(\mu) = L^1(\mu)^*$. In more detail, if $f \in L^\infty(\mu)$ and $g \in L^1(\mu)$ then $fg$ is measurable since $\{x : g(x) \neq 0\}$ is $\sigma$-finite, and it is then integrable since $\{x : |f(x)| > \|f\|_\infty\} \cap \{x : g(x) \neq 0\}$ is null. So $g \mapsto \int fg \, d\mu$ is a well-defined linear functional on $L^1(\mu)$, and its norm is easily seen to be $\|f\|_\infty$. That every element of $L^1(\mu)^*$ arises in this way follows in general from the decomposability of the Carathéodory extension of $\mu$. (See Hewitt-Ross [66, Theorem 12.18], or Folland [45, Exercises 6.23–26].)

In the case of Haar measure $\lambda$ on a locally compact group $G$ (the one we shall be mainly concerned with), this can be established quite simply by the following argument, whose details we leave to the reader. In view of Propositions 2.4 and 2.22, and with the notation used there, it is easy to see that $E \subset G$ is locally Borel $\iff$ $E \cap yH$ is Borel for every $y \in Y$; $E$ is locally null $\iff$ $\lambda(E \cap yH) = 0$ for every $y \in Y$; and $f : G \to \mathbb{C}$ is locally measurable $\iff$ $f|yH$ is measurable for every $y \in Y$. If $\Phi \in L^1(G, \lambda)^*$ then $\Phi|L^1(yH, \lambda)$ is given by a bounded measurable function $f_y$ on $yH$ since $\lambda$ is $\sigma$-finite on $yH$. Define $f : G \to \mathbb{C}$ by $f = f_y$ on $yH$. Then $f$ is locally measurable, $\|f\|_\infty = \sup_{y \in Y} \|f_y\|_\infty = \|\Phi\|$, and $\Phi(g) = \int fg$ for every $g \in L^1(\lambda)$.

Henceforth $L^\infty(\mu)$ will always denote the space defined above. Of course, this coincides with the usual $L^\infty(\mu)$ when $\mu$ is $\sigma$-finite.

## 2.4   The Modular Function

Let $G$ be a locally compact group with left Haar measure $\lambda$. We wish to investigate the extent to which $\lambda$ fails to be right-invariant. If, for $x \in G$, we define $\lambda_x(E) = \lambda(Ex)$, then $\lambda_x$ is again a left Haar measure, by the associative law: $y(Ex) = (yE)x$. By the uniqueness theorem 2.20, there is a number $\Delta(x) > 0$ such that $\lambda_x = \Delta(x)\lambda$, and $\Delta(x)$ is independent of the original choice of $\lambda$. The function $\Delta : G \to (0, \infty)$ thus defined is called the **modular function** of $G$. In what follows, we denote by $\mathbb{R}_\times$ the multiplicative group of positive real numbers.

**2.24 Proposition.** $\Delta$ *is a continuous homomorphism from $G$ to $\mathbb{R}_\times$. Moreover, for any $f \in L^1(\lambda)$,*

$$(2.25) \qquad \int R_y f \, d\lambda = \Delta(y^{-1}) \int f \, d\lambda.$$

*Proof.* For any $x, y \in G$ and $E \subset G$,

$$D(xy)\lambda(E) = \lambda(Exy) = \Delta(y)\lambda(Ex) = \Delta(y)\Delta(x)\lambda(E),$$

so $\Delta$ is a homomorphism from $G$ to $\mathbb{R}_\times$. Moreover, since $\chi_E(xy) = \chi_{Ey^{-1}}(x)$, we have

$$\int \chi_E(xy) \, d\lambda(x) = \lambda(Ey^{-1}) = \Delta(y^{-1})\lambda(E) = \Delta(y^{-1}) \int \chi_E(x) \, d\lambda(x).$$

This proves (2.25) for $f = \chi_E$, and the general case follows by approximating $f$ by simple functions. Finally, since $y \mapsto R_y f$ is continuous from $G$ to $C_c(G)$ (Proposition 2.6), $y \mapsto \int R_y f \, d\lambda$ is continuous from $G$ to $\mathbb{C}$, so the continuity of $\Delta$ follows from (2.25).          $\square$

If we set $y_0 = y^{-1}$ in (2.25) and make the substitution $x \to xy_0$, we obtain

$$\Delta(y_0) \int f(x) \, d\lambda(x) = \int f(xy_0^{-1}) \, d\lambda(x) = \int f(x) d\lambda(xy_0),$$

which yields the following convenient abbreviated form of (2.25):

$$(2.26) \qquad\qquad d\lambda(xy_0) = \Delta(y_0) \, d\lambda(x).$$

$G$ is called **unimodular** if $\Delta \equiv 1$, that is, if left Haar measure is also right Haar measure. Unimodularity is a useful property that makes life simpler in a number of respects. Obviously Abelian groups and discrete groups are unimodular, but many others are too. Here are some classes of examples.

**2.27 Proposition.** *If $K$ is any compact subgroup of $G$ then $\Delta|K \equiv 1$.*

*Proof.* $\Delta(K)$ is a compact subgroup of $\mathbb{R}_\times$, hence equal to $\{1\}$.          $\square$

**2.28 Corollary.** *If $G$ is compact, then $G$ is unimodular.*

Let $[G, G]$ denote the smallest closed subgroup of $G$ containing all elements of the form $[x, y] = xyx^{-1}y^{-1}$. $[G, G]$ is called the **commutator subgroup** of $G$; it is normal since $z[x, y]z^{-1} = [zxz^{-1}, zyz^{-1}]$.

**2.29 Proposition.** *If $G/[G, G]$ is compact, then $G$ is unimodular.*

*Proof.* Since $\mathbb{R}_\times$ is Abelian we have $\Delta([x, y]) = [\Delta(x), \Delta(y)] = 1$, so $\Delta$ must annihilate $[G, G]$ and hence factor through $G/[G, G]$. It then follows as in Proposition 2.27 that $\Delta(G) = \{1\}$.          $\square$

As a consequence of Proposition 2.29, one can see that every connected semi-simple Lie group $G$ is unimodular. Indeed, the Lie algebra $\mathfrak{g}$ of such a group is a direct sum of simple algebras, and it follows easily that $[\mathfrak{g}, \mathfrak{g}] = \mathfrak{g}$. But $[\mathfrak{g}, \mathfrak{g}]$ is the Lie algebra of $[G, G]$, so $G = [G, G]$ since $G$ is connected.

More generally, one has the following result, whose proof may be found in Helgason [64]. (See Lemma 1.2 of Chapter 10 of [64], together with Proposition 2.31 below.)

**2.30 Proposition.** *If $G$ is a connected Lie group and $\mathrm{Ad}$ denotes the adjoint action of $G$ on its Lie algebra, then $\Delta(x) = \det \mathrm{Ad}(x^{-1})$.*

It follows easily from Proposition 2.30 that every connected nilpotent Lie group is unimodular. Thus, Lie groups that are close to being Abelian (i.e., nilpotent) or far from being Abelian (i.e., semi-simple) are unimodular. The simplest example of a non-unimodular group is the $ax + b$ group, which is solvable but not nilpotent; see the list of examples following Proposition 2.21.

To each left Haar measure $\lambda$ is associated the right Haar measure $\rho$ defined by $\rho(E) = \lambda(E^{-1})$. The modular function can be used to relate $\lambda$ to $\rho$:

**2.31 Proposition.** $\lambda$ *and* $\rho$ *are strongly equivalent, and*

$$d\rho(x) = \Delta(x^{-1}) \, d\lambda(x).$$

*Proof.* By (2.25), if $f \in C_c(G)$,

$$\int R_y f(x) \Delta(x^{-1}) \, d\lambda(x) = \Delta(y) \int f(xy) \Delta((xy)^{-1}) \, d\lambda(x)$$

$$= \int f(x) \Delta(x^{-1}) \, d\lambda(x).$$

Thus the functional $f \mapsto \int f(x) \Delta(x^{-1}) \, d\lambda(x)$ is right-invariant, so its associated Radon measure is a right Haar measure and hence equals $c\rho$ for some $c > 0$. By Proposition 2.23, $c \, d\rho(x) = \Delta(x^{-1}) \, d\lambda(x)$, so we need only show that $c = 1$. If $c \neq 1$, we can pick a compact symmetric neighborhood $U$ of 1 in $G$ such that $|\Delta(x^{-1}) - 1| \leq \frac{1}{2}|c - 1|$ on $U$. But then $\lambda(U) = \rho(U)$, so

$$|c - 1|\lambda(U) = \left| c\rho(U) - \lambda(U) \right| = \left| \int_U [\Delta(x^{-1}) - 1] \, d\lambda(x) \right| \leq \frac{1}{2}|c - 1|\lambda(U),$$

a contradiction. $\qquad\qquad\qquad\square$

The formula $d\rho(x) = \Delta(x^{-1}) \, d\lambda(x)$ can be restated in the following ways, convenient for making substitutions in integrals:

$$(2.32) \qquad d\lambda(x^{-1}) = \Delta(x^{-1})d\lambda(x), \qquad d\rho(x^{-1}) = \Delta(x) \, d\rho(x).$$

When $G$ is not unimodular, the function $\Delta$ is unbounded, so the spaces $L^p(\lambda)$ and $L^p(\rho)$ $(1 \leq p < \infty)$ are not the same. There are two ways of passing from one to the other. Namely, let us define

$$\check{f}(x) = f(x^{-1}), \qquad M_p f(x) = \Delta(x)^{1/p} f(x).$$

Then $f \mapsto \check{f}$ and $f \mapsto M_p f$ are isometric isomorphisms from $L^p(\lambda)$ to $L^p(\rho)$, because $d\lambda(x) = d\rho(x^{-1}) = \Delta(x) \, d\rho(x)$. By composing these maps, we get an interesting isometric linear isomorphism of $L^p(\lambda)$ onto itself:

$$(2.33) \qquad M_p^{-1} \check{f}(x) = (M_p f)\check{}(x) = \Delta(x)^{-1/p} f(x^{-1}).$$

## 2.5 Convolutions

*From now on we shall assume that each locally compact group $G$ is equipped with a fixed left Haar measure. Denoting this measure by $\lambda$ for the moment, we shall henceforth write $dx$ for $d\lambda(x)$, $\int f$ for $\int f\,d\lambda$, $|E|$ for $\lambda(E)$, and $L^p$ or $L^p(G)$ for $L^p(\lambda)$. The symbol $\lambda$ is hereby freed to be used for other purposes.*

Let $G$ be a locally compact group, and let $M(G)$ be the space of complex Radon measures on $G$. We define the convolution of two measures $\mu, \nu \in M(G)$ as follows. The map $I(\phi) = \iint \phi(xy)\,d\mu(x)\,d\nu(y)$ is clearly a linear functional on $C_0(G)$ satisfying $|I(\phi)| \leq \|\phi\|_{\sup}\|\mu\|\,\|\nu\|$. Hence it is given by a measure $\mu * \nu \in M(G)$ with $\|\mu * \nu\| \leq \|\mu\|\,\|\nu\|$, called the **convolution** of $\mu$ and $\nu$:

$$(2.34) \qquad \int \phi\,d(\mu * \nu) = \iint \phi(xy)\,d\mu(x)\,d\nu(y).$$

(The order of integration in the double integral is immaterial. What matters is that the variables of integration for $\mu$ and $\nu$ are the first and second factors, respectively, of the product that forms the argument of $\phi$.)

Convolution is associative: if $\mu, \nu, \sigma \in M(G)$ and $\phi \in C_c(G)$,

$$\int \phi\,d[\mu * (\nu * \sigma)] = \iint \phi(xy)\,d\mu(x)\,d(\nu * \sigma)(y)$$
$$= \iiint \phi(xyz)\,d\mu(x)\,d\nu(y)\,d\sigma(z)$$
$$= \iint \phi(yz)\,d(\mu * \nu)(y)\,d\sigma(z)$$
$$= \int \phi\,d[(\mu * \nu) * \sigma].$$

Moreover, convolution is commutative if and only if $G$ is Abelian. Indeed, if $G$ is Abelian we have $\phi(xy) = \phi(yx)$, whence it follows from (2.34) that $\mu * \nu = \nu * \mu$. On the other hand, if $\delta_x \in M(G)$ denotes the point mass at $x \in G$, we have

$$\int \phi\,d(\delta_x * \delta_y) = \iint \phi(uv)\,d\delta_x(u)\,d\delta_y(v) = \phi(xy) = \int \phi\,d\delta_{xy},$$

in other words, $\delta_x * \delta_y = \delta_{xy}$. Thus $\delta_x * \delta_y = \delta_y * \delta_x$ if and only if $xy = yx$.

The estimate $\|\mu * \nu\| \leq \|\mu\|\,\|\nu\|$ implies that convolution makes $M(G)$

into a Banach algebra, called the **measure algebra** of $G$. $M(G)$ has a mulitiplicative identity, namely the point mass $\delta = \delta_1$ at 1:

$$\int \phi \, d(\delta * \mu) = \iint \phi(xy) \, d\delta(x) \, d\mu(y) = \int \phi(y) \, d\mu(y) = \int \phi \, d\mu,$$

and similarly $\mu * \delta = \mu$. $M(G)$ also has a canonical involution $\mu \mapsto \mu^*$ defined by

$$(2.35) \qquad \mu^*(E) = \overline{\mu(E^{-1})}, \quad \text{or} \quad \int \phi \, d\mu^* = \int \phi(x^{-1}) \, d\overline{\mu}(x).$$

This is indeed an involution, because

$$\int \phi \, d(\mu * \nu)^* = \int \phi(x^{-1}) \, d\overline{(\mu * \nu)} = \iint \phi((xy)^{-1}) \, d\overline{\mu}(x) \, d\overline{\nu}(y)$$

$$= \iint \phi(y^{-1} x^{-1}) \, d\overline{\mu}(x) \, d\overline{\nu}(y) = \iint \phi(yx) \, d\mu^*(x) \, d\nu^*(y)$$

$$= \int \phi \, d(\nu^* * \mu^*),$$

so that $(\mu * \nu)^* = \nu^* * \mu^*$.

In many respects, the algebra $M(G)$ is too big and complicated to work with easily, and it is preferable to restrict attention to the space $L^1(G)$, which is a subspace of $M(G)$ if we identify the function $f$ with the measure $f(x) \, dx$. If $f, g \in L^1$, the **convolution** of $f$ and $g$ is the function defined by

$$f * g(x) = \int f(y) g(y^{-1} x) \, dy.$$

An application of Fubini's theorem shows that the integral is absolutely convergent for almost every $x$ and that $\|f * g\|_1 \leq \|f\|_1 \|g\|_1$, for

$$\iint |f(y) g(y^{-1} x)| \, dx \, dy = \iint |f(y) g(x)| \, dx \, dy = \|f\|_1 \|g\|_1$$

by the left invariance of the measure $dx$. This definition of convolution agrees with the preceding one when we identify $f \in L^1$ with the measure $f(x) \, dx$, because

$$\iint \phi(yx) f(y) g(x) \, dx \, dy = \iint \phi(x) f(y) g(y^{-1} x) \, dx \, dy$$

$$= \int \phi(x) (f * g)(x) \, dx.$$

The integral defining $f * g(x)$ can be expressed in several different forms:

(2.36)

$$f * g(x) = \int f(y)g(y^{-1}x)\, dy$$

$$= \int f(xy)g(y^{-1})\, dy$$

$$= \int f(y^{-1})g(yx)\Delta(y^{-1})\, dy$$

$$= \int f(xy^{-1})g(y)\Delta(y^{-1})\, dy.$$

The equality of the integrals follows from the substitutions $y \to xy$ and $y \to y^{-1}$, according to (2.32). To remember how to arrange the variables $x$ and $y$ in them, it may be useful to keep the following two rules in mind. (i) The variable $y$ of integration occurs as $y$ in one factor and as $y^{-1}$ in the other. (ii) The two occurrences of the variable of integration are adjacent to each other, not separated by the variable $x$ at which the convolution is evaluated. When $G$ is unimodular, the factor of $\Delta(y^{-1})$ disappears, in which case these two rules always suffice to give the right answer.

The involution on $M(G)$, restricted to $L^1(G)$, is defined by the relation $f^*(x)\, dx = \overline{f(x^{-1})}\, d(x^{-1})$, so by (2.32) it is given by

(2.37)        $$f^*(x) = \Delta(x^{-1})\overline{f(x^{-1})}.$$

(Cf. (2.33).) With the convolution product (2.36) and the involution (2.37), $L^1(G)$ becomes a Banach $*$-algebra, called the $L^1$ **group algebra** of $G$. (This algebra coincides with the algebraists' "group algebra" when $G$ is a finite group.)

Observe that

(2.38)        $$f * g = \int f(y) L_y g \, dy = \int g(y^{-1}) R_y f \, dy.$$

These equalities can be interpreted in two senses: either pointwise, in which case they are restatements of the first two equalities in (2.36), or in the sense of vector-valued integrals, where the expressions on the right are considered as integrals of $L^1$-valued functions of $y$: see Appendix 4. Thus, $f * g$ is a generalized linear combination of left translates of $g$, or of right translates of $f$. Since left translations commute with right translations (this is the associative law), it follows easily that convolutions have the following behavior under translations:

(2.39)        $$L_z(f * g) = (L_z f) * g, \qquad R_z(f * g) = f * (R_z g).$$

Convolution can be extended from $L^1$ to other $L^p$ spaces. Specifically, we have the following results:

**2.40 Proposition.** *Suppose $1 \leq p \leq \infty$, $f \in L^1(G)$, and $g \in L^p(G)$.*

a. *The integrals in (2.36) converge absolutely for almost every $x$, and we have $f * g \in L^p(G)$ and $\|f * g\|_p \leq \|f\|_1 \|g\|_p$.*

b. *If $G$ is unimodular, the same conclusions hold with $f * g$ replaced by $g * f$.*

c. *If $G$ is not unimodular, we still have $g * f \in L^p(G)$ when $f$ has compact support.*

*Proof.* To prove (a), we apply Minkowski's inequality for integrals to the first integral in (2.36), obtaining

$$\|f * g\|_p = \left\| \int f(y) L_y g(\cdot) \, dy \right\|_p \leq \int |f(y)| \, \|L_y g\|_p \, dy = \|f\|_1 \|g\|_p,$$

since the $L^p$ norm is left-invariant. (The a.e. convergence of the integral for $f * g$ is implicit in this.) If $G$ is unimodular, we apply Minkowski's inequality to the fourth integral in (2.36) with $f$ and $g$ switched,

$$\|g * f\|_p = \left\| \int R_{y^{-1}} g(\cdot) f(y) \, dy \right\|_p \leq \int \|R_{y^{-1}} g\|_p |f(y)| \, dy = \|g\|_p \|f\|_1,$$

which proves (b). If $K = \operatorname{supp} f$ is compact, a similar argument works in the non-unimodular case:

$$\|g * f\|_p = \left\| \int R_{y^{-1}} g(\cdot) f(y) \Delta(y^{-1}) \, dy \right\|_p$$
$$\leq \int \|R_{y^{-1}} g\|_p |f(y)| \Delta(y^{-1}) \, dy = \|g\|_p \int_K |f(y)| \Delta(y)^{(1/p)-1} \, dy$$
$$\leq C \|g\|_p \|f\|_1,$$

where $C = \sup_K \Delta(y)^{(1/p)-1}$. $\qquad\qquad\square$

**2.41 Proposition.** *Suppose $G$ is unimodular. If $f \in L^p(G)$ and $g \in L^q(G)$ where $1 < p, q < \infty$ and $p^{-1} + q^{-1} = 1$, then $f * g \in C_0(G)$ and $\|f * g\|_{\sup} \leq \|f\|_p \|g\|_q$.*

*Proof.* The fact that $|f * g(x)| \leq \|f\|_p \|g\|_q$ for all $x \in G$ follows from Hölder's inequality and the invariance of Haar integrals under translations and inversions. If $f, g \in C_c(G)$, it is easy to check that $f * g \in C_c(G)$. But $C_c(G)$ is dense in $L^p(G)$, and if $f_n \to f$ in $L^p$ and $g_n \to g \in L^q$ then $f_n * g_n \to f * g$ and $g_n * f_n \to g * f$ uniformly; the result follows. $\qquad\square$

This result is not true as stated in the extreme cases $p = 1$ or $\infty$, because then $f * g$ need not vanish at infinity. However, it is still continuous, and even more can be said. To prepare for this result, and for other purposes, we need to establish the continuity of translations on $L^p$.

**2.42 Proposition.** *If $1 \leq p < \infty$ and $f \in L^p(G)$ then $\|L_y f - f\|_p$ and $\|R_y f - f\|_p$ tend to zero as $y \to 1$.*

*Proof.* Fix a compact neighborhood $V$ of 1. First, if $g \in C_c(G)$, let $K = (\operatorname{supp} g)V^{-1} \cup V(\operatorname{supp} g)$. Then $K$ is compact, and $L_y g$ and $R_y g$ are supported in $K$ when $y \in V$. Hence, $\|L_y g - g\|_p \leq |K|^{1/p}\|L_y g - g\|_\infty \to 0$ as $y \to 1$ by Proposition 2.6, and likewise $\|R_y g - g\|_p \to 0$.

Now suppose $f \in L^p(G)$. We have $\|L_y f\|_p = \|f\|_p$ and $\|R_y f\|_p = \Delta(y)^{-1/p}\|f\|_p \leq C\|f\|_p$ for $y \in V$. Given $\epsilon > 0$ we can choose $g \in C_c(G)$ such that $\|f - g\|_p \leq \epsilon$, and then

$$\|R_y f - f\|_p \leq \|R_y(f-g)\|_p + \|R_y g - g\|_p + \|g - f\|_p \leq (C+1)\epsilon + \|R_y g - g\|_p,$$

and the last term tends to zero as $y \to 1$; similarly for $L_y f$. $\qquad\square$

**2.43 Proposition.** *If $f \in L^1(G)$ and $g \in L^\infty(G)$, then $f * g$ is left uniformly continuous and $g * f$ is right uniformly continuous.*

*Proof.* We have $L_z(f * g) - f * g = (L_z f - f) * g$ and $R_z(g * f) - g * f = g * (R_z f - f)$ by (2.39); therefore,

$$\|L_z(f * g) - f * g\|_{\sup} \leq \|L_z f - f\|_1 \|g\|_\infty,$$
$$\|R_z(f * g) - f * g\|_{\sup} \leq \|g\|_\infty \|R_z f - f\|_1,$$

and the quantities on the right tend to 0 as $z \to 1$ by Proposition 2.42. $\qquad\square$

When $G$ is discrete, the function $\delta$ defined by $\delta(1) = 1$ and $\delta(x) = 0$ for $x \neq 1$ is the identity element in the algebra $L^1(G)$: $f * \delta = \delta * f = f$ for any $f$. When $G$ is not discrete there is no function with this property. There is a *measure*, of course — the point mass at the origin — but in many situations one needs to remain in the realm of functions, so one uses an "approximate identity" instead.

**2.44 Proposition.** *Let $\mathfrak{U}$ be a neighborhood base at 1 in $G$. For each $U \in \mathfrak{U}$, let $\psi_U$ be a function such that*
*(i) $\operatorname{supp} \psi_U$ is compact and contained in $U$,*
*(ii) $\psi_U \geq 0$ and $\int \psi_U = 1$.*
*Then $\|\psi_U * f - f\|_p \to 0$ as $U \to \{1\}$ if $1 \leq p < \infty$ and $f \in L^p$, or if $p = \infty$ and $f$ is left uniformly continuous. If, in addition,*
*(iii) $\psi_U(x^{-1}) = \psi_U(x)$ for all $x$,*

then $\|f * \psi_U - f\|_p \to 0$ as $U \to \{1\}$ if $1 \leq p < \infty$ and $f \in L^p$, or if $p = \infty$ and $f$ is right uniformly continuous.

*Proof.* Since $\int \psi_U = 1$, we have

$$\psi_U * f(x) - f(x) = \int \psi_U(y)[L_y f(x) - f(x)] \, dy,$$

so by Minkowski's inequality for integrals,

$$\|\psi_U * f - f\|_p \leq \int \|L_y f - f\|_p \psi_U(y) \, dy \leq \sup_{y \in U} \|L_y f - f\|_p.$$

Hence $\|f * \psi_U - f\|_p \to 0$ by Proposition 2.42, or by the left uniform continuity of $f$ if $p = \infty$. The second assertion follows in the same way, since under condition (iii),

$$f * \psi_U(x) - f(x) = \int f(xy)\psi_U(y^{-1}) \, dy - f(x) \int \psi_U(y) \, dy$$

$$= \int [R_y f(x) - f(x)]\psi_U(y) \, dy. \qquad \square$$

A family $\{\psi_U\}$ of functions satisfying conditions (i)–(iii) of Proposition 2.44 is called an **approximate identity**. Approximate identities exist in great abundance. For example, we could take the sets $U$ to be compact and symmetric and then take $\psi_U = |U|^{-1}\chi_U$, or we could use Urysohn's lemma to obtain continuous $\psi_U$'s. Sometimes we shall informally say, "Let $g$ be an approximate identity"; what we mean is, "Let $g = \psi_U$ and let $U \to \{1\}$."

As a first application of approximate identities, we derive an important characterization of the closed ideals in $L^1(G)$.

**2.45 Theorem.** *Let $\mathcal{I}$ be a closed subspace of $L^1(G)$. Then $\mathcal{I}$ is a left ideal if and only if it is closed under left translations, and $\mathcal{I}$ is a right ideal if and only if it is closed under right translations.*

*Proof.* Suppose $\mathcal{I}$ is a left ideal. If $f \in \mathcal{I}$, $x \in G$, and $\{\psi_U\}$ is an approximate identity, we have $L_x(\psi_U * f) = (L_x\psi_U) * f \in \mathcal{I}$ and hence $L_x f = \lim L_x(\psi_U * f) \in \mathcal{I}$. On the other hand, suppose $\mathcal{I}$ is closed under left translations. If $f \in \mathcal{I}$ and $g \in L^1$, we have $g * f = \int g(y)L_y f \, dy$ by (2.38), so $g * f$ is in the closed linear span of the functions $L_y f$ (see Appendix 4) and hence in $\mathcal{I}$. $\qquad \square$

Finally, we mention a couple of other extensions of the notion of convolution. First, if $\mu \in M(G)$ and $f \in L^p(G)$ one can define the function $\mu * f$ by

$$\mu * f(x) = \int f(y^{-1}x) \, d\mu(y).$$

The argument used to prove Proposition 2.40(a) shows that $\mu * f \in L^p$ and that $\|\mu * f\|_p \leq \|\mu\| \|f\|_p$. When $G$ is unimodular or when $p = 1$ one can define $f * \mu$ similarly. In particular, $L^1(G)$ is a two-sided ideal in $M(G)$.

Second, suppose $G$ is unimodular. We have $L^p * L^1 \subset L^p$ and $L^p * L^{p'} \subset L^\infty$ where $p'$ is the conjugate exponent to $p$, by Propositions 2.40 and 2.41. An application of the Riesz-Thorin interpolation theorem then shows that $L^p * L^q \subset L^r$ and that $\|f * g\|_r \leq \|f\|_p \|g\|_q$ whenever $p^{-1} + q^{-1} = r^{-1} + 1$.

## 2.6   Homogeneous Spaces

Let $G$ be a locally compact group and $S$ a locally compact Hausdorff space. A (left) **action** of $G$ on $S$ is a continuous map $(x, s) \mapsto xs$ from $G \times S$ to $S$ such that (i) $s \mapsto xs$ is a homeomorphism of $S$ for each $x \in G$, and (ii) $x(ys) = (xy)s$ for all $x, y \in G$ and $s \in S$. A space $S$ equipped with an action of $G$ is called a $G$-**space**. A $G$-space is called **transitive** if for every $s, t \in S$ there exists $x \in G$ such that $xs = t$.

The standard examples of transitive $G$-spaces are the quotient spaces $G/H$ (where $H$ is a closed subgroup of $G$), on which $G$ acts by left multiplication. In fact, these are close to being the only examples. If $S$ is a transitive $G$-space, pick $s_0 \in S$, define $\phi : G \to S$ by $\phi(x) = xs_0$, and let $H = \{x \in G : xs_0 = s_0\}$. Then $H$ is a closed subgroup of $G$ and $\phi$ is a continuous surjection of $G$ onto $S$ that is constant on the left cosets of $H$. Hence $\phi$ induces a continuous bijection $\Phi : G/H \to S$ such that $\Phi \circ q = \phi$, where $q : G \to G/H$ is the natural quotient map. The only additional thing needed to identify $S$ with $G/H$ is the continuity of $\Phi^{-1}$. This is not always the case — for example, consider $G = \mathbb{R}$ with the discrete topology, acting by translations on $\mathbb{R}$ with the usual topology. But it is valid if $G$ is $\sigma$-compact.

**2.46 Proposition.** *With notation as above, if $G$ is $\sigma$-compact then $\Phi$ is a homeomorphism.*

*Proof.* It suffices to show that $\phi$ maps open sets in $G$ to open sets in $S$. Suppose $U$ is open in $G$ and $x_0 \in U$; pick a compact symmetric neighborhood $V$ of 1 such that $x_0VV \subset U$. Since $G$ is $\sigma$-compact, there is a countable set $\{y_n\} \subset G$ such that the sets $y_nV$ cover $G$. Then $S = \bigcup_1^\infty \phi(y_nV)$. The sets $\phi(y_nV)$ are all homeomorphic to $\phi(V)$ since $s \mapsto y_ns$ is a homeomorphism of $S$, and they are compact and hence closed. By the Baire category theorem for locally compact Hausdorff spaces (Folland [45, Exercise 5.28[1]]), $\phi(V)$ must have an interior point, say $\phi(x_1)$ $(x_1 \in V)$. But then $\phi(x_0)$ is an interior point of $\phi(x_0 x_1^{-1} V)$, and $x_0 x_1^{-1} V \subset x_0 V V \subset U$, so $\phi(x_0)$ is an interior point of $\phi(U)$. Thus $\phi(U)$ is open. $\qquad\square$

We shall use the term **homogeneous space** to mean a transitive $G$-space $S$ that is isomorphic to a quotient space $G/H$ — that is, one for which the map $\Phi$ described above is a homeomorphism — and we shall generally identify $S$ with $G/H$. This identification depends on the choice of a base point $s_0 \in S$. If we choose a different base point $s_0' = x_0 s_0$, the only effect is to replace $H$ with $H' = x_0 H x_0^{-1}$; and the map $x \mapsto x_0 x x_0^{-1}$ induces a $G$-equivariant homeomorphism between $G/H$ and $G/H'$.

Henceforth we consider homogeneous spaces $G/H$, where $G$ is an arbitrary locally compact group and $H$ is an arbitrary closed subgroup. The question we wish to address is whether there is a $G$-invariant Radon measure on $G/H$, that is, a Radon measure $\mu$ such that $\mu(xE) = \mu(E)$ for every $x \in G$. The answer is not always affirmative. For example, the real line $\mathbb{R}$ is a homogeneous space of the group of affine transformations $x \mapsto ax + b$ of $\mathbb{R}$. The only measure on $\mathbb{R}$ (up to scalar multiples) that is invariant under translations $x \mapsto x + b$ is Lebesgue measure, but it is not invariant under the dilations $x \mapsto ax$. However, we shall obtain a necessary and sufficient condition for the existence of an invariant measure, and a good substitute result for the cases where the condition fails.

In what follows, $G$ is a locally compact group with left Haar measure $dx$, $H$ is a closed subgroup of $G$ with left Haar measure $d\xi$, $q : G \to G/H$ is the canonical quotient map $q(x) = xH$, and $\Delta_G$ and $\Delta_H$ are the modular functions of $G$ and $H$. We define a map $P : C_c(G) \to C_c(G/H)$ by

$$(2.47) \qquad Pf(xH) = \int_H f(x\xi) \, d\xi.$$

This is well-defined by the left-invariance of $d\xi$: if $y = x\eta$ with $\eta \in H$ then $\int f(y\xi) \, d\xi = \int f(x\xi) \, d\xi$. $Pf$ is obviously continuous, and $\text{supp}(Pf) \subset q(\text{supp } f)$. Moreover, if $\phi \in C(G/H)$ we have

$$P[(\phi \circ q) \cdot f] = \phi \cdot Pf.$$

We now show that $P$ maps $C_c(G)$ onto $C_c(G/H)$, obtaining along the way a couple of lemmas that will be needed later.

**2.48 Lemma.** *If $E \subset G/H$ is compact, there exists a compact $K \subset G$ with $q(K) = E$.*

---

[1]This is Exercise 5.32 in the first edition of [45].

*Proof.* Pick an open neighborhood $V$ of 1 in $G$ with compact closure. Since $q$ is an open map, the sets $q(xV)$ $(x \in G)$ are an open cover of $E$, so there is a finite subcover $q(x_j V)$ $(j = 1, \ldots, n)$. Let $K = q^{-1}(E) \cap \bigcup_1^n x_j \overline{V}$. Since $q^{-1}(E)$ is closed, $K$ is compact, and $q(K) = E$.   □

**2.49 Lemma.** *If $F \subset G/H$ is compact, there exists $f \geq 0$ in $C_c(G)$ such that $Pf = 1$ on $F$.*

*Proof.* Let $E$ be a compact neighborhood of $F$ in $G/H$, and use Lemma 2.48 to obtain a compact $K \subset G$ such that $q(K) = E$. Choose nonnegative $g \in C_c(G)$ with $g > 0$ on $K$ and $\phi \in C_c(G/H)$ supported in $E$ such that $\phi = 1$ on $F$, and set

$$ f = \frac{\phi \circ q}{Pg \circ q} g, $$

with the understanding that the fraction is zero wherever the numerator is zero. $f$ is continuous since $Pg > 0$ on $\text{supp}\, \phi$, its support is contained in $\text{supp}\, g$, and $Pf = (\phi/Pg)Pg = \phi$.   □

**2.50 Proposition.** *If $\phi \in C_c(G/H)$, there exists $f \in C_c(G)$ such that $Pf = \phi$ and $q(\text{supp}\, f) = \text{supp}\, \phi$, and also such that $f \geq 0$ if $\phi \geq 0$.*

*Proof.* If $\phi \in C_c(G/H)$, by Lemma 2.49 there exists $g \geq 0$ in $C_c(G)$ such that $Pg = 1$ on $\text{supp}\, \phi$. Let $f = (\phi \circ q)g$. We have $Pf = \phi(Pg) = \phi$, and the other properties of $f$ are obvious.   □

We are now in a position to settle the question of the existence and uniqueness of invariant measures on $G/H$.

**2.51 Theorem.** *Suppose $G$ is a locally compact group and $H$ is a closed subgroup. There is a $G$-invariant Radon measure $\mu$ on $G/H$ if and only if $\Delta_G | H = \Delta_H$. In this case, $\mu$ is unique up to a constant factor, and if this factor is suitably chosen we have*

$$ (2.52) \qquad \int_G f(x)\, dx = \int_{G/H} Pf\, d\mu = \int_{G/H} \int_H f(x\xi)\, d\xi\, d\mu(xH) $$

*for $f \in C_c(G)$.*

*Proof.* Suppose a $G$-invariant measure $\mu$ exists. Then $f \mapsto \int Pf\, d\mu$ is a nonzero left invariant positive linear functional on $C_c(G)$, so $\int Pf\, d\mu = c \int f(x)\, dx$ for some $c > 0$ by the uniqueness of Haar measure on $G$. In view of Proposition 2.50, this formula completely determines $\mu$, so $\mu$ is unique up to the arbitrary constant factor in Haar measure. Replacing

$\mu$ by $c^{-1}\mu$, we may assume $c = 1$, so that (2.52) holds. This being the case, if $\eta \in H$ and $f \in C_c(G)$ we have

$$\Delta_G(\eta) \int_G f(x)\,dx = \int_G f(x\eta^{-1})\,dx$$

$$= \int_{G/H} \int_H f(x\xi\eta^{-1})\,d\xi\,d\mu(xH)$$

$$= \Delta_H(\eta) \int_{G/H} \int_H f(x\xi)\,d\xi\,d\mu(xH)$$

$$= \Delta_H(\eta) \int f(x)\,dx,$$

so that $\Delta_G(\eta) = \Delta_H(\eta)$.

Conversely, suppose $\Delta_G|H = \Delta_H$. We claim that if $f \in C_c(G)$ and $Pf = 0$ then $\int f(x)\,dx = 0$. Indeed, by Lemma 2.49 there exists $\phi \in C_c(G)$ such that $P\phi = 1$ on $q(\text{supp } f)$. We have

$$0 = Pf(xH) = \int f(x\xi)\,d\xi = \int f(x\xi^{-1})\Delta_H(\xi^{-1})\,d\xi$$

$$= \int f(x\xi^{-1})\Delta_G(\xi^{-1})\,d\xi,$$

so

$$0 = \int_G \int_H \phi(x)f(x\xi^{-1})\Delta_G(\xi^{-1})\,d\xi\,dx$$

$$= \int_H \int_G \phi(x)f(x\xi^{-1})\Delta_G(\xi^{-1})\,dx\,d\xi = \int_H \int_G \phi(x\xi)f(x)\,dx\,d\xi$$

$$= \int_G P\phi(xH)f(x)\,dx = \int_G f(x)\,dx.$$

This means that if $Pf = Pg$ then $\int_G f = \int_G g$. It then follows from Proposition 2.50 that the map $Pf \mapsto \int_G f$ is a well-defined $G$-invariant positive linear functional on $C_c(G/H)$. The associated Radon measure is then the desired measure $\mu$. $\qquad\square$

**2.53 Corollary.** *If $H$ is compact, $G/H$ admits a $G$-invariant Radon measure.*

*Proof.* By Proposition 2.27, $\Delta_G|H = \Delta_H = 1$. $\qquad\square$

When no $G$-invariant measure exists, a weaker but still useful result is available. Suppose $\mu$ is a Radon measure on $G/H$. For $x \in G$ we define the translate $\mu_x$ of $\mu$ by

$$\mu_x(E) = \mu(xE).$$

$\mu$ is said to be **quasi-invariant** if the measures $\mu_x$ are all equivalent (i.e., mutually absolutely continuous), and we shall call $\mu$ **strongly quasi-invariant** if there is a continuous function $\lambda : G \times (G/H) \to (0, \infty)$ such that $d\mu_x(p) = \lambda(x, p) \, d\mu(p)$ for all $x \in G$ and $p \in G/H$. Thus, strong quasi-invariance means not only that the measures $\mu_x$ are all strongly equivalent but that the Radon-Nikodym derivative $(d\mu_x/d\mu)(p)$ is jointly continuous in $x$ and $p$.

If $G$ and $H$ are Lie groups, so that $G/H$ is a smooth manifold on which $G$ acts by diffeomorphisms, the construction of strongly quasi-invariant measures is easy, and the derivatives $d\mu_x/d\mu$ can be taken to be smooth. Indeed, the Riemannian volume density associated to any Riemannian metric on $G/H$ will do the job; if $G/H$ is orientable, so will the volume density given by any nonvanishing smooth $n$-form $(n = \dim(G/H))$.

We shall now prove that strongly quasi-invariant measures exist on an arbitrary homogeneous space and show how to construct all of them by modifying the proof of Theorem 2.51. The ideas in this proof will be needed again in Chapter 6, even for the case of Lie groups. First, we need a couple of technical lemmas.

**2.54 Lemma.** *Let $V$ be a symmetric open neighborhood of $1$ in $G$ with compact closure. There exists a set $A \subset G$ such that:*
a. *for every $x \in G$ there exists $a \in A$ such that $xH \cap Va \ne \varnothing$;*
b. *if $K \subset G$ is compact, there are only finitely many $a \in A$ such that $KH \cap \overline{V}a \ne \varnothing$.*

*Proof.* By Zorn's lemma there is a maximal set $A \subset G$ such that if $a, b \in A$ then $a \notin VbH$. (The condition $a \notin VbH$ is symmetric in $a$ and $b$ since $V = V^{-1}$ and $H = H^{-1}$.) For any $x \in G$, $xH$ intersects some $Va$, for otherwise $x \notin VaH$ for all $a$, contradicting maximality. Also, if $K \subset G$ is compact and $KH \cap \overline{V}a \ne \varnothing$ for infinitely many $a$, there exist $a_1, a_2, \ldots \in A$ (all distinct) and $h_1, h_2, \ldots \in H$ such that $a_j h_j \in \overline{V}K$ for all $j$. Since $\overline{V}K$ is compact, the sequence $a_j h_j$ has a cluster point $z$. Pick a symmetric neighborhood $W$ of $1$ such that $WW \subset V$. Then there exist two distinct integers $j$ and $k$ such that $a_j h_j \in Wz$ and $a_k h_k \in Wz$, and hence $a_j h_j \in V a_k h_k$. But then $a_j \in V a_k H$, contradicting the definition of $A$. $\qquad\square$

**2.55 Lemma.** *There exists a continuous $f : G \to [0, \infty)$ such that:*
i. *$\{y : f(y) > 0\} \cap xH \ne \varnothing$ for all $x \in G$;*
ii. *$(\operatorname{supp} f) \cap KH$ is compact for every compact $K \subset G$.*

*Proof.* Pick $g \in C_c^+(G)$ with $g(x) = g(x^{-1})$ and $g(1) > 0$, let $V = \{x : g(x) > 0\}$, choose $A \subset G$ as in Lemma 2.54 for this $V$, and set $f(x) = \sum_{a \in A} g(xa^{-1})$. By (ii) of Lemma 2.54, for $x$ in any compact set there are only finitely many nonzero terms in this sum, so $f$ is well-defined and continuous. Moreover, since $\operatorname{supp} f = \overline{\bigcup_a Va} \subset \bigcup_a \overline{V}a$, for any compact $K \subset G$ $(\operatorname{supp} f) \cap KH$ is contained in a finite union of $\overline{V}a$'s, which is compact. Finally, by (i) of Lemma 2.54, $\{y : f(y) > 0\} = \bigcup_a Va$ intersects every coset $xH$. $\qquad\square$

A **rho-function** for the pair $(G, H)$ is a continuous function $\rho : G \to (0, \infty)$ such that

$$\rho(x\xi) = \frac{\Delta_H(\xi)}{\Delta_G(\xi)}\rho(x) \qquad (x \in G,\ \xi \in H).$$

**2.56 Proposition.** *For any locally compact group $G$ and any closed subgroup $H$, $(G, H)$ admits a rho-function.*

*Proof.* Let $f$ be as in Lemma 2.55, and set

$$\rho(x) = \int_H \frac{\Delta_G(\eta)}{\Delta_H(\eta)} f(x\eta)\, d\eta.$$

The properties of $f$ easily imply that the integral converges for each $x$ and defines a positive continuous function on $G$. Moreover,

$$\rho(x\xi) = \int \frac{\Delta_G(\eta)}{\Delta_H(\eta)} f(x\xi\eta)\, d\eta = \int \frac{\Delta_G(\xi^{-1}\eta)}{\Delta_H(\xi^{-1}\eta)} f(x\eta)\, d\eta = \frac{\Delta_H(\xi)}{\Delta_G(\xi)}\rho(x).$$

$\qquad\square$

In conjunction with Proposition 2.56, the two theorems below give the existence of strongly quasi-invariant measures on $G/H$ and a characterization of all of them.

**2.57 Lemma.** *If $f \in C_c(G)$ and $Pf = 0$ then $\int f\rho = 0$ for any rho-function $\rho$.*

*Proof.* This argument is much the same as in Theorem 2.51. We have

$$0 = \int_H f(x\xi)\, d\xi = \int_H f(x\xi^{-1})\Delta_H(\xi^{-1})\, d\xi$$

for all $x \in G$. By Lemma 2.49 there exists $\phi \geq 0$ in $C_c(G)$ such that

$P\phi = 1$ on $q(\operatorname{supp} f)$, and then

$$
\begin{aligned}
0 &= \int_G \int_H \rho(x)\phi(x)f(x\xi^{-1})\Delta_H(\xi^{-1})\,d\xi\,dx \\
&= \int_H \int_G \rho(x\xi)\phi(x\xi)f(x)\Delta_H(\xi^{-1})\Delta_G(\xi)\,dx\,d\xi \\
&= \int_G \int_H \rho(x)\phi(x\xi)f(x)\,d\xi\,dx \\
&= \int_G f(x)\rho(x)P\phi(q(x))\,dx \\
&= \int_G f(x)\rho(x)\,dx.
\end{aligned}
$$

(The purpose of $\phi$ is to make the integrals absolutely convergent.)   □

**2.58 Theorem.** *Given any rho-function $\rho$ for the pair $(G,H)$, there is a strongly quasi-invariant measure $\mu$ on $G/H$ such that*

$$
(2.59) \qquad \int_{G/H} Pf\,d\mu = \int_G f(x)\rho(x)\,dx \qquad (f \in C_c(G)).
$$

$\mu$ *also satisfies*

$$
\frac{d\mu_x}{d\mu}(yH) = \frac{\rho(xy)}{\rho(y)} \qquad (x,y \in G).
$$

*Proof.* By Proposition 2.50 and Lemma 2.57, the map $Pf \mapsto \int f\rho$ is a well-defined positive linear functional on $C_c(G/H)$, so it defines a Radon measure $\mu$ on $G/H$. Next, the functional equation for rho-functions implies that the quotient $\rho(xy)/\rho(y)$ depends only on the coset $yH$, so it defines a continuous function $\lambda : G \times (G/H) \to (0,\infty)$ by $\lambda(x,q(y)) = \rho(xy)/\rho(y)$. Since the map $P$ commutes with the left action of $G$, for any $x \in G$ and $f \in C_c(G)$ we have

$$
\begin{aligned}
\int_{G/H} Pf(p)\,d\mu_x(p) &= \int_{G/H} Pf(x^{-1}p)\,d\mu(p) \\
&= \int_G f(x^{-1}y)\rho(y)\,dy = \int_G f(y)\rho(xy)\,dy = \int_G f(y)\lambda(x,q(y))\rho(y)\,dy \\
&= \int_{G/H} Pf(p)\lambda(x,p)\,d\mu(p),
\end{aligned}
$$

since $P[f\lambda(x,q(\cdot))] = (Pf)\lambda(x,\cdot)$. This proves the strong quasi-invariance and gives the desired formula for $d\mu_x/d\mu$.   □

**2.60 Proposition.** *If $\mu$ is a quasi-invariant measure on $G/H$, then $\mu(U) > 0$ for every nonempty open set $U$.*

*Proof.* Same as in Proposition 2.19. □

**2.61 Theorem.** *Every strongly quasi-invariant measure on $G/H$ arises from a rho-function as in (2.59), and all such measures are strongly equivalent.*

*Proof.* Suppose $\mu$ is strongly quasi-invariant, so that $(d\mu_x/d\mu)(p) = \lambda(x,p)$ where $\lambda$ is positive and continuous on $G \times (G/H)$. For $x, y \in G$, since $\mu_{xy} = (\mu_x)_y$, the chain rule for Radon-Nikodym derivatives implies that

$$(2.62) \qquad \lambda(xy, p) = \lambda(x, yp)\lambda(y, p)$$

for locally a.e. $p$. But both sides are continuous in $p$, so the set where they disagree is open, and it follows from Proposition 2.60 that (2.62) is valid everywhere.

If $f \in C_c(G)$ and $y \in G$, we have

$$\int_{G/H} \int_H f(y^{-1}x\xi)\lambda(x\xi, H)^{-1}\, d\xi\, d\mu(xH)$$

$$= \int_{G/H} \int_H f(x\xi)\lambda(yx\xi, H)^{-1}\lambda(y, xH)\, d\xi\, d\mu(xH)$$

$$= \int_{G/H} \int_H f(x\xi)\lambda(x\xi, H)^{-1}\, d\xi\, d\mu(xH),$$

since

$$\lambda(yx\xi, H) = \lambda(y, x\xi H)\lambda(x\xi, H) = \lambda(y, xH)\lambda(x\xi, H)$$

by (2.62). Hence $f \mapsto \int_{G/H} \int_H f(x\xi)\lambda(x\xi, H)^{-1}\, d\xi\, d\mu(xH)$ is a left-invariant positive linear functional on $C_c(G)$, so there is a constant $c > 0$ such that

$$(2.63) \qquad \int_{G/H} \int_H f(x\xi)\lambda(x\xi, H)^{-1}\, d\xi\, d\mu(xH) = c\int_G f(x)\, dx.$$

Let $\rho(x) = c\lambda(x, H)$. Replacing $f$ by $f\lambda(\cdot, H)$ in (2.63), we see that

$$\int_{G/H} \int_H f(x\xi)\, d\xi\, d\mu(xH) = \int_G f(x)\rho(x)\, dx,$$

so that (2.59) holds. Moreover, if $\eta \in H$,

$$
\begin{aligned}
\int_G f(x)\rho(x\eta)\,dx &= \Delta_G(\eta)^{-1} \int f(x\eta^{-1})\rho(x)\,dx \\
&= \Delta_G(\eta)^{-1} \int_{G/H} \int_H f(x\xi\eta^{-1})\,d\xi\,d\mu(xH) \\
&= \Delta_G(\eta)^{-1}\Delta_H(\eta) \int_{G/H} \int_H f(x\xi)\,d\xi\,d\mu(xH) \\
&= \Delta_G(\eta)^{-1}\Delta_H(\eta) \int_G f(x)\rho(x)\,dx.
\end{aligned}
$$

This being so for all $f$, we have $\rho(x\eta) = \Delta_G(\eta)^{-1}\Delta_H(\eta)\rho(x)$. Since $\rho$ is continuous and positive, it is a rho-function.

Finally, suppose $\mu$ and $\mu'$ are strongly quasi-invariant measures with associated rho-functions $\rho$ and $\rho'$. By the functional equation for rho-functions, the quotient $\rho'(y)/\rho(y)$ depends only on the coset of $y$ and hence defines a positive continuous function $\phi$ on $G/H$. For $f \in C_c(G)$ we have $P(f\rho'/\rho) = (Pf)\phi$, so

$$
\int_{G/H} Pf\,d\mu' = \int_G f\rho' = \int_G f(\rho'/\rho)\rho = \int_{G/H} (Pf)\phi\,d\mu.
$$

Hence $d\mu'/d\mu = \phi$.    □

Theorem 2.61 implies that all strongly quasi-invariant measures on $G/H$ have the same null sets. (In fact, *all* quasi-invariant measures on $G/H$ have the same null sets; see Bourbaki [18, §VII.2.5].) We conclude this section by relating these null sets — or, more generally, locally null sets — to the locally null sets in $G$. We begin with a result related to Proposition 2.22.

**2.64 Lemma.** *Let $\mu$ be a strongly quasi-invariant measure on $G/H$.*
a. *If $E \subset G/H$ is Borel and $\mu(E) < \infty$, then $E$ is contained in a $\sigma$-compact set.*
b. *If $F \subset G/H$ is locally Borel, then $F$ is locally $\mu$-null if and only if $F \cap K$ is $\mu$-null for every compact $K \subset G/H$.*

*Proof.* Let $G_0$ be an open, closed, and $\sigma$-compact subgroup of $G$ as in Proposition 2.4. For $y \in G$, let $O_y$ be the $G_0$-orbit of $yH$ in $G/H$, that is, $O_y = q(G_0 y) = \{xyH : x \in G_0\}$. Since $q$ maps open sets and compact sets to open sets and compact sets, respectively, $O_y$ is open and $\sigma$-compact. Let $Y$ be a subset of $G$ such that the orbits $O_y$, $y \in Y$, are disjoint and comprise $G/H$. If $\mu(E) < \infty$, there is an open $U \supset E$ with $\mu(U) < \infty$. By Proposition 2.60, for each $y \in Y$ either $\mu(U \cap O_y) > 0$ or

$U \cap O_y = \varnothing$, so there is a countable set $Z \subset Y$ such that $U \cap O_y = \varnothing$ for $y \notin Z$. Hence $E \subset U \subset \bigcup_{z \in Z} O_z$, which proves (a). (b) follows immediately, since if $\mu(F \cap K) = 0$ for $K$ compact, then $\mu(F \cap E) = 0$ for $E$ $\sigma$-compact. $\qquad\square$

In view of Theorem 2.61, Lemma 2.64(b) implies:

**2.65 Corollary.** *All strongly quasi-invariant measures on $G/H$ have the same locally null sets.*

For the remainder of this discussion we fix a strongly quasi-invariant measure $\mu$ on $G/H$ with associated rho-function $\rho$. We observe that the definition (2.47) of $Pf$ makes sense when $f$ is any Borel measurable function from $G$ to $[0, \infty]$ and yields a function $Pf : G/H \to [0, \infty]$.

**2.66 Lemma.** *Suppose $f : G \to [0, \infty]$ is lower semi-continuous. Then $Pf$ is lower semi-continuous, and formula (2.59) holds for $f$.*

*Proof.* If $X$ is a locally compact Hausdorff space, let us call a family $\Phi$ of functions from $X$ to $[0, \infty]$ *directed* if for every $\phi, \psi \in \Phi$ there is a $\chi \in \Phi$ with $\chi \geq \max(\phi, \psi)$. We then have the following facts (Folland [45, Props. 7.11 and 7.12]): A function $f : X \to [0, \infty]$ is lower semi-continuous if and only if $f = \sup_{\phi \in \Phi} \phi$ (pointwise) for some directed family $\Phi \subset C_c(G)$, and in this case, $\int f \, d\nu = \sup_{\phi \in \Phi} \int \phi \, d\nu$ for any Radon measure $\nu$ on $X$.

Now suppose $f : G \to [0, \infty]$ is lower semi-continuous, and let $\Phi$ be a directed family in $C_c(G)$ such that $f = \sup_{\phi \in \Phi} \phi$. For each $x \in G$ we have

$$Pf(xH) = \int_H f(x\xi) \, d\xi = \sup_{\phi \in \Phi} \int_H \phi(x\xi) \, d\xi = \sup_{\phi \in \Phi} P\phi(xH).$$

But $\{P\phi : \phi \in \Phi\}$ is also directed since $P(\max(\phi, \psi)) \geq \max(P\phi, P\psi)$, and $\{\phi\rho : \phi \in \Phi\}$ is directed with $\sup_{\phi \in \Phi} \phi\rho = f\rho$, so $Pf$ is lower semi-continuous and we have

$$\int_{G/H} Pf \, d\mu = \sup_{\phi \in \Phi} \int_{G/H} P\phi \, d\mu = \sup_{\phi \in \Phi} \int_G \phi(x)\rho(x) \, dx = \int_G f(x)\rho(x) \, dx.$$

$\qquad\square$

**2.67 Lemma.** *For every compact $K \subset G$ there is a constant $C_K > 0$ such that $|U| \leq C_K \mu(q(U))$ for every open $U \subset K$.*

*Proof.* Pick $f \geq 0$ in $C_c(G)$ with $f = 1$ on $K$, and let

$$c_K = \left[\inf\{\rho(x) : x \in \operatorname{supp} f\}\right]^{-1}, \qquad C_K = \|Pf\|_{\sup} c_K.$$

Since $q^{-1}(q(U)) \supset U$, we have

$$|U| = \int_G f(x)\chi_U(x)\,dx \le c_K \int_G f(x)\chi_{q(U)}(q(x))\rho(x)\,dx.$$

But $f(\chi_{q(U)} \circ q)$ is lower semi-continuous since $U$ is open, so by Lemma 2.66,

$$|U| \le c_K \int_{G/H} Pf \cdot \chi_{q(U)}\,d\mu \le C_K\mu(q(U)). \qquad \square$$

**2.68 Theorem.** *A set $E \subset G/H$ is locally null (with respect to any quasi-invariant measure $\mu$) if and only if $q^{-1}(E)$ is locally null in $G$ (with respect to Haar measure).*

*Proof.* Suppose $E$ is locally null in $G/H$ and $K$ is compact in $G$; we wish to show that $|K \cap q^{-1}(E)| = 0$. Choose an open set $V \supset K$ with compact closure. Since $q(K) \cap E$ is null, for any $\epsilon > 0$ there is an open set $W$ in $G/H$ such that $W \supset q(K) \cap E$ and $\mu(W) < \epsilon/C_{\overline{V}}$, where $C_{\overline{V}}$ is as in Lemma 2.65. Let $U = V \cap q^{-1}(W)$. Then $U$ is open, $U \supset K \cap q^{-1}(E)$, and by Lemma 2.67, $|U| \le C_{\overline{V}}\mu(W) < \epsilon$. Since $\epsilon$ is arbitrary, $|K \cap q^{-1}(E)| = 0$.

Conversely, suppose $q^{-1}(E)$ is locally null in $G$ and $K$ is compact in $G/H$; we wish to show that $\mu(E \cap K) = 0$. By Lemma 2.49 we can choose $f \ge 0$ in $C_c(G)$ such that $Pf = 1$ on $K$. Let $A = q^{-1}(E \cap K) \cap (\operatorname{supp} f)$; then $|A| = 0$, so for any $\epsilon > 0$ there is an open set $U \supset A$ such that $|U| < \epsilon/\|f\rho\|_{\sup}$. Then $f\chi_U\rho$ is lower semi-continuous, so by Lemma 2.66,

$$\int_{G/H} P(f\chi_U)\,d\mu = \int_G f(x)\chi_U(x)\rho(x)\,dx < \epsilon.$$

But by the construction of $U$ we have $f\chi_U \ge f(\chi_{E \cap K} \circ q)$, so $P(f\chi_U) \ge Pf \cdot \chi_{E \cap K} = \chi_{E \cap K}$. It follows that $\mu(E \cap K) < \epsilon$, and since $\epsilon$ is arbitrary, $\mu(E \cap K) = 0$. $\qquad \square$

## 2.7  Notes and References

The theory of topological groups was developed in the 1930s. For second countable groups, the existence of Haar measure was first proved by Haar [57] and the uniqueness was first proved by von Neumann [144]. The first systematic treatment of analysis on locally compact groups using Haar measure was given by Weil [146], who showed that the countability assumptions were unnecessary. Weil also proved a theorem to the effect that essentially the only measurable groups possessing an invariant measure are the locally compact groups; see [146] for the precise statement. Our proofs of the existence and uniqueness of Haar measure follow Weil [146] and Loomis [84], respectively. H. Cartan [23] devised an argument that yields existence and uniqueness simultaneously and avoids the use of the axiom of choice, which enters our existence proof in the form of Tychonoff's theorem.

For more detailed information about topological groups and Haar measure, as well as extensive historical references, the reader may consult Hewitt and Ross [66]. Further examples of formulas for Haar measure on specific groups or classes of groups may be found in Helgason [64], Hewitt and Ross [66], and Bourbaki [18]. For more about the $p$-adic numbers and other local fields, see Koblitz [80] and Taibleson [132].

Proposition 2.40(a) and Proposition 2.43 can be restated succinctly as $L^1(G) * L^p(G) \subset L^p(G)$, $L^1(G) * L^\infty(G) \subset C_{lu}(G)$, and $L^\infty(G) * L^1(G) \subset C_{ru}(G)$, where $C_{lu}(G)$ (resp. $C_{ru}(G)$) is the space of bounded left (resp. right) uniformly continuous functions on $G$. It is a remarkable and perhaps surprising fact that all these inclusions (except the first one for $p = \infty$) are actually equalities:

**2.69 Theorem.** *On any locally compact group $G$ we have $L^1(G) * L^p(G) = L^p(G)$ for $1 \leq p < \infty$. Moreover, $L^1(G) * L^\infty(G) = L^1(G) * C_{lu}(G) = C_{lu}(G)$ and $L^\infty(G) * L^1(G) = C_{ru}(G) * L^1(G) = C_{ru}(G)$.*

This theorem has a rather complicated history. The fact that $L^1(G) * L^1(G) = L^1(G)$ was first proved by Salem for $G = \mathbb{T}$ and then by Rudin for $G = \mathbb{R}^n \times \mathbb{T}^m$ $(m, n \geq 0)$; for general $G$ it is a special case of a theorem of Cohen concerning factorization in Banach algebras with approximate identities. Finally, Hewitt and (independently) Curtis and Figà-Talamanca generalized Cohen's theorem to modules over such algebras, which yields the full results stated above. The proofs and references can be found in Hewitt and Ross [67, §32]; see also Ross [121] for an engaging and more detailed account of the history of this and related results. (Note, however, that these authors reverse the meanings of $C_{lu}$ and $C_{ru}$.)

The existence of quasi-invariant measures on homogeneous spaces $G/H$ (with merely measurable rho-functions) was first proved by Mackey [88] under the assumption that $G$ is second countable. Bruhat [22] and Loomis [85] showed how to obtain strongly quasi-invariant measures with no countability hypotheses. Our insistence on strongly quasi-invariant measures is motivated by our development of the theory of induced representations in Chapter 6.

Formula (2.59), rewritten as

$$(2.70) \qquad \int_G f(x)\rho(x)\,dx = \int_{G/H} \int_H f(x\xi)\,d\xi\,d\mu(xH),$$

is clearly a sort of Fubini theorem relating Haar measure $d\xi$ on $H$, the quasi-invariant measure $\mu$ on $G/H$, and the weighted Haar measure $\rho(x)\,dx$ on $G$. We have proved this formula for $f \in C_c(G)$ (Theorem 2.58) and for $f$ nonnegative and lower semi-continuous (Lemma 2.66). Although we shall have no need of further results here, it is evidently of interest to know if (2.70) remains valid for more general measurable functions $f$. The answer is affirmative for any $f \in L^1(G)$ and for any $f \geq 0$ that vanishes outside a $\sigma$-finite set; the details can be found in Reiter [117] and Bourbaki [18].

When $G$ is second countable, these assertions actually do reduce to Fubini's theorem. Indeed, in this case there is a Borel set $Y \subset G$ that meets each coset of $H$ in exactly one point. (See Mackey [88, Lemma (1.1)], or Baggett and Ramsay [5, Lemma 7]. The latter paper deals with Polish spaces; every second countable locally compact Hausdorff space is Polish.) $q|Y$ is then a measurable bijection from $Y$ to $G/H$, the map $(y, \xi) \mapsto y\xi$ is a measurable bijection from $Y \times H$ to $G$. In short, $G$ can be identified as a measurable space with $G/H \times H$, and when this identification is made, (2.70) says that the measure $\rho(x)\,dx$ on $G$ is the product of the measure $\mu$ on $G/H$ and Haar measure on $H$.

# 3

## Basic Representation Theory

In this chapter we present the basic concepts in the theory of unitary representations of locally compact groups and derive a few fundamental results: Schur's lemma, the correspondence between unitary representations of $G$ and $*$-representations of $L^1(G)$, and the Gelfand-Raikov existence theorem for irreducible representations. The main tool in proving the latter theorem is the connection between cyclic representations and functions of positive type, an extremely fertile idea that will play a role in a number of places later in the book.

## 3.1 Unitary Representations

Let $G$ be a locally compact group. A **unitary representation** of $G$ is a homomorphism $\pi$ from $G$ into the group $U(\mathcal{H}_\pi)$ of unitary operators on some nonzero Hilbert space $\mathcal{H}_\pi$ that is continuous with respect to the strong operator topology — that is, a map $\pi : G \to U(\mathcal{H}_\pi)$ that satisfies $\pi(xy) = \pi(x)\pi(y)$ and $\pi(x^{-1}) = \pi(x)^{-1} = \pi(x)^*$, and for which $x \mapsto \pi(x)u$ is continuous from $G$ to $\mathcal{H}_\pi$ for any $u \in \mathcal{H}_\pi$. $\mathcal{H}_\pi$ is called the **representation space** of $\pi$, and its dimension is called the **dimension** or **degree** of $\pi$.

More generally, one can consider nonunitary representations of $G$, that is, continuous homomorphisms from $G$ to the group of invertible continuous linear operators on some topological vector space. However, *in this book we shall consider only unitary representations*. Accordingly, although we shall sometimes add the word "unitary" for emphasis, when we say "representation" we shall always mean "unitary representation" unless the contrary is explicitly stated.

We do not require a representation $\pi$ to be continuous in the norm topology of $\mathcal{L}(\mathcal{H}_\pi)$; norm continuity is too restrictive a condition to be of much interest. However, it is worth noting that strong continuity is implied by the apparently less restrictive condition of weak continuity, namely, that $x \mapsto \langle \pi(x)u, v \rangle$ should be continuous from $G$ to $\mathbb{C}$ for each

$u, v \in \mathcal{H}_\pi$. This is because the weak and strong operator topologies coincide on $U(\mathcal{H}_\pi)$. Indeed, if $\{T_\alpha\}$ is a net of unitary operators that converges weakly to $T$, then for any $u \in \mathcal{H}_\pi$,

$$\|(T_\alpha - T)u\|^2 = \|T_\alpha u\|^2 - 2\operatorname{Re}\langle T_\alpha u, Tu \rangle + \|Tu\|^2$$
$$= 2\|u\|^2 - 2\operatorname{Re}\langle T_\alpha u, Tu \rangle.$$

The last term converges to $2\|Tu\|^2 = 2\|u\|^2$, so $\|(T_\alpha - T)u\| \to 0$.

Unitary representations are likely to be found whenever the group $G$ acts on a locally compact Hausdorff space $S$. In this case $G$ also acts on functions on $S$, by

$$[\pi(x)f](s) = f(x^{-1}s).$$

If $S$ has a $G$-invariant Radon measure $\mu$, then $\pi$ defines a unitary representation on $L^2(\mu)$. (The continuity of $\pi$ results from the argument that proves Proposition 2.42.) More generally, suppose $G$ acts on $S$, and $S$ admits a strongly quasi-invariant measure $\mu$; that is, a Radon measure $\mu$ such that $d\mu(xs) = \phi(x, s)\, d\mu(s)$ for some positive continuous function $\phi$. Then one can modify $\pi$ to obtain a unitary representation of $G$ on $L^2(\mu)$, as follows:

$$[\widetilde{\pi}(x)f](s) = \phi(x, x^{-1}s)^{-1/2} f(x^{-1}s).$$

This is easily seen to be unitary:

$$\int \phi(x, x^{-1}s)^{-1}|f(x^{-1}s)|^2\, d\mu(s) = \int \phi(x, s)^{-1}|f(s)|^2\, d\mu(xs)$$
$$= \int |f(s)|^2\, d\mu(s).$$

We have $\widetilde{\pi}(xy) = \widetilde{\pi}(x)\widetilde{\pi}(y)$ because the chain rule for Radon-Nikodym derivatives gives $\phi(xy, s) = \phi(x, ys)\phi(y, s)$, and the continuity of $\widetilde{\pi}$ follows as above.

The most basic example of these constructions arises from the action of $G$ on itself by left or right translations. Left translations yield the **left regular representation** $\pi_L$ of $G$ on $L^2(G)$; this is nothing but the representation

$$[\pi_L(x)f](y) = L_x f(y) = f(x^{-1}y)$$

defined by (2.5). The right translation operators $R_x$ likewise define a unitary representation $\pi_R$ on $L^2(G, \rho)$ where $\rho$ is *right* Haar measure on $G$, or they can be made into a unitary representation $\widetilde{\pi}_R$ on $L^2(G)$ (with *left* Haar measure) by the device described above:

$$[\pi_R(x)f](y) = R_x f(y) = f(yx),$$
$$[\widetilde{\pi}_R(x)f](y) = \Delta(x)^{1/2} R_x f(y) = \Delta(x)^{1/2} f(yx).$$

$\pi_R$ and $\widetilde{\pi}_R$ are both called the **right regular representation** of $G$.

Any unitary representation $\pi$ of $G$ on $\mathcal{H}_\pi$ determines another representation $\overline{\pi}$ on the dual space $\mathcal{H}'_\pi$ of $\mathcal{H}_\pi$, namely $\overline{\pi}(x) = \pi(x^{-1})'$ where the prime denotes the transpose. This may seem a little confusing, as one normally considers a Hilbert space to be identical to its dual; but the identification of $\mathcal{H}_\pi$ with $\mathcal{H}'_\pi$ is *antilinear*, and if we do not make it we are led to consider the transpose rather than the adjoint of an operator. Thus, if we choose an orthonormal basis for $\mathcal{H}_\pi$, so that $\pi(x)$ is represented by a matrix $M(x)$, then the matrix for $\overline{\pi}(x)$ is the inverse transpose of $M(x)$, and since $\pi$ is unitary this is nothing but the complex conjugate of $M(x)$. $\overline{\pi}$ is called the **contragredient** of $\pi$. In some cases — for example, when there is a basis for $\mathcal{H}_\pi$ with respect to which the matrices $M(x)$ are all real — $\overline{\pi}$ is equivalent to $\pi$ in a sense to be described shortly, but in general it is not.

We now introduce some standard terminology associated to unitary representations. If $\pi_1$ and $\pi_2$ are unitary representations of $G$, an **intertwining operator** for $\pi_1$ and $\pi_2$ is a bounded linear map $T : \mathcal{H}_{\pi_1} \to \mathcal{H}_{\pi_2}$ such that $T\pi_1(x) = \pi_2(x)T$ for all $x \in G$. The set of all such operators is denoted by $\mathcal{C}(\pi_1, \pi_2)$:

$$\mathcal{C}(\pi_1, \pi_2) = \left\{ T : \mathcal{H}_{\pi_1} \to \mathcal{H}_{\pi_2} : T\pi_1(x) = \pi_2(x)T \text{ for all } x \in G \right\}.$$

$\pi_1$ and $\pi_2$ are (**unitarily**) **equivalent** if $\mathcal{C}(\pi_1, \pi_2)$ contains a unitary operator $U$, so that $\pi_2(x) = U\pi_1(x)U^{-1}$. We shall not consider any other equivalence relation among representations here, so we shall generally omit the adverb "unitarily."

> *Example*: The right regular representations $\pi_R$ and $\widetilde{\pi}_R$ on $L^2(G, \rho)$ and $L^2(G, \lambda)$ are equivalent, and $f \mapsto \Delta^{1/2} f$ is an intertwining operator. Moreover, $\pi_R$ is equivalent to the left regular representation $\pi_L$, and $Uf(x) = f(x^{-1})$ is an intertwining operator.

We set

$$\mathcal{C}(\pi) = \mathcal{C}(\pi, \pi).$$

This is the space of bounded operators on $\mathcal{H}_\pi$ that commute with $\pi(x)$ for every $x \in G$; it is called the **commutant** or **centralizer** of $\pi$. $\mathcal{C}(\pi)$ is obviously an algebra that is closed under weak limits; it is also closed under taking adjoints, for if $T \in \mathcal{C}(\pi)$ then $T^*\pi(x) = [\pi(x^{-1})T]^* = [T\pi(x^{-1})]^* = \pi(x)T^*$. In short, $\mathcal{C}(\pi)$ is a von Neumann algebra.

Suppose $\mathcal{M}$ is a closed subspace of $\mathcal{H}_\pi$. $\mathcal{M}$ is called an **invariant subspace** for $\pi$ if $\pi(x)\mathcal{M} \subset \mathcal{M}$ for all $x \in G$. If $\mathcal{M}$ is invariant and $\neq \{0\}$, the restriction of $\pi$ to $\mathcal{M}$,

$$\pi^{\mathcal{M}}(x) = \pi(x)|\mathcal{M},$$

defines a representation of $G$ on $\mathcal{M}$, called a **subrepresentation** of $\pi$. We shall consistently use the notation $\pi^{\mathcal{M}}$ for the subrepresentation of $\pi$ on $\mathcal{M}$. If $\pi$ admits an invariant subspace that is nontrivial (i.e., $\neq \{0\}$ or $\mathcal{H}_\pi$) then $\pi$ is called **reducible**, otherwise $\pi$ is **irreducible**.

If $\{\pi_i\}_{i \in I}$ is a family of unitary representations, their **direct sum** $\bigoplus \pi_i$ is the representation $\pi$ on $\mathcal{H} = \bigoplus \mathcal{H}_{\pi_i}$ defined by $\pi(x)(\sum v_i) = \sum \pi_i(x) v_i$ $(v_i \in \mathcal{H}_{\pi_i})$. (See Appendix 1 for more about direct sums of Hilbert spaces.) In this case the $\mathcal{H}_{\pi_i}$'s, as subspaces of $\mathcal{H}$, are invariant under $\pi$, and each $\pi_i$ is a subrepresentation of $\pi$. In fact, subrepresentations always arise as summands in direct sums:

**3.1 Proposition.** *If $\mathcal{M}$ is invariant under $\pi$, then so is $\mathcal{M}^\perp$.*

*Proof.* If $u \in \mathcal{M}$ and $v \in \mathcal{M}^\perp$, $\langle \pi(x)v, u \rangle = \langle v, \pi(x^{-1})u \rangle = 0$, so $\pi(x)v \in \mathcal{M}^\perp$. $\qquad\square$

**3.2 Corollary.** *If $\pi$ has a nontrivial invariant subspace $\mathcal{M}$, then $\pi$ is the direct sum of $\pi^{\mathcal{M}}$ and $\pi^{\mathcal{M}^\perp}$.*

We remark that this result is false for nonunitary representations. For example, $\pi(t) = \left(\begin{smallmatrix} 1 & t \\ 0 & 1 \end{smallmatrix}\right)$ defines a representation of $\mathbb{R}$ on $\mathbb{C}^2$, and the only nontrivial invariant subspace is the one spanned by $(1, 0)$.

If $\pi$ is a unitary representation of $G$ and $u \in \mathcal{H}_\pi$, the closed linear span $\mathcal{M}_u$ of $\{\pi(x)u : x \in G\}$ in $\mathcal{H}_\pi$ is called the **cyclic subspace** generated by $u$. Clearly $\mathcal{M}_u$ is invariant under $\pi$. If $\mathcal{M}_u = \mathcal{H}_\pi$, $u$ is called a **cyclic vector** for $\pi$. $\pi$ is called a **cyclic representation** if it has a cyclic vector.

**3.3 Proposition.** *Every unitary representation is a direct sum of cyclic representations.*

*Proof.* Let $\pi$ be a representation on $\mathcal{H}_\pi$. By Zorn's lemma, there is a maximal collection $\{\mathcal{M}_\alpha\}_{\alpha \in A}$ of mutually orthogonal cyclic subspaces of $\mathcal{H}_\pi$. If there were a nonzero $u \in \mathcal{H}_\pi$ orthogonal to all the $\mathcal{M}_\alpha$'s, the cyclic subspace generated by $u$ would also be orthogonal to the $\mathcal{M}_\alpha$'s by Proposition 3.1, contradicting maximality. Hence $\mathcal{H}_\pi = \bigoplus \mathcal{M}_\alpha$, and $\pi = \bigoplus \pi^{\mathcal{M}_\alpha}$. $\qquad\square$

We now give some results relating operators in $\mathcal{C}(\pi)$ to reducibility properties of $\pi$. The first of these is very simple, but the next is one of the fundamental theorems of the subject.

**3.4 Proposition.** *Let $\mathcal{M}$ be a closed subspace of $\mathcal{H}_\pi$ and let $P$ be the orthogonal projection onto $\mathcal{M}$. Then $\mathcal{M}$ is invariant under $\pi$ if and only if $P \in \mathcal{C}(\pi)$.*

*Proof.* If $P \in \mathcal{C}(\pi)$ and $v \in \mathcal{M}$, then $\pi(x)v = \pi(x)Pv = P\pi(x)v \in \mathcal{M}$, so $\mathcal{M}$ is invariant. Conversely, if $\mathcal{M}$ is invariant we have $\pi(x)Pv = \pi(x)v = P\pi(x)v$ for $v \in \mathcal{M}$ and $\pi(x)Pv = 0 = P\pi(x)v$ for $v \in \mathcal{M}^{\perp}$ (by Proposition 3.1). Hence $\pi(x)P = P\pi(x)$. □

**3.5 Theorem** (Schur's Lemma).
a. *A unitary representation $\pi$ of $G$ is irreducible if and only if $\mathcal{C}(\pi)$ contains only scalar multiples of the identity.*
b. *Suppose $\pi_1$ and $\pi_2$ are irreducible unitary representations of $G$. If $\pi_1$ and $\pi_2$ are equivalent then $\mathcal{C}(\pi_1, \pi_2)$ is one-dimensional; otherwise, $\mathcal{C}(\pi_1, \pi_2) = \{0\}$.*

*Proof.* (a) If $\pi$ is reducible, $\mathcal{C}(\pi)$ contains nontrivial projections, by Proposition 3.4. Conversely, suppose $T \in \mathcal{C}(\pi)$ and $T \neq cI$. Then $A = \frac{1}{2}(T + T^*)$ and $B = \frac{1}{2i}(T - T^*)$ are in $\mathcal{C}(\pi)$, and at least one of them — say, $A$ — is not a multiple of $I$. $A$ is self-adjoint, so every operator that commutes with $A$, and in particular every $\pi(x)$, commutes with all the projections $\chi_E(A)$, $E \subset \mathbb{R}$ (Theorem 1.51(c)). Thus $\mathcal{C}(\pi)$ contains nontrivial projections, and so $\pi$ is reducible by Proposition 3.4 again.

(b) If $T \in \mathcal{C}(\pi_1, \pi_2)$ then $T^* \in \mathcal{C}(\pi_2, \pi_1)$ because

$$T^*\pi_2(x) = [\pi_2(x^{-1})T]^* = [T\pi_1(x^{-1})]^* = \pi_1(x)T^*.$$

It follows that $T^*T \in \mathcal{C}(\pi_1)$ and $TT^* \in \mathcal{C}(\pi_2)$, so $T^*T = cI$ and $TT^* = cI$. Hence, either $T = 0$ or $c^{-1/2}T$ is unitary. This shows that $\mathcal{C}(\pi_1, \pi_2) = \{0\}$ precisely when $\pi_1$ and $\pi_2$ are inequivalent, and that $\mathcal{C}(\pi_1, \pi_2)$ consists of scalar multiples of unitary operators. If $T_1, T_2 \in \mathcal{C}(\pi_1, \pi_2)$ then $T_2^{-1}T_1 = T_2^*T_1 \in \mathcal{C}(\pi_1)$, so $T_2^{-1}T_1 = cI$ and $T_1 = cT_2$, so $\dim \mathcal{C}(\pi_1, \pi_2) = 1$. □

**3.6 Corollary.** *If $G$ is Abelian, then every irreducible representation of $G$ is one-dimensional.*

*Proof.* If $\pi$ is a representation of $G$, the operators $\pi(x)$ all commute with one another and so belong to $\mathcal{C}(\pi)$. If $\pi$ is irreducible, we therefore have $\pi(x) = c_x I$ for each $x \in G$. But then every one-dimensional subspace of $\mathcal{H}_\pi$ is invariant, so $\dim \mathcal{H}_\pi = 1$. □

The irreducible unitary representations of a locally compact group $G$ are the basic building blocks of the harmonic analysis associated to $G$. Actually, this assertion is a bit premature at this point, for it may not be obvious that a given group $G$ has any irreducible representations except the trivial one-dimensional representation $\pi_0(x) \equiv I$. But, in fact, $G$ always has enough irreducible representations to separate points: this is the Gelfand-Raikov theorem, the final major result of this chapter. Once one has this assurance, the basic questions of harmonic analysis on $G$ are the following.

   i. Describe all the irreducible unitary representations of $G$, up to equivalence.

  ii. Determine how arbitrary unitary representations of $G$ can be built out of irreducible ones.

 iii. Given a specific unitary representation of $G$ such as the regular representation, show concretely how to build it out of irreducible ones.

The answer to question (i) will, of course, depend strongly on the nature of $G$. We shall discuss a number of examples in later chapters, and in Chapter 6 we shall develop a general technique that can be used to classify the irreducible representations of many groups.

As to question (ii), one might hope that every representation would be a direct sum of irreducible subrepresentations. This is the case if $G$ is compact, as we shall see in §5.1, but not otherwise. For example, consider the regular representation of $\mathbb{R}$ on $L^2(\mathbb{R})$, $[\pi(x)f](t) = f(t-x)$. This representation has *no* irreducible subrepresentations. If there were one, the space on which it acts would be one-dimensional by Corollary 3.6, hence of the form $\{cf : c \in \mathbb{C}\}$ for some $f \neq 0 \in L^2$. But then for each $x \in \mathbb{R}$ we would have $f(t-x) = c_x f(t)$ for some $c_x \in \mathbb{C}$ of modulus 1, so $|f(t)|$ would be constant. This is impossible for $f \in L^2$ unless $f = 0$.

Instead, the Fourier inversion formula exhibits $\pi$ as a direct *integral* of irreducible representations. Namely, for each $\xi \in \mathbb{R}$, $\pi_\xi(x) = e^{-2\pi i \xi x}$ (acting by scalar multiplication on $\mathbb{C}$) is an irreducible representation of $\mathbb{R}$. If $f \in L^2(\mathbb{R})$, we have $f(t) = \int e^{2\pi i \xi t} \widehat{f}(\xi)\, d\xi$ (with the integral suitably interpreted) where $\widehat{f}$ is the Fourier transform of $f$, and hence

$$[\pi(x)f](t) = \int e^{2\pi i \xi(t-x)} \widehat{f}(\xi)\, d\xi = \int \pi_\xi(x)[e^{2\pi i \xi t} \widehat{f}(\xi)]\, d\xi.$$

There is a general theory of direct integrals of Hilbert spaces, including direct sums as special cases, that allows one construct direct integrals of unitary representations in the abstract setting. It can then be shown that every unitary representation is a direct integral of irreducible ones. There remains the question of uniqueness of direct integral decompositions, which is more delicate than one might suspect at first. We shall address these issues for Abelian groups in §4.4 and for compact groups in §5.1, and we shall sketch the results for non-compact, non-Abelian groups in §7.4.

Our concern with question (iii) will be largely limited to the regular representation. Its answer in this case, for a given group $G$, is called the *Plancherel theorem* for $G$. We shall prove the Plancherel theorem for Abelian groups in §4.2 and for compact groups (where it is part of the Peter-Weyl theorem) in §5.2, and we shall discuss the Plancherel theorem for non-compact, non-Abelian groups in §7.5.

## 3.2 Representations of a Group and Its Group Algebra

If $G$ is a locally compact group, we recall from §2.5 that $L^1(G)$ is a Banach $*$-algebra under the convolution product and the involution $f^*(x) = \Delta(x^{-1})\overline{f(x^{-1})}$, where $\Delta$ is the modular function of $G$. In this section we show that there is a one-to-one correspondence between the unitary representations of $G$ and the nondegenerate $*$-representations of $L^1(G)$.

Any unitary representation $\pi$ of $G$ determines a representation of $L^1(G)$, still denoted by $\pi$, in the following way. If $f \in L^1(G)$, we define the bounded operator $\pi(f)$ on $\mathcal{H}_\pi$ by

$$\pi(f) = \int f(x)\pi(x)\,dx.$$

We interpret this operator-valued integral in the weak sense, as explained in Appendix 4. That is, for any $u \in \mathcal{H}_\pi$ we define $\pi(f)u$ by specifying its inner product with an arbitrary $v \in \mathcal{H}_\pi$, and the latter is given by

$$(3.7) \qquad \langle \pi(f)u, v \rangle = \int f(x)\langle \pi(x)u, v \rangle\,dx.$$

Since $\langle \pi(x)u, v \rangle$ is a bounded continuous function of $x \in G$, the integral on the right is the ordinary integral of a function in $L^1(G)$. It is obvious from this formula that $\langle \pi(f)u, v \rangle$ depends linearly on $u$ and antilinearly on $v$ and that $|\langle \pi(f)u, v \rangle| \le \|f\|_1 \|u\| \|v\|$, so $\pi(f)$ is indeed a bounded linear operator on $\mathcal{H}_\pi$ with norm $\|\pi(f)\| \le \|f\|_1$.

*Example.* Let $\pi_L$ be the left regular representation of $G$, $\pi_L(x) = L_x$. By (2.38), $\pi_L(f)$ is convolution with $f$ on the left:

$$(3.8) \qquad [\pi_L(f)g] = \int f(y)L_y g\,dy = f * g.$$

**3.9 Theorem.** *Let $\pi$ be a unitary representation of $G$. The map $f \mapsto \pi(f)$ is a nondegenerate $*$-representation of $L^1(G)$ on $\mathcal{H}_\pi$. Moreover, for $x \in G$ and $f \in L^1(G)$,*

$$(3.10) \qquad \pi(x)\pi(f) = \pi(L_x f), \qquad \pi(f)\pi(x) = \Delta(x^{-1})\pi(R_{x^{-1}}f).$$

*Proof.* The correspondence $f \mapsto \pi(f)$ is obviously linear. Formally, we have

$$\pi(f * g) = \iint f(y)g(y^{-1}x)\pi(x)\,dy\,dx = \iint f(y)g(x)\pi(yx)\,dx\,dy$$

$$= \iint f(y)g(x)\pi(y)\pi(x)\,dx\,dy = \pi(f)\pi(g),$$

$$\pi(f^*) = \int \Delta(x^{-1})\overline{f(x^{-1})}\pi(x)\,dx = \int \overline{f(x)}\pi(x^{-1})\,dx$$

$$= \int [f(x)\pi(x)]^*\,dx = \pi(f)^*,$$

$$\pi(x)\pi(f) = \int f(y)\pi(x)\pi(y)\,dy = \int f(y)\pi(xy)\,dy$$

$$= \int f(x^{-1}y)\pi(y)\,dy = \pi(L_x f),$$

$$\pi(f)\pi(x) = \int f(y)\pi(y)\pi(x)\,dy = \int f(y)\pi(yx)\,dy$$

$$= \Delta(x^{-1})\int f(yx^{-1})\pi(y)\,dy = \Delta(x^{-1})\pi(R_{x^{-1}}f).$$

Each of these calculations is justified by applying the operators to $u \in \mathcal{H}_\pi$ and taking the inner product with $v \in \mathcal{H}_\pi$ as in (3.7). In a couple of places we use the fact that integration of vector-valued functions commutes with application of bounded linear maps (see Appendix 4) to bring an operator $\pi(x)$ into or out of an integral; details are left to the reader.

This shows that $\pi$ is a $*$-homomorphism and proves (3.9). To see that $\pi$ is nondegenerate, suppose $u \neq 0 \in \mathcal{H}_\pi$. Pick a compact neighborhood $V$ of 1 in $G$ such that $\|\pi(x)u - u\| < \|u\|$ for $x \in V$, and set $f = |V|^{-1}\chi_V$. Then

$$\|\pi(f)u - u\| = \frac{1}{|V|}\left\|\int_V [\pi(x)u - u]\,dx\right\| < \|u\|,$$

and in particular $\pi(f)u \neq 0$. $\qquad\square$

**3.11 Theorem.** *Suppose $\pi$ is a nondegenerate $*$-representation of $L^1(G)$ on the Hilbert space $\mathcal{H}$. Then $\pi$ arises from a unique unitary representation of $G$ on $\mathcal{H}$ according to (3.7).*

*Proof.* The idea is that $\pi(x)$ should be the limit of $\pi(f)$ as $f$ approaches the $\delta$-function at $x$. Thus, let $\{\psi_U\}$ be an approximate identity in $L^1$, as in Proposition 2.44. If $f \in L^1$, we have $\psi_U * f \to f$ in $L^1$, so $(L_x\psi_U) * f = L_x(\psi_U * f) \to L_x f$ in $L^1$ for any $x \in G$, and hence $\pi(L_x\psi_U)\pi(f)v \to \pi(L_x f)v$ for all $v \in \mathcal{H}$.

Let $\mathcal{D}$ be the (finite) linear span of $\{\pi(f)v : f \in L^1,\ v \in \mathcal{H}\}$. Then $\mathcal{D}$ is a dense subspace of $\mathcal{H}$, for if $u \perp \mathcal{D}$ then $0 = \langle u, \pi(f)v \rangle = \langle \pi(f^*)u, v \rangle$ for all $v$ and $f$, whence $u = 0$ since $\pi$ is nondegenerate. The preceding calculations show that the operators $\pi(L_x \psi_U)$ converge strongly on $\mathcal{D}$ to an operator $\widetilde{\pi}(x) : \mathcal{D} \to \mathcal{D}$ such that $\widetilde{\pi}(x)\pi(f)v = \pi(L_x f)v$. $\widetilde{\pi}(x)$ is well-defined because

$$\sum \pi(f_j)v_j = 0 \implies \sum \pi(L_x f_j)v_j = \lim \sum \pi(L_x \psi_U)\pi(f_j)v_j = 0.$$

Moreover, the operators $\pi(L_x \psi_U)$ satisfy $\|\pi(L_x \psi_U)\| \leq \|L_x \psi_U\|_1 = 1$, by Proposition 1.24(b). Therefore, the operator $\widetilde{\pi}(x)$ extends uniquely to $\mathcal{H}$ in such a way that $\|\widetilde{\pi}(x)\| \leq 1$ and $\widetilde{\pi}(x)\pi(f) = \pi(L_x f)$.

We claim that $\widetilde{\pi}$ is a unitary representation of $G$. First,

$$\widetilde{\pi}(xy)\pi(f) = \pi(L_{xy}f) = \pi(L_x L_y f) = \widetilde{\pi}(x)\pi(L_y f) = \widetilde{\pi}(x)\widetilde{\pi}(y)\pi(f),$$

so $\widetilde{\pi}(xy) = \widetilde{\pi}(x)\widetilde{\pi}(y)$ on $\mathcal{D}$ and hence on $\mathcal{H}$. Next, $\widetilde{\pi}(1) = I$, so $\widetilde{\pi}$ is a homomorphism from $G$ to the group of invertible operators on $\mathcal{H}$. Since

$$\|u\| = \|\widetilde{\pi}(x^{-1})\widetilde{\pi}(x)u\| \leq \|\widetilde{\pi}(x)u\| \leq \|u\| \qquad (u \in \mathcal{H}),$$

$\widetilde{\pi}(x)$ is an isometry and hence a unitary operator. Finally, if $x_\alpha \to x$ in $G$ then $L_{x_\alpha}f \to L_x f$ in $L^1$ for any $f \in L^1$, so $\widetilde{\pi}(x_\alpha)\pi(f) = \pi(L_{x_\alpha}f) \to \pi(L_x f) = \widetilde{\pi}(x)\pi(f)$ strongly. Thus $\widetilde{\pi}(x_\alpha) \to \widetilde{\pi}(x)$ strongly on $\mathcal{D}$, and since $\|\widetilde{\pi}(x_\alpha)\| = 1$ for all $\alpha$, a simple $\epsilon/3$-argument shows that $\widetilde{\pi}(x_\alpha) \to \widetilde{\pi}(x)$ strongly on $\mathcal{H}$, so $\widetilde{\pi}$ is continuous.

It remains to show that $\pi(f) = \widetilde{\pi}(f)$ for $f \in L^1$, where $\widetilde{\pi}(f)$ arises from $\widetilde{\pi}$ by (3.7). But if $f, g \in L^1$ we have $f * g = \int f(y)L_y g\, dy$ by (2.38), where we interpret the integral as the integral of an $L^1$-valued function of $y$. Since $\pi$ is a bounded linear map from $L^1$ to $\mathcal{L}(\mathcal{H})$, it commutes with integration (see Appendix 4), so

$$\pi(f)\pi(g) = \pi(f * g) = \int f(y)\pi(L_y g)\, dy = \int f(y)\widetilde{\pi}(y)\pi(g)\, dy$$

$$= \left[ \int f(y)\widetilde{\pi}(y)\, dy \right] \pi(g) = \widetilde{\pi}(f)\pi(g).$$

Thus $\widetilde{\pi}(f) = \pi(f)$ on $\mathcal{D}$ and hence on $\mathcal{H}$.

Finally, suppose $\widehat{\pi}$ is another unitary representation of $G$ such that $\widehat{\pi}(f) = \pi(f)$ for $f \in L^1(G)$. It follows from (3.7) that $\langle \widehat{\pi}(x)u, v \rangle = \langle \widetilde{\pi}(x)u, v \rangle$ for all $x \in G$ and all $u, v \in \mathcal{H}$, and hence $\widehat{\pi}(x) = \widetilde{\pi}(x)$ for all $x \in G$. $\qquad \square$

Let $\pi$ be a unitary representation of $G$. If $G$ is discrete, the associated representation of $L^1(G)$ includes the representation of $G$ itself, since

$\pi(x) = \pi(\delta_x)$ where $\delta_x$ is the function whose value at $x$ is 1 and whose value elsewhere is 0. On the other hand, if $G$ is not discrete and $\pi$ is infinite-dimensional, the families

$$\pi(G) = \{\pi(x) : x \in G\} \quad \text{and} \quad \pi(L^1(G)) = \{\pi(f) : f \in L^1(G)\}$$

are quite different. (For one thing, the operators $\pi(f)$ are rarely invertible.) In fact, the C* algebras generated by these two families frequently have trivial intersection. However, they are related:

**3.12 Theorem.** *Let $\pi$ be a unitary representation of $G$.*

a. *The von Neumann algebras generated by $\pi(G)$ and $\pi(L^1(G))$ are identical.*

b. *$T \in \mathcal{L}(\mathcal{H}_\pi)$ belongs to $\mathcal{C}(\pi)$ if and only if $T\pi(f) = \pi(f)T$ for every $f \in L^1(G)$.*

c. *A closed subspace $\mathcal{M}$ of $\mathcal{H}_\pi$ is invariant under $\pi$ if and only if $\pi(f)\mathcal{M} \subset \mathcal{M}$ for every $f \in L^1(G)$.*

*Proof.* (a) First, we claim that if $g \in C_c(G)$, $\pi(g)$ is the strong limit of Riemann sums $\Sigma_E = \sum g(x_j)\pi(x_j)|E_j|$, where $E = \{E_j\}$ denotes a finite partition of supp $g$ and $x_j \in E_j$. Indeed, given $\epsilon > 0$ and $u_1, \ldots, u_n \in \mathcal{H}_\pi$, by using the uniform continuity of the maps $x \mapsto g(x)\pi(x)u_m$ it is easy to find a partition $E = \{E_j\}$ of supp $g$ such that $\|g(x)\pi(x)u_m - g(y)\pi(y)u_m\| < \epsilon$ for $m = 1, \ldots, n$ when $x$ and $y$ lie in the same $E_j$, so that $\|\Sigma_E u_m - \pi(g)u_m\| < \epsilon|\operatorname{supp} g|$ for $m = 1, \ldots, n$. Thus every strong neighborhood of $\pi(g)$ contains sums $\Sigma_E$.

Now, if $f \in L^1(G)$, $f$ is the $L^1$ limit of functions in $C_c(G)$, so $\pi(f)$ is the norm limit of operators $\pi(g)$ with $g \in C_c(G)$. These in turn are strong limits of Riemann sums, and these sums are in the algebra generated by $\pi(G)$. On the other hand, the proof of Theorem 3.11 shows that $\pi(x)$ is the strong limit of $\pi(L_x\psi_U)$ as $U \to \{1\}$. Hence the algebras generated by $\pi(G)$ and $\pi(L^1(G))$ have the same strong closure, and *a fortiori* the same weak closure.

(b) If $T \in \mathcal{C}(\pi)$ then $T$ commutes with every element of the von Neumann algebra generated by $\pi(G)$, and in particular with every $\pi(f)$; and vice versa.

(c) This follows from (b) together with Proposition 3.4 (and its analogue for representations of $L^1(G)$, which has the same proof). $\qquad\square$

## 3.3  Functions of Positive Type

A **function of positive type** on a locally compact group $G$ is a function $\phi \in L^\infty(G)$ that defines a positive linear functional on the Banach $*$-algebra $L^1(G)$, i.e., that satisfies

$$\int (f^* * f)\phi \geq 0 \text{ for all } f \in L^1(G).$$

We have

$$\int (f^* * f)\phi = \iint \Delta(y^{-1})\overline{f(y^{-1})}f(y^{-1}x)\phi(x)\,dy\,dx$$
$$= \iint \overline{f(y)}f(yx)\phi(x)\,dy\,dx,$$

so reversing the order of integration and substituting $y^{-1}x$ for $x$ shows that $\phi$ is of positive type if and only if

$$(3.13) \qquad \iint f(x)\overline{f(y)}\phi(y^{-1}x)\,dy\,dx \geq 0 \qquad (f \in L^1(G)).$$

We shall prove below (Corollary 3.21) that any function of positive type agrees locally a.e. with a continuous function, and from that point on all functions of positive type will be implicitly assumed to be continuous. We shall set

$$\mathcal{P} = \mathcal{P}(G) = \text{the set of all continuous functions of positive type on } G.$$

**3.14 Proposition.** *If $\phi$ is of positive type then so is $\overline{\phi}$.*

*Proof.* By examination of (3.13) one sees that $\int (f^* * f)\overline{\phi}$ is the complex conjugate of $\int [(\overline{f})^* * \overline{f}]\phi$. The last integral is nonnegative for all $f \in L^1$, hence so is the first. $\qquad\square$

There is a beautiful connection between functions of positive type and unitary representations. The first ingredient is the following result.

**3.15 Proposition.** *If $\pi$ is a unitary representation of $G$ and $u \in \mathcal{H}_\pi$, let $\phi(x) = \langle \pi(x)u, u \rangle$. Then $\phi \in \mathcal{P}$.*

*Proof.* $\phi$ is clearly continuous. Also, $\phi(y^{-1}x) = \langle \pi(y^{-1})\pi(x)u, u \rangle = \langle \pi(x)u, \pi(y)u \rangle$, so if $f \in L^1$,

$$\iint f(x)\overline{f(y)}\phi(y^{-1}x)\,dx\,dy = \iint \langle f(x)\pi(x)u, f(y)\pi(y)u \rangle\,dx\,dy$$
$$= \|\pi(f)u\|^2 \geq 0. \qquad\square$$

**3.16 Corollary.** *If $f \in L^2(G)$, let $\widetilde{f}(x) = \overline{f(x^{-1})}$; then $f * \widetilde{f} \in \mathcal{P}$.*

*Proof.* Let $\pi$ be the left regular representation: then

$$\langle \pi(x)f, f \rangle = \int f(x^{-1}y)\overline{f(y)}\,dy = \overline{f * \widetilde{f}(x)}.$$

Hence $f * \widetilde{f} \in \mathcal{P}$ in view of Proposition 3.14. $\qquad\square$

We now show that every nonzero function of positive type arises from a unitary representation as in Proposition 3.15. To begin with, if $\phi \neq 0$ is of positive type, it defines a positive semi-definite Hermitian form on $L^1(G)$ by

$$(3.17) \qquad \langle f, g \rangle_\phi = \int (g^* * f)\phi = \iint f(x)\overline{g(y)}\phi(y^{-1}x)\,dx\,dy,$$

which clearly satisfies

$$(3.18) \qquad |\langle f, g \rangle_\phi| \leq \|\phi\|_\infty \|f\|_1 \|g\|_1.$$

Let $\mathcal{N} = \{ f \in L^1 : \langle f, f \rangle_\phi = 0 \}$. By the Schwarz inequality (see Appendix 1), $f \in \mathcal{N}$ if and only if $\langle f, g \rangle_\phi = 0$ for all $g \in L^1$. The form $\langle \cdot, \cdot \rangle_\phi$ therefore induces an inner product on the quotient space $L^1/\mathcal{N}$, still denoted by $\langle \cdot, \cdot \rangle_\phi$. We denote the Hilbert space completion of $L^1/\mathcal{N}$ by $\mathcal{H}_\phi$, and we denote the image of $f \in L^1$ in $L^1/\mathcal{N} \subset \mathcal{H}_\phi$ by $\widetilde{f}$. By (3.18),

$$\|\widetilde{f}\|_{\mathcal{H}_\phi} \leq \|\phi\|_\infty^{1/2} \|f\|_1.$$

Now, if $f, g \in L^1$ and $x \in G$,

$$\langle L_x f, L_x g \rangle_\phi = \iint f(x^{-1}y)\overline{g(x^{-1}z)}\phi(z^{-1}y)\,dy\,dz$$

$$= \iint f(y)\overline{g(z)}\phi((xz)^{-1}(xy))\,dy\,dz = \langle f, g \rangle_\phi.$$

In particular, $L_x(\mathcal{N}) \subset \mathcal{N}$, so the operators $L_x$ yield a unitary representation $\pi_\phi$ of $G$ on $\mathcal{H}_\phi$ that is determined by

$$(3.19) \qquad \pi_\phi(x)\widetilde{f} = (L_x f)^{\sim} \qquad (f \in L^1(G)).$$

In view of (3.8), it is easily verified that the corresponding representation of $L^1(G)$ on $\mathcal{H}_\phi$ is given by $\pi_\phi(f)\widetilde{g} = (f * g)^{\sim}$.

**3.20 Theorem.** *Given a function $\phi$ of positive type on $G$, let $\mathcal{H}_\phi$ be the Hilbert space determined as above by the Hermitian form (3.17), let $f \mapsto \widetilde{f}$ be the canonical map from $L^1(G)$ to $\mathcal{H}_\phi$, and let $\pi_\phi$ be the unitary representation of $G$ on $\mathcal{H}_\phi$ defined by (3.19). There is a cyclic vector $\epsilon$ for $\pi_\phi$ such that $\pi_\phi(f)\epsilon = \widetilde{f}$ for all $f \in L^1$ and $\phi(x) = \langle \pi_\phi(x)\epsilon, \epsilon \rangle$ locally a.e.*

*Proof.* Let $\{\psi_U\}$ be an approximate identity. Then $\{\psi_U^*\}$ satisfies conditions (i) and (ii) of Proposition 2.44,[1] so for any $f \in L^1$, $\langle \widetilde{f}, \widetilde{\psi}_U \rangle_\phi = \int (\psi_U^* * f)\phi \to \int f\phi$. Also, $\|\widetilde{\psi}_U\|_{\mathcal{H}_\phi} \leq \|\phi\|_\infty^{1/2} \|\psi_U\|_1 = \|\phi\|_\infty^{1/2}$. It follows easily that $\lim \langle v, \psi_U \rangle_\phi$ exists for all $v \in \mathcal{H}_\phi$, and hence that $\widetilde{\psi}_U$ converges weakly in $\mathcal{H}_\phi$ to an element $\epsilon$ such that $\langle \widetilde{f}, \epsilon \rangle_\phi = \int f\phi$ for all $f \in L^1$.

If $f, g \in L^1$ and $y \in G$, we have

$$\langle \widetilde{g}, \pi_\phi(y)\epsilon \rangle_\phi = \langle \pi_\phi(y^{-1})\widetilde{g}, \epsilon \rangle_\phi = \langle (L_{y^{-1}}g)\widetilde{\;}, \epsilon \rangle_\phi$$
$$= \int g(yx)\phi(x)\,dx = \int g(x)\phi(y^{-1}x)\,dx,$$

and hence

$$\langle \widetilde{g}, \widetilde{f} \rangle_\phi = \int \langle \widetilde{g}, \pi_\phi(y)\epsilon \rangle_\phi \overline{f(y)}\,dy = \langle \widetilde{g}, \pi_\phi(f)\epsilon \rangle_\phi.$$

It follows that $\widetilde{f} = \pi_\phi(f)\epsilon$ for all $f \in L^1$. It also follows that if $\langle \widetilde{g}, \pi_\phi(y)\epsilon \rangle = 0$ for all $y$ then $\widetilde{g} = 0$, so the linear span of $\{\pi_\phi(y)\epsilon : y \in G\}$ is dense in $\mathcal{H}_\phi$ and $\epsilon$ is a cyclic vector. Moreover,

$$\int \langle \epsilon, \pi_\phi(y)\epsilon \rangle \overline{f(y)}\,dy = \lim \int \langle \widetilde{\psi}_U, \pi_\phi(y)\epsilon \rangle_\phi \overline{f(y)}\,dy = \lim \langle \widetilde{\psi}_U, \widetilde{f} \rangle_\phi$$
$$= \langle \epsilon, \widetilde{f} \rangle_\phi = \overline{\langle \widetilde{f}, \epsilon \rangle_\phi} = \int \overline{\phi(y)f(y)}\,dy$$

for every $f \in L^1$, and hence

$$\langle \pi_\phi(y)\epsilon, \epsilon \rangle = \overline{\langle \epsilon, \pi_\phi(y)\epsilon \rangle} = \phi(y) \text{ locally a.e.} \qquad \square$$

**3.21 Corollary.** *Every function of positive type agrees locally a.e. with a continuous function.*

**3.22 Corollary.** *If $\phi \in \mathcal{P}$ then $\|\phi\|_\infty = \phi(1)$ and $\phi(x^{-1}) = \overline{\phi(x)}$.*

*Proof.* We have $\phi(x) = \langle \pi(x)u, u \rangle$ for some $\pi$ and $u$, so $|\phi(x)| = |\langle \pi(x)u, u \rangle| \leq \|u\|^2 = \phi(1)$ and $\phi(x^{-1}) = \langle \pi(x^{-1})u, u \rangle = \langle u, \pi(x)u \rangle = \overline{\phi(x)}$. $\qquad \square$

Proposition 3.15 and Theorem 3.20 establish a correspondence between cyclic representations and functions of positive type. (The representation $\pi$ in Proposition 3.15 is not assumed cyclic, but $\langle \pi(x)u, u \rangle$ clearly depends only on the subrepresentation of $\pi$ on the cyclic subspace generated by $u$.) The picture is completed by the following uniqueness theorem.

---

[1] It satisfies condition (iii) only when $G$ is unimodular.

**3.23 Proposition.** *Suppose $\pi$ and $\rho$ are cyclic representations of $G$ with cyclic vectors $u$ and $v$. If $\langle \pi(x)u, u \rangle = \langle \rho(x)v, v \rangle$ for all $x \in G$, then $\pi$ and $\rho$ are unitarily equivalent; more precisely, there is a unitary $T \in \mathcal{C}(\pi, \rho)$ such that $Tu = v$.*

*Proof.* For any $x, y \in G$ we have

$$\langle \pi(x)u, \pi(y)u \rangle = \langle \pi(y^{-1}x)u, u \rangle = \langle \rho(y^{-1}x)v, v \rangle = \langle \rho(x)v, \rho(y)v \rangle.$$

It follows that if we define $T[\pi(x)u] = \rho(x)v$, then $T$ extends by linearity to an isometry from the span of $\{\pi(x)u : x \in G\}$ to the span of $\{\rho(x)v : x \in G\}$, and it then extends by continuity to a unitary map from $\mathcal{H}_\pi$ to $\mathcal{H}_\rho$. Since $\rho(y)T[\pi(x)u] = \rho(yx)v = T[\pi(y)\pi(x)u]$ we have $\rho(y)T = T\pi(y)$, so $T \in \mathcal{C}(\pi, \rho)$. $\qquad\square$

**3.24 Corollary.** *If $\pi$ is a cyclic representation of $G$ with cyclic vector $u$ and $\phi(x) = \langle \pi(x)u, u \rangle$, then $\pi$ is unitarily equivalent to the representation $\pi_\phi$ defined by (3.19).*

*Remark:* The proof of Proposition 3.23 shows that if $\pi$ and $\rho$ are arbitrary unitary representations of $G$, $u \in \mathcal{H}_\pi$, $v \in \mathcal{H}_\rho$, and $\langle \pi(x)u, u \rangle = \langle \rho(x)v, v \rangle$ for all $x$, there exists $T \in \mathcal{C}(\pi, \rho)$ such that $Tu = v$. Namely, if $\mathcal{M}$ is the cyclic subspace generated by $u$, the proof yields an isometric $T \in \mathcal{C}(\pi^\mathcal{M}, \rho)$, and one can extend $T$ from $\mathcal{M}$ to $\mathcal{H}_\pi$ by setting $T = 0$ on $\mathcal{M}^\perp$.

The set $\mathcal{P}$ of continuous functions of positive type is a convex cone. We single out some subsets of $\mathcal{P}$ for special attention:

$$\mathcal{P}_1 = \{\phi \in \mathcal{P} : \|\phi\|_\infty = 1\} = \{\phi \in \mathcal{P} : \phi(1) = 1\},$$
$$\mathcal{P}_0 = \{\phi \in \mathcal{P} : \|\phi\|_\infty \leq 1\} = \{\phi \in \mathcal{P} : 0 \leq \phi(1) \leq 1\}.$$

(The equalities on the right follow from Corollary 3.22.) $\mathcal{P}_1$ and $\mathcal{P}_0$ are bounded convex sets, and we set

$$\mathcal{E}(\mathcal{P}_j) = \text{the set of extreme points of } \mathcal{P}_j, \qquad (j = 0, 1).$$

The extreme points of $\mathcal{P}_1$ are of particular interest for the following reason.

**3.25 Theorem.** *If $\phi \in \mathcal{P}_1$, then $\phi \in \mathcal{E}(\mathcal{P}_1)$ if and only if the representation $\pi_\phi$ of Theorem 3.20 is irreducible.*

*Proof.* Suppose $\pi_\phi$ is reducible, say $\mathcal{H}_\phi = \mathcal{M} \oplus \mathcal{M}^\perp$ where $\mathcal{M}$ is nontrivial and invariant under $\pi_\phi$. Let $\epsilon \in \mathcal{H}_\phi$ be as in Theorem 3.20. Since $\epsilon$ is a cyclic vector for $\pi_\phi$, it cannot belong to $\mathcal{M}$ or $\mathcal{M}^\perp$, so $\epsilon = \alpha u + \beta v$ where

$u$ and $v$ are unit vectors in $\mathcal{M}$ and $\mathcal{M}^\perp$, respectively, and $\alpha, \beta > 0$. Let $\psi_1(x) = \langle \pi_\phi(x)u, u \rangle_\phi$ and $\psi_2(x) = \langle \pi_\phi(x)v, v \rangle_\phi$. Then $\psi_1, \psi_2 \in \mathcal{P}_1$, and

$$\psi_1(x) - \psi_2(x) = \langle \pi_\phi(x)\epsilon, \alpha^{-1}u - \beta^{-1}v \rangle_\phi.$$

Since $\epsilon$ is a cyclic vector and $\alpha^{-1}u - \beta^{-1}v \neq 0$, there are $x$'s (an open set of them) for which this quantity is nonzero. Hence $\psi_1$ and $\psi_2$ are unequal, and since $\psi_1(1) = \psi_2(1) = 1$ they are actually linearly independent. Moreover,

$$\phi(x) = \langle \pi_\phi(x)\epsilon, \epsilon \rangle_\phi = \alpha^2 \langle \pi_\phi(x)u, u \rangle_\phi + \beta^2 \langle \pi_\phi(x)v, v \rangle_\phi$$
$$= \alpha^2 \psi_1(x) + \beta^2 \psi_2(x),$$

and $\alpha^2 + \beta^2 = \phi(1) = 1$. Hence $\phi$ is not extreme.

On the other hand, suppose $\pi_\phi$ is irreducible, but that $\phi = \psi + \psi'$ with $\psi, \psi' \in \mathcal{P}$. Then for any $f, g \in L^1$, with the notations of (3.17)–(3.19), we have

$$\langle f, f \rangle_\psi = \langle f, f \rangle_\phi - \langle f, f \rangle_{\psi'} \leq \langle f, f \rangle_\phi$$

and hence

$$|\langle f, g \rangle_\psi| \leq \langle f, f \rangle_\psi^{1/2} \langle g, g \rangle_\psi^{1/2} \leq \langle f, f \rangle_\phi^{1/2} \langle g, g \rangle_\phi^{1/2}.$$

Thus the map $(f, g) \mapsto \langle f, g \rangle_\psi$ induces a bounded Hermitian form on $\mathcal{H}_\phi$, so there is a bounded self-adjoint operator $T$ on $\mathcal{H}_\phi$ such that $\langle f, g \rangle_\psi = \langle T\tilde{f}, \tilde{g} \rangle_\phi$ for all $f, g \in L^1$. (Recall that $\tilde{f}$ is the image of $f$ in $\mathcal{H}_\phi$.) By (3.19), if $x \in G$ and $f, g \in L^1$ we have

$$\langle T\pi_\phi(x)\tilde{f}, \tilde{g} \rangle_\phi = \langle T(L_x f)\tilde{\ }, \tilde{g} \rangle_\phi = \langle L_x f, g \rangle_\psi = \langle f, L_{x^{-1}}g \rangle_\psi$$
$$= \langle T\tilde{f}, (L_{x^{-1}}g)\tilde{\ } \rangle_\phi = \langle T\tilde{f}, \pi_\phi(x^{-1})\tilde{g} \rangle_\phi = \langle \pi_\phi(x)T\tilde{f}, \tilde{g} \rangle_\phi.$$

Therefore, $T \in \mathcal{C}(\pi_\phi)$, so by Schur's lemma, $T = cI$ and $\langle f, g \rangle_\psi = c\langle f, g \rangle_\phi$ for all $f, g$. In view of (3.17), this implies that $\psi = c\phi$ and hence $\psi' = (1 - c)\phi$, so $\phi$ is extreme. $\qquad\square$

The condition $\int (f^* * f)\phi \geq 0$ on $\phi$ is clearly preserved under weak* limits, so $\mathcal{P}_0$ is a weak* closed subset of the closed unit ball in $L^\infty$. By Alaoglu's theorem, $\mathcal{P}_0$ is weak* compact, and then by the Krein-Milman theorem (see Rudin [123]), $\mathcal{P}_0$ is the weak* closure of the convex hull of its extreme points. $\mathcal{P}_1$ is in general *not* weak* closed (unless $G$ is discrete, in which case $\phi \mapsto \phi(1)$ is a bounded linear functional on $L^\infty$). Nonetheless, the conclusion of the Krein-Milman theorem holds for it too.

**3.26 Lemma.** $\mathcal{E}(\mathcal{P}_0) = \mathcal{E}(\mathcal{P}_1) \cup \{0\}$.

*Proof.* Suppose $\phi_1, \phi_2 \in \mathcal{P}_0$, $c_1, c_2 \geq 0$, and $c_1 + c_2 = 1$. If $c_1\phi_1 + c_2\phi_2 = 0$ then $c_1\phi_1(1) + c_2\phi_2(1) = 0$, which implies that $\phi_1(0) = \phi_2(0) = 0$ and hence $\phi_1 = \phi_2 = 0$ by Corollary 3.22. Thus 0 is extreme. On the other hand, if $c_1\phi_1 + c_2\phi_2 = \phi$ where $\phi \in \mathcal{P}_1$ then $c_1\phi_1(1) + c_2\phi_2(1) = 1$, which implies that $\phi_1(1) = \phi_2(1) = 1$ and hence $\phi_1, \phi_2 \in \mathcal{P}_1$. Thus if $\phi$ is extreme in $\mathcal{P}_1$ it is extreme in $\mathcal{P}_0$. Finally, no element $\phi$ of $\mathcal{P}_0 \setminus (\mathcal{P}_1 \cup \{0\})$ is extreme, since it is interior to the line segment joining 0 to $\phi/\phi(0)$. $\square$

**3.27 Theorem.** *The convex hull of $\mathcal{E}(\mathcal{P}_1)$ is weak\* dense in $\mathcal{P}_1$.*

*Proof.* Suppose $\phi_0 \in \mathcal{P}_1$. By Lemma 3.26 and the preceding remarks, $\phi_0$ is the weak\* limit of a net of functions $\phi_\alpha$ of the form $c_1\psi_1 + \cdots + c_n\psi_n + c_{n+1}0$, where $\psi_1, \ldots, \psi_n \in \mathcal{E}(\mathcal{P}_1)$, $c_1, \ldots, c_{n+1} \geq 0$, and $\sum c_j = 1$. Since $\|\phi_0\|_\infty = 1$, $\|\phi_\alpha\|_\infty \leq 1$, and $\{f \in L^\infty : \|f\|_\infty \leq 1 - \epsilon\}$ is weak\* closed, we must have $\lim \phi_\alpha(1) = \lim \|\phi_\alpha\|_\infty = 1$. But then, if we set $\phi'_\alpha = \phi_\alpha/\phi_\alpha(1)$, we have

$$\phi'_\alpha = \frac{1}{\phi_\alpha(1)} \sum_1^n c_j\psi_j, \qquad \frac{1}{\phi_\alpha(1)} \sum_1^n c_j = \frac{\phi_\alpha(1)}{\phi_\alpha(1)} = 1.$$

Thus $\phi'_\alpha$ is in the convex hull of $\mathcal{E}(\mathcal{P}_1)$ and $\phi_0 = \lim \phi'_\alpha$. $\square$

Our next goal is to establish the remarkable fact that the weak\* topology that $\mathcal{P}_1$ inherits as a subset of $L^\infty$ coincides with the topology of uniform convergence on compact subsets of $G$, or for short, the **topology of compact convergence** on $G$. In this topology, a neighborhood base at the function $\phi_0$ is provided by the sets

$$N(\phi_0; \epsilon, K) = \{\phi : |\phi(x) - \phi_0(x)| < \epsilon \text{ for } x \in K\},$$

where $\epsilon$ ranges over positive numbers and $K$ ranges over compact sets in $G$. The coincidence of these two topologies on $\mathcal{P}_1$ is the more remarkable when one observes that they are, in general, *not* the same on $\mathcal{P}_0$. For example, the functions $\phi_\xi(x) = e^{i\xi x}$ belong to $\mathcal{P}_1(\mathbb{R})$. They converge to 0 in the weak\* topology as $\xi \to \infty$ (this is a restatement of the Riemann-Lebesgue lemma), but they have no limit in the topology of compact convergence on $\mathbb{R}$ (or in the weak\* topology *restricted* to $\mathcal{P}_1$). The proof is based on the following general lemma.

**3.28 Lemma.** *Suppose $X$ is a Banach space and $B$ is a norm-bounded subset of $X^*$. On $B$, the weak\* topology coincides with the topology of compact convergence on $X$.*

*Proof.* The weak\* topology is the topology of pointwise convergence on $\mathfrak{X}$, so it is no stronger than the topology of compact convergence. On the other hand, if $\lambda_0 \in B$, $\epsilon > 0$, and $K \subset \mathfrak{X}$ is compact, let $C = \sup\{\|\lambda\| : \lambda \in B\}$ and $\delta = \epsilon/3C$. Then there exist $\xi_1, \dots, \xi_n \in K$ such that the balls $B(\delta, \xi_j)$ cover $K$. If $\lambda \in B$ and $\xi \in K$ then $\|\xi - \xi_j\| < \delta$ for some $j$, so that

$$|\lambda(\xi) - \lambda_0(\xi)| < |\lambda(\xi - \xi_j)| + |(\lambda - \lambda_0)(\xi_j)| + |\lambda_0(\xi_j - \xi)| < \frac{2\epsilon}{3} + |(\lambda - \lambda_0)(\xi_j)|,$$

so the weak\* neighborhood $\bigcap_1^n \{\lambda : |(\lambda - \lambda_0)(\xi_j)| < \epsilon/3\}$ of $\lambda_0$ is contained in the neighborhood $N(\lambda_0; \epsilon, K)$ for the topology of compact convergence. $\qquad\square$

**3.29 Lemma.** *Suppose $\phi_0 \in \mathcal{P}_1$ and $f \in L^1(G)$. For every $\epsilon > 0$ and every compact $K \subset G$ there is a weak\* neighborhood $\Phi$ of $\phi_0$ in $\mathcal{P}_1$ such that $|f * \phi(x) - f * \phi_0(x)| < \epsilon$ for all $\phi \in \Phi$ and $x \in K$.*

*Proof.* By Corollary 3.22 we have $f * \phi(x) = \int f(xy)\phi(y^{-1})\, dy = \int (L_{x^{-1}}f)\overline{\phi}$. Since $x \mapsto L_{x^{-1}}f$ is continuous from $G$ to $L^1$ (Proposition 2.42), $\{L_{x^{-1}}f : x \in K\}$ is compact in $L^1$, and we can apply Lemma 3.28. $\qquad\square$

**3.30 Lemma.** *If $\phi \in \mathcal{P}_1$, $|\phi(x) - \phi(y)|^2 \leq 2 - 2\operatorname{Re}\phi(yx^{-1})$.*

*Proof.* By Theorem 3.20 we have $\phi(x) = \langle \pi(x)u, u \rangle$ for some unitary representation $\pi$ and some unit vector $u \in \mathcal{H}_\pi$, so

$$|\phi(x) - \phi(y)|^2 = \left|\langle [\pi(x) - \pi(y)]u, u \rangle\right|^2 = \left|\langle u, [\pi(x^{-1}) - \pi(y^{-1})]u \rangle\right|^2$$
$$\leq \|\pi(x^{-1})u - \pi(y^{-1})u\|^2 = 2 - 2\operatorname{Re}\langle \pi(x^{-1})u, \pi(y^{-1})u \rangle$$
$$= 2 - 2\operatorname{Re}\langle \pi(yx^{-1})u, u \rangle = 2 - 2\operatorname{Re}\phi(yx^{-1}). \quad \square$$

**3.31 Theorem.** *On $\mathcal{P}_1$, the weak\* topology coincides with the topology of compact convergence on $G$.*

*Proof.* If $f \in L^1(G)$ and $\epsilon > 0$, there is a compact $K \subset G$ such that $\int_{G \setminus K} |f| < \frac{1}{4}\epsilon$. If $\phi, \phi_0 \in \mathcal{P}_1$ and $|\phi - \phi_0| < \epsilon/2\|f\|_1$ on $K$ then

$$\left| \int (f\phi - f\phi_0) \right| \leq \int_K |f||\phi - \phi_0| + \int_{G \setminus K} |f||\phi - \phi_0| < \tfrac{1}{2}\epsilon + \tfrac{1}{2}\epsilon = \epsilon,$$

so compact convergence on $G$ implies weak\* convergence.

Conversely, suppose $\phi_0 \in \mathcal{P}_1$, $\epsilon > 0$, and $K \subset G$ is compact. We wish to find a weak\* neighborhood $\Phi$ of $\phi_0$ in $\mathcal{P}_1$ such that $|\phi - \phi_0| < \epsilon$ on $K$

when $\phi \in \Phi$. First, if $\eta > 0$ there is a compact neighborhood $V$ of 1 in $G$ such that $|\phi_0(x) - 1| < \eta$ for $x \in V$. Let

$$\Phi_1 = \left\{ \phi \in \mathcal{P}_1 : \left| \int_V (\phi - \phi_0) \right| < \eta |V| \right\}.$$

$\Phi_1$ is a weak* neighborhood of $\phi_0$ since $\chi_V \in L^1$. If $\phi \in \Phi_1$ then

$$(3.32) \qquad \left| \int_V (1 - \phi) \right| \leq \left| \int_V (1 - \phi_0) \right| + \left| \int_V (\phi_0 - \phi) \right| < 2\eta |V|.$$

Also, if $\phi \in \Phi_1$ and $x \in G$, we have

$$\left| \chi_V * \phi(x) - |V|\phi(x) \right| = \left| \int_V [\phi(y^{-1}x) - \phi(x)] \, dy \right| \leq \int_V |\phi(y^{-1}x) - \phi(x)| \, dy.$$

By Lemma 3.30, the Schwarz inequality, and (3.32), this is bounded by

$$\int_V [2 - 2\operatorname{Re}\phi(y)]^{1/2} \, dy \leq \left( \int_V [2 - 2\operatorname{Re}\phi(y)] \, dy \right)^{1/2} |V|^{1/2} < 2|V|\sqrt{\eta}.$$

By Lemma 3.29, there is a weak* neighborhood $\Phi_2$ of $\phi_0$ in $\mathcal{P}_1$ such that $|\chi_V * \phi(x) - \chi_V * \phi_0(x)| < \eta |V|$ for $\phi \in \Phi_2$ and $x \in K$. Hence, if $\phi \in \Phi_1 \cap \Phi_2$ and $x \in K$, $|\phi(x) - \phi_0(x)|$ is bounded by

$$\frac{1}{|V|} \Big[ \big| \, |V|\phi(x) - \chi_V * \phi(x) \big| + \big| \chi_V * (\phi - \phi_0)(x) \big| + \big| \chi_V * \phi_0(x) - |V|\phi_0(x) \big| \Big]$$

$$\leq \frac{1}{|V|} \left( 2|V|\sqrt{\eta} + |V|\eta + 2|V|\sqrt{\eta} \right) = \eta + 4\sqrt{\eta}.$$

Therefore, if we choose $\eta$ so that $\eta + 4\sqrt{\eta} < \epsilon$ and take $\Phi = \Phi_1 \cap \Phi_2$, we are done. $\qquad \square$

We are almost ready to prove our final major theorem. First, we need one more simple result that will also be useful in the next chapter.

**3.33 Proposition.** *The linear span of $C_c(G) \cap \mathcal{P}(G)$ includes all functions of the form $f * g$ with $f, g \in C_c(G)$. It is dense in $C_c(G)$ in the uniform norm, and dense in $L^p(G)$ $(1 \leq p < \infty)$ in the $L^p$ norm.*

*Proof.* By Corollary 3.16, $C_c(G) \cap \mathcal{P}(G)$ includes all functions of the form $f * \widetilde{f}$ with $f \in C_c(G)$, where $\widetilde{f}(x) = \overline{f(x^{-1})}$. By polarization (see Appendix 1), its linear span includes all functions of the form $f * \widetilde{h}$ with $f, h \in C_c(G)$ and hence all functions of the form $f * g$ with $f, g \in C_c(G)$ (take $h = \widetilde{g}$). $\{f * g : f, g \in C_c(G)\}$ is dense in $C_c(G)$ in the uniform norm or the $L^p$ norm because $g$ can be taken to be an approximate identity, and $C_c(G)$ is itself dense in $L^p$. $\qquad \square$

**3.34 Theorem** (The Gelfand-Raikov Theorem). *If $G$ is any locally compact group, the irreducible unitary representations of $G$ separate points on $G$. That is, if $x$ and $y$ are distinct points of $G$, there is an irreducible representation $\pi$ such that $\pi(x) \neq \pi(y)$.*

*Proof.* If $x \neq y$ there exists $f \in C_c(G)$ such that $f(x) \neq f(y)$, and by Proposition 3.33 we can take $f$ to be a linear combination of functions of positive type. By Theorems 3.27 and 3.31, there is a linear combination $g$ of extreme points of $\mathcal{P}_1$ that approximates $f$ on the compact set $\{x, y\}$ closely enough so that $g(x) \neq g(y)$. Hence there must be an extreme point $\phi$ of $\mathcal{P}_1$ such that $\phi(x) \neq \phi(y)$. The associated representation $\pi_\phi$ of Theorem 3.20 is irreducible by Theorem 3.25, and it satisfies

$$\langle \pi_\phi(x)\epsilon, \epsilon \rangle = \phi(x) \neq \phi(y) = \langle \pi_\phi(y)\epsilon, \epsilon \rangle,$$

whence $\pi_\phi(x) \neq \pi_\phi(y)$. $\qquad\qquad\qquad\qquad\qquad\qquad\qquad\qquad\qquad$ $\square$

We shall give a simpler proof of the Gelfand-Raikov theorem for the case of compact groups in §5.2; see Theorem 5.11 and the remarks preceding it.

It should be noted that when $G$ is neither compact nor Abelian, the irreducible representations of $G$ may be infinite-dimensional, and the finite-dimensional ones usually do not separate points on $G$. We shall see some examples of this phenomenon — the $ax + b$ group, the Heisenberg groups, and $SL(2, \mathbb{R})$ — in §6.7 and §7.6.

We conclude this section by discussing a commonly-used variant of the notion of functions of positive type. A function $\phi : G \to \mathbb{C}$ is called **positive definite** if

$$\sum_{i,j=1}^{n} c_i \bar{c}_j \phi(x_j^{-1} x_i) \geq 0 \text{ for all } c_1, \ldots, c_n \in \mathbb{C} \text{ and } x_1, \ldots, x_n \in G.$$

(Here $n$ is an arbitrary positive integer.) When $n = 2$, $x_1 = x$, and $x_2 = 1$, this condition says that the matrix

$$\begin{pmatrix} \phi(1) & \phi(x) \\ \phi(x^{-1}) & \phi(1) \end{pmatrix}$$

is positive semi-definite. Therefore $\phi(x^{-1}) = \overline{\phi(x)}$ and $\phi(1)^2 - \phi(x)\phi(x^{-1}) \geq 0$, which implies that $|\phi(x)| \leq \phi(1)$ for all $x$. In particular, positive definite functions are bounded.

Positive definite functions need not be continuous: for example, $\phi(x) = 1$ when $x = 1$, $\phi(x) = 0$ otherwise. They need not even be measurable: if $G = \mathbb{R}$ and $\psi$ is any automorphism of $\mathbb{R}$ as a vector space over $\mathbb{Q}$ then $e^{i\psi}$ is positive definite. However, the continuous positive definite functions are nothing but the functions of positive type.

**3.35 Proposition.** *If $\phi$ is a bounded continuous function on $G$, the following are equivalent:*

  *i. $\phi$ is of positive type.*
  *ii. $\phi$ is positive definite.*
  *iii. $\int (f^* * f)\phi \geq 0$ for all $f \in C_c(G)$.*

*Proof.* (i) $\Rightarrow$ (ii): Let $\{\psi_U\}$ be an approximate identity. Given $c_1, \ldots, c_n \in \mathbb{C}$ and $x_1, \ldots, x_n \in G$, let $f_U = \sum_1^n c_j L_{x_j} \psi_U$. Then

$$0 \leq \int (f_U^* * f_U)\phi = \sum_{i,j=1}^n c_i \bar{c}_j \iint \psi_U(x_i^{-1}y)\psi_U(x_j^{-1}z)\phi(z^{-1}y)\, dy\, dz.$$

Since $\phi$ is continuous, the sum on the right approaches $\sum c_i \bar{c}_j \phi(x_j^{-1}x_i)$ as $U \to \{1\}$, so the latter sum is nonnegative.

  (ii) $\Rightarrow$ (iii): If $f \in C_c(G)$, the function $F(x,y) = f(x)\overline{f(y)}\phi(y^{-1}x)$ is in $C_c(G \times G)$, hence is uniformly continuous. Let $K = \operatorname{supp} f$, so that $\operatorname{supp} F \subset K \times K$. Given $\epsilon > 0$, we can cover $K \times K$ by finitely many open sets of the form $U \times U$ such that the variation of $F$ on each such set is less than $\epsilon$. By discarding the overlaps, we can obtain a partition of $K$ into disjoint sets $E_1, \ldots, E_n$ and points $x_j \in E_j$ such that $|F(x,y) - F(x_i, x_j)| < \epsilon$ when $(x,y) \in E_i \times E_j$. But then

$$\int (f^* * f)\phi = \iint F(x,y)\, dx\, dy = \sum_{i,j} \iint_{E_i \times E_j} F(x,y)\, dx\, dy$$

$$\text{(3.36)} \qquad = \sum_{i,j} F(x_i, x_j)|E_i||E_j| + R$$

$$= \sum_{i,j} f(x_i)|E_i|\overline{f(x_j)|E_j|}\phi(x_j^{-1}x_i) + R,$$

where

$$|R| = \left| \sum_{i,j} \iint_{E_i \times E_j} [F(x,y) - F(x_i, x_j)]\, dx\, dy \right| < \epsilon |K|^2.$$

The last sum in 3.36 is nonnegative, and $\epsilon$ is arbitrary, so $\int (f^* * f)\phi \geq 0$.

  (iii) $\Rightarrow$ (i): If $f \in L^1$ there is a sequence $\{f_n\} \subset C_c(G)$ such that $f_n \to f$ in $L^1$. Then $f_n^* * f_n \to f^* * f$ in $L^1$, so $\int (f^* * f)\phi = \lim \int (f_n^* * f_n)\phi \geq 0$. $\qquad \square$

## 3.4   Notes and References

The theory of representations of finite groups was developed by Frobenius, Schur, Burnside, and others beginning in the 1890s. Representations of arbitrary compact groups, and finite-dimensional (possibly nonunitary) representations of the classical matrix groups, were studied by Weyl and others beginning in the 1920s; see Peter and Weyl [112] and Weyl [147]. The theory of (possibly infinite-dimensional) unitary representations of locally compact groups was initiated in the 1940s, although a few special cases had been known earlier. At that time, various researchers began looking at both abstract representation theory for general groups and concrete representation theory for specific groups, particularly those arising in quantum mechanics.

Simon [129] has a good treatment of representations of finite groups from an analyst's perspective. Other treatises in which various aspects of the theory of unitary representations are developed include Dixmier [32], Fell and Doran [40], [41], Hewitt and Ross [66], [67], Kaniuth and Taylor [71], Kirillov [77], and Mackey [98]; see also the survey article of Mackey [94]. Discussions of the connections between representation theory and other parts of mathematics and physics can be found in Mackey [96], [99], [100].

The material in §3.3 was first developed in the fundamental paper of Gelfand and Raikov [50].

The term "positive definite function" is frequently used synonymously with "function of positive type." I find it unsatisfactory because functions of positive type are generally not positive themselves, and the positive linear functionals they define are generally not positive definite. I have therefore adapted the usual term in French, *fonction de type positif*, as the standard terminology for this book.

# 4

## Analysis on Locally Compact Abelian Groups

The central idea of this chapter is the Fourier transform on locally compact Abelian groups, which provides a single theory that includes Fourier series and integrals on $\mathbb{R}^n$, Walsh expansions, finite (or discrete) Fourier transforms, and many other things as special cases. The fundamental results about Fourier analysis on locally compact Abelian groups are developed in §§4.1–3. The rest of the chapter is devoted to three separate topics — classification of unitary representations (§4.4), the circle of ideas centering on Wiener's general Tauberian theorem (§§4.5–6), and the Bohr compactification (§4.7) —that can be read independently of one another.

*Throughout this chapter, G will denote a locally compact Abelian group.* Here left and right translations are the same thing, so we have our choice of notation:

$$L_y f(x) = f(y^{-1}x) = f(xy^{-1}), \quad R_y f(x) = L_{y^{-1}} f(x) = f(xy) = f(yx).$$

Also, $G$ is unimodular and convolution is commutative:

$$f * g(x) = g * f(x) = \int f(xy^{-1})g(y)\,dy = \int f(y)g(y^{-1}x)\,dy.$$

We shall continue to write the group operation as multiplication. One must keep in mind that in many of the common Abelian groups, such as $\mathbb{R}$ and $\mathbb{Z}$, the group law is addition; the notation must be adjusted accordingly.

## 4.1 The Dual Group

Let $G$ be a locally compact Abelian group. By Corollary 3.6, the irreducible representations of $G$ are all one-dimensional. Thus, for each such representation $\pi$ we can take $\mathcal{H}_\pi = \mathbb{C}$, and then $\pi(x)(z) = \xi(x)z$ ($z \in \mathbb{C}$)

where $\xi$ is a continuous homomorphism of $G$ into the circle group $\mathbb{T}$. Such homomorphisms are called (unitary) **characters** of $G$, and the set of all characters of $G$ is denoted by $\widehat{G}$. Since $\xi(x) = \langle\pi(x)1, 1\rangle$, Proposition 3.15 shows that $\widehat{G}$ is contained in $\mathcal{P}_1(G)$, the set of functions of positive type on $G$ of norm 1. In fact, by Theorem 3.25, $\widehat{G}$ is the set of extreme points of $\mathcal{P}_1(G)$. For reasons of symmetry (which will become more cogent in §4.3) we shall use the notation

$$\langle x, \xi \rangle = \xi(x) \qquad (x \in G, \ \xi \in \widehat{G}).$$

According to Theorem 3.9, each $\xi \in \widehat{G}$ determines a nondegenerate $*$-representation of $L^1(G)$ on $\mathbb{C}$ by

$$(4.1) \qquad \xi(f) = \int \langle x, \xi \rangle f(x)\, dx.$$

Here we identify $\mathcal{L}(\mathbb{C})$ with $\mathbb{C}$; with this identification, such a representation is a multiplicative functional on $L^1(G)$ as defined in §1.2. Conversely, every multiplicative functional $\Phi$ is given by integration against a character. This is not quite an instance of Theorem 3.11, because we do not assume that $\Phi(f^*) = \overline{\Phi(f)}$; rather, we have the following simple argument.

First, $\Phi \in (L^1)^*$ is given by integration against some $\phi \in L^\infty$. Thus, for any $f, g \in L^1$,

$$\Phi(f)\int \phi(x)g(x)\, dx = \Phi(f)\Phi(g) = \Phi(f * g)$$

$$= \iint \phi(y)f(yx^{-1})g(x)\, dx\, dy = \int \Phi(L_x f)g(x)\, dx,$$

so that

$$(4.2) \qquad \Phi(f)\phi(x) = \Phi(L_x f)$$

locally a.e. Fix $f \in L^1$ with $\Phi(f) \neq 0$; we can redefine $\phi(x)$ to be $\Phi(L_x f)/\Phi(f)$ for every $x$, and then $\phi$ is continuous and (4.2) holds everywhere. Moreover, by (4.2), first with $x$ replaced by $xy$ and then with $f$ replaced by $L_y f$,

$$\phi(xy)\Phi(f) = \Phi(L_{xy}f) = \Phi(L_x L_y f) = \phi(x)\phi(L_y f) = \phi(x)\phi(y)\Phi(f),$$

so $\phi(xy) = \phi(x)\phi(y)$. Finally, $\phi(x^n) = \phi(x)^n$ for every $n$, and $\phi$ is bounded, which necessitates $|\phi(x)| = 1$, that is, $\phi : G \to \mathbb{T}$.

In short, we have:

**4.3 Theorem.** $\widehat{G}$ *can be identified with the spectrum of* $L^1(G)$ *via* (4.1).

$\widehat{G}$ is clearly an Abelian group under pointwise multiplication; its identity element is the constant function 1, and

$$\langle x, \xi^{-1} \rangle = \langle x^{-1}, \xi \rangle = \overline{\langle x, \xi \rangle}.$$

We shall give $\widehat{G}$ the topology of compact convergence on $G$, under which the group operations are obviously continuous. By Theorem 3.31, this topology coincides with the weak* topology that $\widehat{G}$ inherits as a subset of $L^\infty$. But $\widehat{G} \cup \{0\}$ is the set of all homomorphisms from $L^1$ to $\mathbb{C}$, which is a subset of the closed unit ball of $L^\infty$ (by Proposition 1.10(c)) and is clearly weak* closed, and hence is weak* compact by Alaoglu's theorem. Therefore, $\widehat{G}$ is locally compact. In short, $\widehat{G}$ is a locally compact Abelian group, called the **dual group** of $G$.

When $G$ is compact or discrete, we can say more about the structure of $\widehat{G}$. First, if $G$ is compact, $\widehat{G} \subset L^\infty(G) \subset L^p(G)$ for all $p \geq 1$, and we have:

**4.4 Proposition.** *If $G$ is compact and Haar measure is normalized so that $|G| = 1$, then $\widehat{G}$ is an orthonormal set in $L^2(G)$.*

*Proof.* If $\xi \in \widehat{G}$ then $|\xi|^2 \equiv 1$, so clearly $\|\xi\|_2 = 1$. If $\xi \neq \eta$ there is an $x_0 \in G$ such that $\langle x_0, \xi\eta^{-1} \rangle \neq 1$, and we then have

$$\int \xi\overline{\eta} = \int \langle x, \xi\eta^{-1} \rangle \, dx = \langle x_0, \xi\eta^{-1} \rangle \int \langle x_0^{-1}x, \xi\eta^{-1} \rangle \, dx$$

$$= \langle x_0, \xi\eta^{-1} \rangle \int \langle x, \xi\eta^{-1} \rangle \, dx = \langle x_0, \xi\eta^{-1} \rangle \int \xi\overline{\eta},$$

which implies that $\int \xi\overline{\eta} = 0$. $\qquad\qquad\square$

**4.5 Proposition.** *If $G$ is discrete then $\widehat{G}$ is compact. If $G$ is compact then $\widehat{G}$ is discrete.*

*Proof.* If $G$ is discrete then $L^1(G)$ has a unit — namely, the function $\delta$ that equals 1 at the identity and 0 elsewhere. Hence its spectrum $\widehat{G}$ is compact.

If $G$ is compact then the constant function 1 is in $L^1$, so $\{f \in L^\infty : |\int f| > \frac{1}{2}\}$ is a weak* open set. By Proposition 4.4, for $\xi \in \widehat{G}$ we have $\int \xi = 1$ if $\xi = 1$, $\int \xi = 0$ if $\xi \neq 1$. Thus $\{1\}$ is an open set in $\widehat{G}$, so every singleton set in $\widehat{G}$ is open and $\widehat{G}$ is discrete. $\qquad\square$

The remainder of this section is devoted to the calculation of $\widehat{G}$ for various locally compact Abelian groups $G$. We begin by reviewing the classic simple examples.

**4.6 Theorem.**

a. $\widehat{\mathbb{R}} \cong \mathbb{R}$, with the pairing $\langle x, \xi \rangle = e^{2\pi i \xi x}$.

b. $\widehat{\mathbb{T}} \cong \mathbb{Z}$, with the pairing $\langle \alpha, n \rangle = \alpha^n$.

c. $\widehat{\mathbb{Z}} \cong \mathbb{T}$, with the pairing $\langle n, \alpha \rangle = \alpha^n$.

d. If $Z_k$ is the additive group of integers mod $k$, then $\widehat{Z}_k \cong Z_k$, with the pairing $\langle m, n \rangle = e^{2\pi i mn/k}$.

*Proof.* (a) If $\phi \in \widehat{\mathbb{R}}$ we have $\phi(0) = 1$, so there exists $a > 0$ such that $\int_0^a \phi(t)\, dt \neq 0$. Setting $A = \int_0^a \phi(t)\, dt$, we have

$$A\phi(x) = \int_0^a \phi(x + t)\, dt = \int_x^{a+x} \phi(t)\, dt,$$

so $\phi$ is differentiable and

$$\phi'(x) = A^{-1}[\phi(a + x) - \phi(x)] = c\phi(x) \text{ where } c = A^{-1}[\phi(a) - 1].$$

It follows that $\phi(t) = e^{ct}$, and since $|\phi| = 1$, $c = 2\pi i \xi$ for some $\xi \in \mathbb{R}$.

(b) Since $\mathbb{T} \cong \mathbb{R}/\mathbb{Z}$ via the identification of $x \in \mathbb{R}/\mathbb{Z}$ with $\alpha = e^{2\pi i x} \in \mathbb{T}$, the characters of $\mathbb{T}$ are just the characters of $\mathbb{R}$ that are trivial on $\mathbb{Z}$. The result therefore follows from (a).

(c) If $\phi \in \widehat{\mathbb{Z}}$ then $\alpha = \phi(1) \in \mathbb{T}$, and $\phi(n) = \phi(1)^n = \alpha^n$.

(d) The characters of $Z_k$ are the characters of $\mathbb{Z}$ that are trivial on $k\mathbb{Z}$, hence are of the form $\phi(n) = \alpha^n$ where $\alpha$ is a $k$th root of 1.  $\square$

We can generate more examples from these by taking products.

**4.7 Proposition.** *If* $G_1, \ldots, G_n$ *are locally compact Abelian groups, then*

$$(G_1 \times \cdots \times G_n)^{\widehat{}} \cong \widehat{G}_1 \times \cdots \times \widehat{G}_n.$$

*Proof.* Each $\xi = (\xi_1, \ldots, \xi_n) \in \prod_1^n \widehat{G}_j$ defines a character on $\prod_1^n G_j$ by

$$\langle (x_1, \ldots, x_n), (\xi_1, \ldots, \xi_n) \rangle = \langle x_1, \xi_1 \rangle \cdots \langle x_n, \xi_n \rangle.$$

Moreover, every character $\chi$ on $\prod_1^n G_j$ is of this form, where $\xi_j$ is defined by

$$\langle x_j, \xi_j \rangle = \langle (1, \ldots, 1, x_j, 1, \ldots, 1), \chi \rangle.$$

$\square$

**4.8 Corollary.** $(\mathbb{R}^n)^{\widehat{}} \cong \mathbb{R}^n$, $(\mathbb{T}^n)^{\widehat{}} \cong \mathbb{Z}^n$, $(\mathbb{Z}^n)^{\widehat{}} \cong \mathbb{T}^n$, *and* $\widehat{G} \cong G$ *for any finite Abelian group* $G$.

Proposition 4.7 can be extended to infinite products of compact groups. If $\{H_\alpha\}_{\alpha \in A}$ is any family of groups, we define $\bigoplus_{\alpha \in A} H_\alpha$ to be the set of all $(h_\alpha)_{\alpha \in A} \in \prod_{\alpha \in A} H_\alpha$ such that $h_\alpha = 1$ for all but finitely many $\alpha$.

**4.9 Proposition.** *If $G = \prod_{\alpha \in A} G_\alpha$ where each $G_\alpha$ is a compact Abelian group, then $\widehat{G} \cong \bigoplus_{\alpha \in A} \widehat{G}_\alpha$.*

*Proof.* As in the proof of Proposition 4.7, it is easy to see that every $(\xi_\alpha) \in \bigoplus \widehat{G}_\alpha$ defines a character $\xi$ on $G$ by $\langle x, \xi \rangle = \prod \langle x_\alpha, \xi_\alpha \rangle$ (the product being finite), and that each $\xi \in \widehat{G}$ determines an element $(\xi_\alpha)$ of $\prod \widehat{G}_\alpha$ where $\xi_\alpha$ is the restriction of $\xi$ to the $\alpha$th factor. We need only show that in this situation, $\xi_\alpha = 1$ for all but finitely many $\alpha$.

There is a neighborhood $V$ of 1 in $G$ such that $|\langle x, \xi \rangle - 1| < 1$ for $x \in V$. By definition of the product topology, $V$ includes a set $\prod_\alpha V_\alpha$ where $V_\alpha = G_\alpha$ for all but finitely many $\alpha$. If $V_\alpha = G_\alpha$ then $\xi(V) \supset \xi_\alpha(G_\alpha)$. $\xi_\alpha(G_\alpha)$ is therefore a subgroup of $\mathbb{T}$ contained in $\{\alpha \in \mathbb{T} : |\alpha - 1| < 1\}$, and it therefore equals $\{1\}$; hence $\xi_\alpha = 1$. $\qquad\square$

*Example.* Let $G = (Z_2)^\omega$, the product of countably many copies of the 2-element group $Z_2$, as discussed in §2.2. For each $n$ there is a unique character $\xi_n$ on $G$ that is nontrivial only on the $n$th factor, namely $\langle (a_j)_1^\infty, \xi_n \rangle = (-1)^{a_n}$. The characters on $G$ are then the finite products of the $\xi_n$'s, together with the trivial character 1.

If we identify $G$ with $[0,1]$ as in §2.2, $\xi_n$ becomes the $n$th **Rademacher function** $r_n$, which takes on the values 1 and $-1$ alternately on the intervals $[0, 2^{-n})$, $[2^{-n}, 2 \cdot 2^{-n})$, ..., $[1 - 2^{-n}, 1)$. The finite products of the Rademacher functions are called **Walsh functions**. There is a standard way of well-ordering the Walsh functions, as follows. If $n$ is a nonnegative integer, let $b_k, \ldots, b_1$ be the digits in its binary expansion (i.e., $n = b_k \ldots b_1$ in base 2); then the $n$th Walsh function is $w_n = r_1^{b_1} \cdots r_k^{b_k}$. From Proposition 4.4, or by direct inspection, $\{w_n\}_0^\infty$ is an orthonormal set in $L^2(0,1)$. It is actually an orthonormal basis. This will follow from the Plancherel theorem in the next section, but it is also an elementary exercise to verify it directly. (Hint: show that the linear span of $\{w_k\}_0^{2^n - 1}$ is the set of all step functions on $[0,1]$ that are constant on the intervals $[0, 2^{-n})$, ..., $[1 - 2^{-n}, 1)$.)

For our final example, we compute the dual of the $p$-adic numbers $\mathbb{Q}_p$. We begin by writing down one character $\xi_1$ of $\mathbb{Q}_p$ explicitly. Let us recall from Proposition 2.8 that each $p$-adic number $x$ can be written uniquely as $\sum_{-\infty}^\infty c_j p^j$ where $c_j \in \{0, 1, \ldots, p-1\}$ and $c_j = 0$ for $j \ll 0$; moreover, $x \in \mathbb{Z}_p$ if and only if $c_j = 0$ for $j < 0$. We define

$$\Big\langle \sum_{-\infty}^\infty c_j p^j, \, \xi_1 \Big\rangle = \exp\Big( 2\pi i \sum_{-\infty}^{-1} c_j p^j \Big).$$

More simply, we can write $\langle x, \xi_1 \rangle = e^{2\pi i x}$, where it is understood that the terms with $j \geq 0$ in the series $x = \sum c_j p^j$ contribute nothing because $e^{2\pi i c_j p^j} = 1$. From this it is clear that $\langle x+y, \xi_1 \rangle = \langle x, \xi_1 \rangle \langle y, \xi_1 \rangle$ (recall that $\mathbb{Q}_p$ is a group under addition, not multiplication!) and that $\xi_1$ is constant on cosets of the open subgroup $\mathbb{Z}_p$ and hence continuous. In short, $\xi_1$ is a character whose kernel is $\mathbb{Z}_p$. Next, for $y \in \mathbb{Q}_p$ we define $\xi_y$ by

$$\langle x, \xi_y \rangle = \langle xy, \xi_1 \rangle.$$

$\xi_y$ is also a character, and its kernel is $\{x : |x| \leq |y|^{-1}\}$. Our claim is that every character on $\mathbb{Q}_p$ is of the form $\xi_y$.

**4.10 Lemma.** *If $\xi \in \widehat{\mathbb{Q}}_p$, there is an integer $k$ such that $\xi = 1$ on $B(p^{-k}, 0)$.*

*Proof.* Since $\xi$ is continuous, there is an integer $k$ such that $\xi$ maps $B(p^{-k}, 0)$ into $\{z \in \mathbb{T} : |z - 1| < 1\}$. But $B(p^{-k}, 0)$ is a subgroup of $\mathbb{Q}_p$, so its image under $\xi$ is a subgroup of $\mathbb{T}$; hence it equals $\{1\}$.   $\square$

Any $\xi \in \widehat{\mathbb{Q}}_p$ is completely determined by its values on the numbers $p^j$, $j \in \mathbb{Z}$, and by Lemma 4.10, if $\xi \neq 1$ there is an integer $j_0$ such that $\langle p^j, \xi \rangle = 1$ for $j \geq j_0$, but $\langle p^{j_0-1}, \xi \rangle \neq 1$. Let us analyze the case $j_0 = 0$.

**4.11 Lemma.** *Suppose $\xi \in \widehat{\mathbb{Q}}_p$, $\langle 1, \xi \rangle = 1$, and $\langle p^{-1}, \xi \rangle \neq 1$. There is a sequence $\{c_j\}_0^\infty$ with $c_0 \in \{1, \ldots, p-1\}$ and $c_j \in \{0, \ldots, p-1\}$ for $j \geq 1$ such that $\langle p^{-k}, \xi \rangle = \exp(2\pi i \sum_1^k c_{k-j} p^{-j})$ for $k = 1, 2, 3, \ldots$.*

*Proof.* Let $\omega_k = \langle p^{-k}, \xi \rangle$; then

$$\omega_{k+1}^p = \langle p^{-k-1}, \xi \rangle^p = \langle p \cdot p^{-k-1}, \xi \rangle = \langle p^{-k}, \xi \rangle = \omega_k.$$

Now $\omega_1 \neq 1 = \omega_0$, so $\omega_1 = \exp(2\pi i c_0 p^{-1})$ for some $c_0 \in \{1, \ldots, p-1\}$. Proceeding by induction, suppose $\omega_k = \exp[2\pi i \sum_1^k c_{k-j} p^{-j}]$. Since $\omega_{k+1}$ is a $p$th root of $\omega_k$, there exists $c_k \in \{0, \ldots, p-1\}$ such that

$$\omega_{k+1} = \exp\left[2\pi i \sum_1^k c_{k-j} p^{-j-1}\right] \exp[2\pi i c_k p^{-1}] = \exp\left[\sum_1^{k+1} c_{k+1-j} p^{-j}\right].$$

$\square$

**4.12 Lemma.** *If $\xi \in \widehat{\mathbb{Q}}_p$, $\langle 1, \xi \rangle = 1$, and $\langle p^{-1}, \xi \rangle \neq 1$, there exists $y \in \mathbb{Q}_p$ with $|y| = 1$ such that $\xi = \xi_y$.*

*Proof.* Let $\{c_j\}$ be as in Lemma 4.11 and set $y = \sum_0^\infty c_j p^j$. Then $|y| = 1$

since $c_0 \neq 0$, and for $k \geq 1$,

$$\langle p^{-k}, \xi \rangle = \exp\left[2\pi i \sum_1^k c_{k-j} p^{-j}\right] = \exp\left[2\pi i \sum_{-k}^{-1} c_{j+k} p^j\right]$$

$$= \left\langle \sum_{-k}^{\infty} c_{j+k} p^j, \xi_1 \right\rangle = \langle p^{-k} y, \xi_1 \rangle = \langle p^{-k}, \xi_y \rangle.$$

It follows that $\langle x, \xi \rangle = \langle x, \xi_y \rangle$ for every $x$. □

**4.13 Theorem.** *The map $y \mapsto \xi_y$ from $\mathbb{Q}_p$ to $\widehat{\mathbb{Q}}_p$ is an isomorphism of topological groups.*

*Proof.* The map $y \mapsto \xi_y$ is clearly an injective group homomorphism. Suppose $\xi \in \widehat{\mathbb{Q}}_p$. If $\xi = 1$ then $\xi = \xi_0$. If $\xi \neq 1$, by Lemma 4.10 there is a smallest integer $j$ such that $\langle p^j, \xi \rangle = 1$. Then the character $\eta$ defined by $\langle x, \eta \rangle = \langle p^j x, \xi \rangle$ satisfies the conditions of Lemma 4.12, so $\eta = \xi_z$ for some $z$ with $|z| = 1$. But then $\xi = \xi_y$ where $y = p^{-j} z$.

The map $y \mapsto \xi_y$ is therefore a group isomorphism. To see that it is a homeomorphism, observe that the sets

$$N(j,k) = \left\{ \xi \in \widehat{\mathbb{Q}}_p : |\langle x, \xi \rangle - 1| < j^{-1} \text{ for } |x| \leq p^k \right\} \qquad (j \geq 1, \ k \in \mathbb{Z})$$

are a neighborhood base at 1 for $\widehat{\mathbb{Q}}_p$. But the image of the set $\{x : |x| \leq p^k\}$ under $\xi_1$ is $\{1\}$ if $k \leq 0$ and is the group of $p^k$th roots of 1 if $k > 0$, and hence is contained in $\{z : |z - 1| < j^{-1}\}$ if and only if $k \leq 0$. It follows that $\xi_y \in N(j,k)$ if and only if $|y| \leq p^{-k}$, and we are done. □

## 4.2 The Fourier Transform

Henceforth it will be convenient to employ a slightly different identification of $\widehat{G}$ with the spectrum of $L^1(G)$ than the one given by (4.1). Namely, we shall associate to $\xi \in \widehat{G}$ the functional

$$f \mapsto \overline{\xi}(f) = \xi^{-1}(f) = \int \overline{\langle x, \xi \rangle} f(x) \, dx.$$

The Gelfand transform on $L^1(G)$ then becomes the map from $L^1(G)$ to $C(\widehat{G})$ defined by

$$\mathcal{F}f(\xi) = \widehat{f}(\xi) = \int \overline{\langle x, \xi \rangle} f(x) \, dx.$$

This map is the **Fourier transform** on $G$. (We denote the Fourier transform as an operator by $\mathcal{F}$, but we usually denote the Fourier transform of $f \in L^1(G)$ by $\widehat{f}$ rather than $\mathcal{F}f$.)

**4.14 Proposition.** *The Fourier transform is a norm-decreasing $*$-homomorphism from $L^1(G)$ to $C_0(\widehat{G})$ (or $C(\widehat{G})$ if $\widehat{G}$ is compact). Its range is a dense subspace of $C_0(\widehat{G})$.*

*Proof.* That $\mathcal{F}$ is a norm-decreasing $*$-homomorphism is simple to check directly (cf. the proof of Theorem 3.9). That its range lies in $C_0(\widehat{G})$ is an instance of a general property of Gelfand transforms (Theorem 1.30). That its range is dense follows from the Stone-Weierstrass theorem as in the proof of Proposition 1.14(c), since the fact that $\mathcal{F}$ is a $*$-homomorphism means that $L^1(G)$ is symmetric. $\qquad\square$

Two points are worth emphasizing here. First, the fact that $\mathcal{F}(L^1) \subset C_0(\widehat{G})$ is the abstract form of the Riemann-Lebesgue lemma of classical Fourier analysis. Second, the fact that $\mathcal{F}$ is a $*$-homomorphism rather than just a homomorphism is equivalent to the fact that $L^1$ is a symmetric algebra.

The other basic operational properties of the Fourier transform are as follows:

(4.15)
$$(L_y f)\widehat{\,}(\xi) = \int \overline{\langle x, \xi\rangle} f(y^{-1}x)\, dx = \int \overline{\langle yx, \xi\rangle} f(x)\, dx = \overline{\langle y, \xi\rangle}\widehat{f}(\xi),$$

(4.16) $$\qquad (\eta f)\widehat{\,}(\xi) = \int \overline{\langle x, \xi\rangle}\langle x, \eta\rangle f(x)\, dx = \widehat{f}(\eta^{-1}\xi) = L_\eta \widehat{f}(\xi).$$

The Fourier transform can be extended to complex Radon measures on $G$: if $\mu \in M(G)$, its Fourier transform (sometimes called the Fourier-Stieltjes transform) is the bounded continuous function $\widehat{\mu}$ on $\widehat{G}$ defined by

$$\widehat{\mu}(\xi) = \int \overline{\langle x, \xi\rangle}\, d\mu(x).$$

The formula $(\mu * \nu)\widehat{\,} = \widehat{\mu}\widehat{\nu}$ is still valid in this context:

$$(\mu * \nu)\widehat{\,}(\xi) = \iint \overline{\langle xy, \xi\rangle}\, d\mu(x)\, d\nu(y) = \iint \overline{\langle x, \xi\rangle\langle y, \xi\rangle}\, d\mu(x)\, d\nu(y)$$
$$= \widehat{\mu}(\xi)\widehat{\nu}(\xi).$$

(The formula $\int \phi\, d(\mu * \nu) = \iint \phi(xy)\, d\mu(x)\, d\nu(y)$ that we have used here is true by definition of $\mu * \nu$ for $\phi \in C_0(G)$; it remains true for any bounded continuous $\phi$ since $\mu$ and $\nu$ can be approximated in norm by compactly supported measures.) Thus, $\widehat{G}$ can be regarded as part of the spectrum of the Banach algebra $M(G)$, and $\widehat{\mu}$ is the restriction of the Gelfand transform of $\mu$ to $\widehat{\widehat{G}}$.

Of more interest to us than the Fourier transform on $M(G)$ is a similar construction for measures on $\widehat{G}$. Namely, if $\mu \in M(\widehat{G})$ we define the bounded continuous function $\phi_\mu$ on $G$ by

$$(4.17) \qquad \phi_\mu(x) = \int \langle x, \xi \rangle \, d\mu(\xi).$$

**4.18 Proposition.** *The map $\mu \mapsto \phi_\mu$ is a norm-decreasing linear injection from $M(\widehat{G})$ to the space of bounded continuous functions on $G$ (with the uniform norm).*

*Proof.* The only nontrivial point is the injectivity. If $\phi_\mu = 0$ then

$$0 = \iint f(x)\langle x, \xi \rangle \, d\mu(\xi) \, dx = \int \widehat{f}(\xi^{-1}) \, d\mu(\xi)$$

for any $f \in L^1(G)$. But this implies that $\mu = 0$ since $\mathcal{F}(L^1)$ is dense in $C_0(\widehat{G})$. $\qquad\square$

If $\mu \in M(\widehat{G})$ is positive, then $\phi_\mu$ is a generalized linear combination of characters with positive coefficients, and hence is a function of positive type on $G$. Indeed, if $f \in L^1(G)$,

$$\iint f(x)\overline{f(y)}\phi_\mu(y^{-1}x) \, dx \, dy = \iiint f(x)\overline{f(y)}\langle y, \xi \rangle \langle x, \xi \rangle \, d\mu(\xi) \, dx \, dy$$

$$= \int |\widehat{f}(\xi)|^2 \, d\mu(\xi) \geq 0.$$

The converse of this is one of the fundamental results of the theory.

**4.19 Theorem** (Bochner's Theorem). *If $\phi \in \mathcal{P}(G)$, there is a unique positive $\mu \in M(\widehat{G})$ such that $\phi = \phi_\mu$ as in (4.17).*

*Proof.* The uniqueness of $\mu$ was established in Proposition 4.18. There are two nice proofs of the existence, one using the Krein-Milman theorem and one using Gelfand theory. We shall give them both.

*First proof:* It suffices to assume that $\phi \in \mathcal{P}_0$, i.e., $\phi(1) \leq 1$. Let $M_0$ be the set of positive measures $\mu \in M(\widehat{G})$ such that $\mu(\widehat{G}) \leq 1$; $M_0$ is compact in the weak* topology of $M(\widehat{G})$. If $\{\mu_\alpha\}$ is a net in $M_0$ converging to $\mu$ in this topology, for any $f \in L^1$ we have

$$\int f(x)\phi_{\mu_\alpha}(x) \, dx = \iint f(x)\langle x, \xi \rangle d\mu_\alpha(\xi) \, dx = \int \widehat{f}(\xi^{-1}) d\mu_\alpha(\xi)$$

$$\to \int \widehat{f}(\xi^{-1}) d\mu(\xi) = \iint f(x)\langle x, \xi \rangle \, d\mu(\xi) \, dx = \int f(x)\phi_\mu(x) \, dx.$$

Thus $\phi_{\mu_\alpha} \to \phi_\mu$ in the weak* topology of $\mathcal{P}_0 \subset L^\infty$. In other words, the map $\mu \to \phi_\mu$ is continuous from $M_0$ to $\mathcal{P}_0$, and its range is therefore a compact convex subset of $\mathcal{P}_0$. But the range contains every character $\xi \in \widehat{G}$ (take $\mu$ to be the point mass at $\xi$) as well as 0 (take $\mu = 0$), and these are the extreme points of $\mathcal{P}_0$ by Theorem 3.25 and Lemma 3.26. By the Krein-Milman theorem, then, the range is all of $\mathcal{P}_0$.

*Second proof*: Without loss of generality, we shall assume $\phi(1) = 1$. Applying the Schwarz inequality to the positive Hermitian form $\langle f, g \rangle_\phi = \int \phi(g^* * f)$, we have

$$\left| \int \phi(g^* * f) \right|^2 \leq \int \phi(f^* * f) \int \phi(g^* * g) \qquad (f, g \in L^1).$$

Take $g = \psi_U$, an approximate identity. Then $\psi_U^* * f \to f$ in $L^1$, so $\int \phi(\psi_U^* * f) \to \int \phi f$. Also, $\psi_U^* * \psi_U$ is again an approximate identity (if $\operatorname{supp} \psi_U \subset U$ then $\operatorname{supp} \psi_U^* * \psi_U \subset U^{-1}U$, and $\int(\psi_U^* * \psi_U) = |\int \psi_U|^2 = 1$ by Fubini's theorem), so $\int \phi(\psi_U^* * \psi_U) \to \phi(1) = 1$. Thus,

$$\left| \int \phi f \right|^2 \leq \int \phi(f^* * f).$$

The function $h = f^* * f$ satisfies $h^* = h$, so if we set $h^{(2)} = h * h$, $h^{(3)} = h * h * h$, etc., and apply this estimate successively to $f, h, h^{(2)}, \ldots$, we get

$$\left| \int \phi f \right| \leq \left| \int \phi h \right|^{1/2} \leq \left| \int \phi h^{(2)} \right|^{1/4} \cdots \leq \left| \int \phi h^{(2^n)} \right|^{2^{-n-1}}$$
$$\leq \|h^{(2^n)}\|_1^{2^{-n-1}},$$

since $\|\phi\|_\infty = \phi(1) = 1$. But by Theorems 1.8 and 1.13 or Theorem 1.30,

$$\lim \|h^{(2^n)}\|_1^{2^{-n-1}} = \|\widehat{h}\|_\infty^{1/2} = \| |\widehat{f}|^2 \|_\infty^{1/2} = \|\widehat{f}\|_\infty.$$

Thus the map $f \mapsto \int \phi f$ induces a linear functional $\widehat{f} \mapsto \int \phi f$ on $\mathcal{F}(L^1)$, and since $\mathcal{F}(L^1)$ is dense in $C_0(\widehat{G})$, it extends to a linear functional on $C_0(\widehat{G})$ of norm $\leq 1$. By the Riesz representation theorem, there is a $\check{\mu} \in M(\widehat{G})$ with $\|\check{\mu}\| \leq 1$ such that

$$\int \phi f = \int \widehat{f} \, d\check{\mu} = \iint f(x)\langle x, \xi^{-1}\rangle \, d\check{\mu}(\xi) \, dx.$$

But this means that $\phi(x) = \int \langle x, \xi \rangle \, d\mu(\xi)$ where $d\mu(\xi) = d\check{\mu}(\xi^{-1})$. Finally, $1 = \phi(1) = \mu(\widehat{G}) \leq \|\mu\| \leq 1$, so that $\|\mu\| = \mu(\widehat{G})$ and hence $\mu \geq 0$. $\qquad\square$

We now introduce some function spaces on $G$ that will be useful below:

$$\mathcal{B}(G) = \{\phi_\mu : \mu \in M(\widehat{G})\}, \qquad \mathcal{B}^p = \mathcal{B}^p(G) = \mathcal{B}(G) \cap L^p(G) \quad (p < \infty),$$

where $\phi_\mu$ is defined by (4.17). By Bochner's theorem,

$$\mathcal{B}(G) = \text{the linear span of } \mathcal{P}(G).$$

Proposition 3.33 therefore says that $\mathcal{B}(G)$ contains all functions of the form $f * g$ with $f, g \in C_c(G)$ and that $\mathcal{B}^p$ is dense in $L^p$ for all $p < \infty$. Our next step is to establish the Fourier inversion formula for functions in $\mathcal{B}^1$.

**4.20 Lemma.** *If $K \subset \widehat{G}$ is compact, there exists $f \in C_c(G) \cap \mathcal{P}$ such that $\widehat{f} \geq 0$ on $\widehat{G}$ and $\widehat{f} > 0$ on $K$.*

*Proof.* Pick $h \in C_c(G)$ with $\widehat{h}(1) = \int h = 1$ and set $g = h^* * h$. Then $\widehat{g} = |\widehat{h}|^2$; in particular, $\widehat{g} \geq 0$ and $\widehat{g}(1) = 1$, so there is a neighborhood $V$ of 1 in $\widehat{G}$ such that $\widehat{g} > 0$ on $V$. $K$ can be covered by finitely many translates of $V$, say $K \subset \bigcup_1^n \xi_j V_j$. Let $f = (\sum_1^n \xi_j)g$. Then $\widehat{f}(\xi) = \sum_1^n \widehat{g}(\xi_j^{-1}\xi)$ by (4.16), so $\widehat{f} > 0$ on $K$ and $\widehat{f} \geq 0$ everywhere. Also, $g \in \mathcal{P}$ by Corollary 3.16, and it follows that $f \in \mathcal{P}$ (in fact, $\int f(a^* * a) = \sum_1^n \int g[(\xi_j a)^* * \xi_j a)] \geq 0$ for any $a \in L^1$). $\qquad \square$

The correspondence $\mu \mapsto \phi_\mu$ is a bijection from $M(\widehat{G})$ to $\mathcal{B}(G)$. In the next two arguments we shall denote its inverse by $\phi \mapsto \mu_\phi$. That is, if $\phi \in \mathcal{B}(G)$, $\mu_\phi$ is the measure such that $\phi_{\mu_\phi} = \phi$.

**4.21 Lemma.** *If $f, g \in \mathcal{B}^1$ then $\widehat{f} \, d\mu_g = \widehat{g} \, d\mu_f$.*

*Proof.* If $h \in L^1(G)$ we have

$$\int \widehat{h} \, d\mu_f = \iint \langle x^{-1}, \xi \rangle h(x) \, dx \, d\mu_f(\xi) = \int h(x) f(x^{-1}) \, dx = h * f(1).$$

Replacing $h$ by $h * g$ or $h$ by $h * f$ and $f$ by $g$ in this calculation, we obtain

$$\int \widehat{h}\widehat{g} \, d\mu_f = (h * g) * f(1) = (h * f) * g(1) = \int \widehat{h}\widehat{f} \, d\mu_g.$$

Since $\mathcal{F}(L^1)$ is dense in $C_0(\widehat{G})$ it follows that $\widehat{g} \, d\mu_f = \widehat{f} \, d\mu_g$. $\qquad \square$

**4.22 Theorem** (Fourier Inversion Theorem I). *If $f \in \mathcal{B}^1$ then $\widehat{f} \in L^1(\widehat{G})$, and if Haar measure $d\xi$ on $\widehat{G}$ is suitably normalized relative to the given Haar measure $dx$ on $G$, we have $d\mu_f(\xi) = \widehat{f}(\xi) \, d\xi$; that is,*

$$f(x) = \int \langle x, \xi \rangle \widehat{f}(\xi) \, d\xi.$$

*Proof.* We are going to manufacture a positive linear functional on $C_c(\widehat{G})$. If $\psi \in C_c(\widehat{G})$, by Lemma 4.20 there exists $f \in L^1(G) \cap \mathcal{P}$ such that $\widehat{f} > 0$ on supp $\psi$. Let

$$I(\psi) = \int \frac{\psi}{\widehat{f}} d\mu_f.$$

If $g$ is another such function, by Lemma 4.21 we have

$$\int \frac{\psi}{\widehat{f}} d\mu_f = \int \frac{\psi}{\widehat{f}\widehat{g}} \widehat{g} \, d\mu_f = \int \frac{\psi}{\widehat{f}\widehat{g}} \widehat{f} \, d\mu_g = \int \frac{\psi}{\widehat{g}} d\mu_g,$$

so $I(\psi)$ depends only on $\psi$ and not on the choice of $f$. From this it is easy to check that $I(\psi)$ depends linearly on $\psi$, and it is clear that $I(\psi) \geq 0$ for $\psi \geq 0$ since $\widehat{f} \geq 0$ and $\mu_f \geq 0$. Moreover, if $g \in \mathcal{B}^1$ then by Lemma 4.21,

(4.23) $$I(\widehat{g}\psi) = \int \frac{\psi}{\widehat{f}} \widehat{g} \, d\mu_f = \int \psi \, d\mu_g.$$

There clearly exist $\psi$ and $g$ so that $\int \psi \, d\mu_g \neq 0$, so $I \not\equiv 0$.

$I$ is therefore a nontrivial positive linear functional on $C_c(\widehat{G})$. Moreover, if $\eta \in \widehat{G}$,

$$\int \langle x, \xi \rangle d\mu_f(\eta\xi) = \int \langle x, \eta^{-1}\xi \rangle d\mu_f(\xi) = \overline{\langle x, \eta \rangle} f(x) = (\overline{\eta} f)(x),$$

so $d\mu_f(\eta\xi) = d\mu_{\overline{\eta} f}(\xi)$. Also, $(\overline{\eta} f)\widehat{\ }(\xi) = \widehat{f}(\eta\xi)$ by (4.16), so if we pick $f$ so that $\widehat{f} > 0$ on supp $\psi \cup$ supp $L_\eta \psi$,

$$I(L_\eta\psi) = \int \frac{\psi(\eta^{-1}\xi)}{\widehat{f}(\xi)} d\mu_f(\xi) = \int \frac{\psi(\xi)}{\widehat{f}(\eta\xi)} d\mu_f(\eta\xi)$$

$$= \int \frac{\psi(\xi)}{(\overline{\eta} f)\widehat{\ }(\xi)} d\mu_{\overline{\eta} f}(\xi) = I(\psi).$$

Thus $I$ is translation-invariant. It follows that $I(\psi) = \int \psi(\xi) \, d\xi$ where $d\xi$ is a Haar measure on $\widehat{G}$. Finally, if $f \in \mathcal{B}^1$ and $\psi \in C_c(\widehat{G})$, by (4.23) we have

$$\int \psi(\xi)\widehat{f}(\xi) \, d\xi = I(\psi\widehat{f}) = \int \psi \, d\mu_f,$$

so that $\widehat{f}(\xi) \, d\xi = d\mu_f(\xi)$. It follows that $\widehat{f} \in L^1(\widehat{G})$ and that $f(x) = \int \langle x, \xi \rangle \widehat{f}(\xi) \, d\xi$. $\square$

We shall show in the next section that the Fourier inversion formula remains valid if the condition $f \in \mathcal{B}^1$ is replaced by $f \in L^1$ and $\widehat{f} \in L^1$. For the moment, we have the following simple corollary.

**4.24 Corollary.** *If $f \in L^1(G) \cap \mathcal{P}$ then $\widehat{f} \geq 0$.*

*Proof.* $\widehat{f}(\xi)\,d\xi = d\mu_f(\xi)$, and $\mu_f \geq 0$ by Bochner's theorem. $\qquad\square$

When a Haar measure $dx$ on $G$ is given, the Haar measure $d\xi$ on $\widehat{G}$ that makes Theorem 4.22 true is called the **dual measure** of $dx$. If the dual of $dx$ is $d\xi$, the dual of $c\,dx$ is $c^{-1}\,d\xi$. (Replacing $dx$ by $c\,dx$ has the effect of replacing $\widehat{f}$ by $c\widehat{f}$; hence one must replace $d\xi$ by $c^{-1}\,d\xi$ in the inversion formula to compensate.) *Henceforth, we always take the Haar measure on $\widehat{G}$ to be the dual of the given Haar measure on $G$.*

When doing Fourier analysis on specific groups, it is important to know precisely the normalization of the dual Haar measure, and this is usually accomplished by computing the Fourier transform of a specific function.

> *Example 1.* If we identify $\widehat{\mathbb{R}}$ with $\mathbb{R}$ by the pairing $\langle x, \xi \rangle = e^{2\pi i \xi x}$, then Lebesgue measure is self-dual. This can be seen by considering $g(x) = e^{-\pi x^2}$. We have $\widehat{g}(\xi) = \int e^{-2\pi i \xi x - \pi x^2}\,dx$; differentiation under the integral followed by integration by parts shows that $(\widehat{g})'(\xi) = -2\pi\xi\widehat{g}(\xi)$, and $\widehat{g}(0) = \int g = 1$. Solving the differential equation shows that $\widehat{g} = g$, and since $g$ is even, this means that $g(x) = \int e^{2\pi i \xi x - \pi \xi^2}\,d\xi$. Hence the inversion formula holds with $dx$ and $d\xi$ both equal to Lebesgue measure, thus:
>
> $$\widehat{f}(\xi) = \int f(x)e^{-2\pi i \xi x}\,dx, \qquad f(x) = \int \widehat{f}(\xi)e^{2\pi i x \xi}\,d\xi.$$
>
> If we identified $\widehat{\mathbb{R}}$ with $\mathbb{R}$ via $\langle x, \xi \rangle = e^{i\xi x}$, as is frequently done, the dual of Lebesgue measure $dx$ would be $d\xi/2\pi$. With this pairing, the self-dual normalization of Lebesgue measure is $dx/\sqrt{2\pi}$.

> *Example 2.* If we identify $\widehat{\mathbb{Q}}_p$ with $\mathbb{Q}_p$ as in Theorem 4.13, the Haar measure on $\mathbb{Q}_p$ such that $|\mathbb{Z}_p| = 1$ is self-dual. To see this, let $f$ be the characteristic function of $\mathbb{Z}_p$. The restriction of any character $\xi_y$ on $\mathbb{Q}_p$ to the compact group $\mathbb{Z}_p$ is a character on $\mathbb{Z}_p$. Hence, if $|\mathbb{Z}_p| = 1$ and we identify $\xi_y$ with $y$, by Proposition 4.4 $\widehat{f}(y) = \int_{\mathbb{Z}_p} \overline{\xi}_y$ equals 1 if $\xi_y$ is trivial on $\mathbb{Z}_p$ and 0 otherwise. But by the construction of $\xi_y$, $\xi_y$ is trivial on $\mathbb{Z}_p$ if and only if $y \in \mathbb{Z}_p$. Hence $f$ is its own Fourier transform, and this implies that the Haar measure chosen above is self-dual.

We also have the following general result.

**4.25 Proposition.** *If $G$ is compact and Haar measure is chosen so that $|G| = 1$, the dual measure on $\widehat{G}$ is counting measure. If $G$ is discrete and*

*Haar measure is chosen to be counting measure, the dual measure on $\widehat{G}$ is the one such that $|\widehat{G}| = 1$.*

*Proof.* If $G$ is compact, let $g \equiv 1$. Then $\widehat{g} = \chi_{\{1\}}$ by Proposition 4.4, so $g(x) = \sum_{\xi \in \widehat{G}} \langle x, \xi \rangle \widehat{g}(\xi)$, which proves the first assertion. Similarly, if $G$ is discrete, let $g = \chi_{\{1\}}$. Then $\widehat{g} \equiv 1$, and $g(x) = \int \langle x, \xi \rangle \, d\xi$ when $d\xi$ is chosen so that $|\widehat{G}| = 1$. (This is another application of Proposition 4.4, as $\xi \mapsto \langle x, \xi \rangle$ is a character on $\widehat{G}$ for each $x \in G$.) $\qquad\square$

*Example 3.* The groups $\mathbb{T}$ and $\mathbb{Z}$ are dual to each other; the natural dual measures on them are normalized Lebesgue measure $d\theta/2\pi$ on $\mathbb{T}$ and counting measure on $\mathbb{Z}$. The Fourier inversion theorem for functions on $\mathbb{T}$ reads:

$$\widehat{f}(n) = \int_0^{2\pi} f(\theta) e^{-in\theta} \frac{d\theta}{2\pi}, \qquad f(\theta) = \sum_{-\infty}^{\infty} \widehat{f}(n) e^{in\theta}.$$

*Example 4.* If $G$ is a finite cyclic group $Z_k$, the dual of counting measure is counting measure divided by $k$ (the measure such that $|Z_k| = 1$), and the Fourier inversion theorem reads:

$$\widehat{f}(m) = \sum_0^k f(n) e^{-2\pi i mn/k}, \qquad f(n) = \frac{1}{k} \sum_0^k \widehat{f}(m) e^{2\pi i mn/k}.$$

The factor $1/k$ can be relocated from the second sum to the first one if one wishes.

We now come to the fundamental theorem in the $L^2$ theory of the Fourier transform.

**4.26 Theorem** (The Plancherel Theorem). *The Fourier transform on $L^1(G) \cap L^2(G)$ extends uniquely to a unitary isomorphism from $L^2(G)$ to $L^2(\widehat{G})$.*

*Proof.* If $f \in L^1 \cap L^2$ then $f * f^* \in L^1 \cap \mathcal{P}$ by Corollary 3.16, and $(f * f^*)\widehat{\phantom{a}} = |\widehat{f}|^2$, so by Theorem 4.22,

$$\int |f(x)|^2 \, dx = f * f^*(1) = \int (f * f^*)\widehat{\phantom{a}}(\xi) \, d\xi = \int |\widehat{f}(\xi)|^2 \, d\xi.$$

Thus $f \mapsto \widehat{f}$ is an isometry in the $L^2$ norm, so it extends uniquely to an isometry from $L^2(G)$ into $L^2(\widehat{G})$. To show that it is surjective, suppose $\psi \in L^2(\widehat{G})$ is orthogonal to all $\widehat{f}$ with $f \in L^1(G) \cap L^2(G)$. Then by (4.15),

$$0 = \int \psi \overline{(L_x f)\widehat{\phantom{a}}} = \int \langle x, \xi \rangle \psi(\xi) \overline{\widehat{f}(\xi)} \, d\xi$$

for all $f \in L^1 \cap L^2$ and all $x \in G$. But $\psi \overline{\hat{f}} \in L^1(\widehat{G})$ since $\psi, \hat{f} \in L^2(\widehat{G})$, so $\psi(\xi)\overline{\hat{f}(\xi)}\, d\xi \in M(\widehat{G})$. It then follows from Proposition 4.18 that $\psi \overline{\hat{f}} = 0$ a.e. for all $f \in L^1 \cap L^2$, and Lemma 4.20 then implies that $\psi = 0$ a.e. $\quad\square$

**4.27 Corollary.** *If $G$ is compact and $|G| = 1$, $\widehat{G}$ is an orthonormal basis for $L^2(G)$.*

*Proof.* We have seen in Proposition 4.4 that $\widehat{G}$ is an orthonormal set. If $f \in L^2$ is orthogonal to every $\xi \in \widehat{G}$ then $0 = \int f\bar{\xi} = \hat{f}(\xi)$ for all $\xi$, so $f = 0$ by the Plancherel theorem. $\quad\square$

The Fourier transform has now been defined on $L^1(G) + L^2(G)$. If $1 \le p \le 2$ then $L^p \subset L^1 + L^2$, so we can define $\hat{f}$ for $f \in L^p(G)$, and we have:

**4.28 Theorem** (The Hausdorff-Young Inequality). *Suppose $1 \le p \le 2$ and $p^{-1} + q^{-1} = 1$. If $f \in L^p(G)$ then $\hat{f} \in L^q(\widehat{G})$ and $\|\hat{f}\|_q \le \|f\|_p$.*

*Proof.* This follows by the Riesz-Thorin interpolation theorem from the estimates $\|\hat{f}\|_\infty \le \|f\|_1$ and $\|\hat{f}\|_2 = \|f\|_2$. (See Folland [45] for the Riesz-Thorin theorem.) $\quad\square$

## 4.3 The Pontrjagin Duality Theorem

The elements of $\widehat{G}$ are characters on $G$, but we can equally well regard elements of $G$ as characters on $\widehat{G}$. More precisely, each $x \in G$ defines a character $\Phi(x)$ on $\widehat{G}$ by

$$(4.29) \qquad \langle \xi, \Phi(x) \rangle = \langle x, \xi \rangle.$$

$\Phi$ is clearly a group homomorphism from $G$ to $\widehat{\widehat{G}}$. It is a fundamental fact that $\Phi$ is actually an isomorphism, so that every locally compact Abelian group is "reflexive." Before proving this, we need a couple of technical lemmas.

**4.30 Lemma.** *If $\phi, \psi \in C_c(\widehat{G})$ then $\phi * \psi = \hat{h}$ where $h \in \mathcal{B}^1(G)$. In particular, $\mathcal{F}(\mathcal{B}^1)$ is dense in $L^p(\widehat{G})$ for $p < \infty$.*

*Proof.* Let

$$f(x) = \int \langle x, \xi \rangle \phi(\xi)\, d\xi, \qquad g(x) = \int \langle x, \xi \rangle \psi(\xi)\, d\xi,$$

$$h(x) = \int \langle x, \xi \rangle (\phi * \psi)(\xi)\, d\xi.$$

Then $f, g, h \in \mathcal{B}(G)$ since $\phi, \psi, \phi * \psi \in L^1(\widehat{G})$. Also, if $k \in L^1(G) \cap L^2(G)$,

$$\left| \int f\overline{k} \right| = \left| \iint \langle x, \xi \rangle \phi(\xi) \overline{k(x)} \, dx \, d\xi \right| = \left| \int \phi \overline{\widehat{k}} \right| \leq \|\phi\|_2 \|\widehat{k}\|_2 = \|\phi\|_2 \|k\|_2$$

by the Plancherel theorem. This implies that $f \in L^2(G)$, and likewise $g \in L^2(G)$. Next,

$$h(x) = \iint \langle x, \xi \rangle \phi(\xi\eta^{-1}) \psi(\eta) \, d\eta \, d\xi = \iint \langle x, \xi\eta \rangle \phi(\xi) \psi(\eta) \, d\xi \, d\eta$$
$$= f(x)g(x),$$

so $h \in L^1(G)$. Thus $h \in \mathcal{B}^1$, so by Theorem 4.22, $h(x) = \int \langle x, \xi \rangle \widehat{h}(\xi) \, d\xi$. On comparing this with the definition of $h$ and using Proposition 4.18, we see that $\widehat{h} = \phi * \psi$.  □

**4.31 Lemma.** *Suppose $G$ is a locally compact group and $H$ is a subgroup. If $H$ is locally compact in the relative topology then $H$ is closed.*

*Proof.* If $H$ is locally compact, there is an open neighborhood $U$ of 1 in $G$ such that the closure in $H$ of $U \cap H$ — call it $K$ — is compact in $H$. But then $K$ is also compact and hence closed in $G$, so $K$ is the closure in $G$ of $U \cap H$.

Now suppose $x \in \overline{H}$. Pick a net $\{x_\alpha\}$ in $H$ that converges to $x$, and pick a symmetric neighborhood $V$ of 1 in $G$ such that $VV \subset U$. Then $x^{-1} \in \overline{H}$ since $\overline{H}$ is a subgroup, so $Vx^{-1} \cap H \neq \varnothing$. Pick $y \in Vx^{-1} \cap H$. $x_\alpha$ is eventually in $xV$, so $yx_\alpha$ is eventually in $(Vx^{-1})(xV) = VV \subset U$. Moreover, $yx_\alpha \in H$, and $yx_\alpha \to yx$, so $yx \in K \subset H$. But then $x = y^{-1}(yx) \in H$, so $H$ is closed.  □

**4.32 Theorem** (The Pontrjagin Duality Theorem). *The map $\Phi : G \to \widehat{\widehat{G}}$ defined by (4.29) is an isomorphism of topological groups.*

*Proof.* In the first place, if $\Phi(x_1) = \Phi(x_2)$ then $\langle x_1, \xi \rangle = \langle x_2, \xi \rangle$ for all $\xi \in \widehat{G}$. This implies that $x_1 = x_2$ since characters separate points on $G$ (the Gelfand-Raikov theorem), so $\Phi$ is injective.

Next, suppose $x \in G$ and $\{x_\alpha\}_{\alpha \in A}$ is a net in $G$, and consider the following assertions:

   i.  $x_\alpha \to x$ in $G$.
  ii.  $f(x_\alpha) \to f(x)$ for every $f \in \mathcal{B}^1(G)$.
 iii.  $\int \langle x_\alpha, \xi \rangle \widehat{f}(\xi) \, d\xi \to \int \langle x, \xi \rangle \widehat{f}(\xi) \, d\xi$ for every $f \in \mathcal{B}^1(G)$.
 iv.  $\Phi(x_\alpha) \to \Phi(x)$ in $\widehat{\widehat{G}}$.

(i) implies (ii) trivially. On the other hand, if $x_\alpha \not\to x$, there is a neighborhood $U$ of $x$ and a cofinal $B \subset A$ such that $x_\beta \notin U$ for $\beta \in B$, and by Proposition 3.33 there is an $f \in \mathcal{B}^1$ with $\operatorname{supp} f \subset U$ and $f(x) \neq 0$. Thus $f(x_\alpha) \not\to f(x)$, so (ii) implies (i). (iii) is just a restatement of (ii) in view of Theorem 4.22. Finally, (iii) says that $\int \Phi(x_\alpha)\widehat{f} \to \int \Phi(x)\widehat{f}$ for all $f \in \mathcal{F}(\mathcal{B}^1)$. Since the topology of $\widehat{\widehat{G}}$ is the weak* topology of $L^\infty(\widehat{G})$, $\|\Phi(x_\alpha)\|_\infty = 1$ for all $\alpha$, and $\mathcal{F}(\mathcal{B}^1)$ is dense in $L^1(\widehat{G})$ by Lemma 4.30, it follows easily that (iii) is equivalent to (iv). In short, (i) is equivalent to (iv), which means that $\Phi$ is a homeomorphism of $G$ onto $\Phi(G)$.

It now follows that $\Phi(G)$ is locally compact, and hence, by Lemma 4.31, that $\Phi(G)$ is closed in $\widehat{\widehat{G}}$. Suppose there exists $\mathbf{x} \in \widehat{\widehat{G}} \setminus \Phi(G)$. Pick a symmetric neighborhood $V$ of 1 in $\widehat{\widehat{G}}$ such that $\mathbf{x}VV \cap \Phi(G) = \varnothing$ and pick nonnegative (and nonzero) $\phi, \psi \in C_c(\widehat{\widehat{G}})$ with $\operatorname{supp}\phi \subset \mathbf{x}V$ and $\operatorname{supp}\psi \subset V$. Then $\phi * \psi \neq 0$, $\operatorname{supp}(\phi * \psi) \cap \Phi(G) = \varnothing$, and by Lemma 4.30, $\phi * \psi = \widehat{h}$ where $h \in \mathcal{B}^1(\widehat{G})$. But then

$$0 = \widehat{h}(\Phi(x^{-1})) = \int \langle \xi, \Phi(x) \rangle h(\xi)\, d\xi = \int \langle x, \xi \rangle h(\xi)\, d\xi$$

for all $x \in G$. By Proposition 4.18, this implies that $h = 0$ and hence $\widehat{h} = \phi * \psi = 0$, a contradiction. Conclusion: $\Phi(G) = \widehat{\widehat{G}}$. □

Henceforth we shall identify $\widehat{\widehat{G}}$ with $G$ and omit writing $\Phi$. Accordingly, we may write either $\langle x, \xi \rangle$ or $\langle \xi, x \rangle$ for the pairing between $G$ and $\widehat{G}$. The Pontrjagin duality theorem has several important corollaries:

**4.33 Theorem** (Fourier Inversion Theorem II). *If $f \in L^1(G)$ and $\widehat{f} \in L^1(\widehat{G})$ then $f(x) = (\widehat{f})\widehat{\ }(x^{-1})$ for a.e. $x$; that is,*

$$f(x) = \int \langle x, \xi \rangle \widehat{f}(\xi)\, d\xi \text{ for a.e. } x.$$

*If $f$ is continuous, these relations hold for every $x$.*

*Proof.* Since

$$\widehat{f}(\xi) = \int \overline{\langle x, \xi \rangle} f(x)\, dx = \int \langle x^{-1}, \xi \rangle f(x)\, dx = \int \langle x, \xi \rangle f(x^{-1})\, dx,$$

we have $\widehat{f} \in \mathcal{B}^1(\widehat{G})$ and $d\mu_{\widehat{f}}(x) = f(x^{-1})dx$. By Theorem 4.22, then, $f(x^{-1}) = (\widehat{f})\widehat{\ }(x)$ a.e. $(\widehat{f})\widehat{\ }$ is automatically continuous, so if $f$ is also continuous we have $f(x^{-1}) = (\widehat{f})\widehat{\ }(x)$ for every $x$. □

**4.34 Corollary** (The Fourier Uniqueness Theorem). *If $\mu, \nu \in M(G)$ and $\widehat{\mu} = \widehat{\nu}$ then $\mu = \nu$. In particular, if $f, g \in L^1(G)$ and $\widehat{f} = \widehat{g}$ then $f = g$.*

*Proof.* By Proposition 4.18 (with $G$ and $\widehat{G}$ interchanged), $\mu$ is completely determined by the function $\phi_\mu(\xi) = \widehat{\mu}(\xi^{-1})$. □

**4.35 Corollary.** *$M(G)$ and $L^1(G)$ are semisimple Banach algebras.*

**4.36 Proposition.** *If $\widehat{G}$ is compact then $G$ is discrete. If $\widehat{G}$ is discrete then $G$ is compact.*

*Proof.* Combine Proposition 4.5 with Pontrjagin duality. □

**4.37 Proposition.** *If $f, g \in L^2(G)$ then $(fg)\widehat{\phantom{x}} = \widehat{f} * \widehat{g}$.*

*Proof.* First suppose $f, g \in L^2(G) \cap \mathcal{F}[\mathcal{B}^1(\widehat{G})]$, so that we can write $f(x) = \widehat{\phi}(x^{-1})$ and $g = \widehat{\psi}(x^{-1})$ with $\phi, \psi \in L^2(\widehat{G}) \cap \mathcal{B}^1(\widehat{G})$. Then, as in the proof of Lemma 4.30,

$$(4.38) \qquad (\phi * \psi)\widehat{\phantom{x}}(x^{-1}) = \iint \langle x, \xi \rangle \phi(\xi\eta^{-1})\psi(\eta)\, d\eta\, d\xi = f(x)g(x).$$

Theorem 4.22 applies to $\phi$ and $\psi$, so $\phi = \widehat{f}$ and $\psi = \widehat{g}$. Moreover, $\phi * \psi \in L^1 * L^1 \subset L^1$ and $fg \in L^2 \cdot L^2 = L^1$, so Theorem 4.33 applies to $\phi * \psi$. Combining this with (4.38) yields

$$\widehat{f} * \widehat{g}(\xi) = \phi * \psi(\xi) = (\phi * \psi)\widehat{\phantom{x}}(\xi^{-1}) = (fg)\widehat{\phantom{x}}(\xi).$$

Finally, we remove the assumption that $f, g \in \mathcal{F}[\mathcal{B}^1(\widehat{G})]$. If $f, g \in L^2(G)$, by Lemma 4.30 there are sequences $\{f_n\}, \{g_n\}$ in $L^2(G) \cap \mathcal{F}[\mathcal{B}^1(\widehat{G})]$ that converge to $f$ and $g$ in $L^2$. Then $f_n g_n \to fg$ in $L^1$, so $(f_n g_n)\widehat{\phantom{x}} \to (fg)\widehat{\phantom{x}}$ uniformly; also $\widehat{f_n} * \widehat{g_n} \to \widehat{f} * \widehat{g}$ uniformly, so the desired result follows. □

The Pontrjagin duality theorem leads to a neat duality between subgroups and quotient groups of a locally compact Abelian group. If $H$ is a closed subgroup of $G$, we define

$$H^\perp = \big\{ \xi \in \widehat{G} : \langle x, \xi \rangle = 1 \text{ for all } x \in H \big\}.$$

$H^\perp$ is clearly a closed subgroup of $\widehat{G}$.

**4.39 Proposition.** *$(H^\perp)^\perp = H$ for any closed subgroup $H$ of $G$.*

*Proof.* Obviously $H \subset (H^\perp)^\perp$. To prove the reverse inclusion, let $q : G \to G/H$ be the canonical projection. If $x_0 \notin H$, by the Gelfand-Raikov theorem there is a character $\eta$ on the group $G/H$ such that $\eta(q(x_0)) \neq 1$. Then $\eta \circ q \in H^\perp$ and $(\eta \circ q)(x_0) \neq 1$, so $x_0 \notin (H^\perp)^\perp$. □

**4.40 Theorem.** *Suppose $H$ is a closed subgroup of $G$. Define $\Phi$ : $(G/H)\hat{} \to H^\perp$ and $\Psi : \widehat{G}/H^\perp \to \widehat{H}$ by*

$$\Phi(\eta) = \eta \circ q, \qquad \Psi(\xi H^\perp) = \xi|H,$$

*where $q : G \to G/H$ is the canonical projection. Then $\Phi$ and $\Psi$ are isomorphisms of topological groups.*

*Proof.* $\Phi$ is obviously a group isomorphism from $(G/H)\hat{}$ to $H^\perp$. If $\eta_\alpha \to \eta$ in $(G/H)\hat{}$ and $K \subset G$ is compact, then $\eta_\alpha \to \eta$ uniformly on $q(K)$, so $\eta_\alpha \circ q \to \eta \circ q$ uniformly on $K$; hence $\eta_\alpha \circ q \to \eta \circ q$ in $\widehat{G}$. Conversely, if $\eta_\alpha \circ q \to \eta \circ q$ in $\widehat{G}$ and $F \subset G/H$ is compact, by Lemma 2.48 there exists a compact $K \subset G$ with $q(K) = F$; we have $\eta_\alpha \circ q \to \eta \circ q$ uniformly on $K$, hence $\eta_\alpha \to \eta$ uniformly on $F$, so $\eta_\alpha \to \eta$ in $(G/H)\hat{}$. Therefore $\Phi$ is a homeomorphism.

Replacing $G$ by $\widehat{G}$ and $H$ by $H^\perp$, by Proposition 4.39 we have $(\widehat{G}/H^\perp)\hat{} \cong (H^\perp)^\perp = H$. More specifically, if $x \in H$, the corresponding element $\eta$ of $(\widehat{G}/H^\perp)\hat{}$ is given by

$$(4.41) \qquad \langle \eta, \xi H^\perp \rangle = \langle x, \xi \rangle.$$

Pontrjagin duality then gives an isomorphism $\widehat{G}/H^\perp \cong (\widehat{G}/H^\perp)\hat{}\hat{} \cong \widehat{H}$, which in view of (4.41) is the restriction map $\Psi$. $\qquad\square$

The surjectivity of $\Psi$ yields a sort of Hahn-Banach theorem for locally compact Abelian groups:

**4.42 Corollary.** *If $H$ is a closed subgroup of $G$, every character on $H$ extends to a character of $G$.*

> *Example.* Consider the $p$-adic numbers $\mathbb{Q}_p$ and the $p$-adic integers $\mathbb{Z}_p$. If we identify $\widehat{\mathbb{Q}}_p$ with $\mathbb{Q}_p$ by the correspondence $y \leftrightarrow \xi_y$ of Theorem 4.13, it is easily verified that $\mathbb{Z}_p^\perp = \mathbb{Z}_p$. By Theorem 4.40, then, $\widehat{\mathbb{Z}}_p$ is isomorphic to $\mathbb{Q}_p/\mathbb{Z}_p$. Moreover, the kernel of the character $\xi_1$ is exactly $\mathbb{Z}_p$, so $\mathbb{Q}_p/\mathbb{Z}_p$ is isomorphic as a group to the range of $\xi_1$, namely the union $U_p$ of the groups of $p^k$th roots of unity, $k \geq 1$. Finally, $\widehat{\mathbb{Z}}_p$ is discrete since $\mathbb{Z}_p$ is compact (Proposition 4.5), so $\widehat{\mathbb{Z}}_p$ is isomorphic to the group $U_p$ with the discrete topology.

We conclude this section by giving a general form of the Poisson summation formula. If $H$ is a closed subgroup of $G$, the (suitably normalized) Haar measure on $G/H$ is obviously the invariant measure on $G/H$ called for in Theorem 2.51. Hence the Fubini-type formula (2.52) holds for functions in $C_c(G)$, with $d\mu(xH) = d(xH) =$ Haar measure on $G/H$.

**4.43 Theorem.** *Suppose $H$ is a closed subgroup of $G$. If $f \in C_c(G)$, define $F \in C_c(G/H)$ by $F(xH) = \int_H f(xy)\, dy$. Then $\widehat{F} = \widehat{f}|H^{\perp}$, where we identify $(G/H)\widehat{\phantom{x}}$ with $H^{\perp}$. If also $\widehat{f}|H^{\perp} \in L^1(H^{\perp})$, then (with Haar measures on $H$ and $H^{\perp}$ suitably normalized)*

$$(4.44) \qquad \int_H f(xy)\, dy = \int_{H^{\perp}} \widehat{f}(\xi)\langle x, \xi \rangle \, d\xi.$$

*Proof.* If $\xi \in H^{\perp}$ we have $\langle xy, \xi \rangle = \langle x, \xi \rangle$ for any $y \in H$, so

$$\widehat{F}(\xi) = \int_{G/H} \int_H f(xy)\overline{\langle xy, \xi \rangle}\, dy\, d(xH) = \int_G f(x)\overline{\langle x, \xi \rangle}\, dx = \widehat{f}(\xi).$$

Hence, if $\widehat{f}|H^{\perp} \in L^1(H^{\perp})$ we can apply the inversion theorem 4.33 to $F$ to get (4.44). $\qquad\square$

*Remark.* The hypothesis that $f \in C_c(G)$ can be weakened. Indeed, if $f \in L^1(G)$ then $F$ is well defined a.e. and belongs to $L^1(G/H)$; we have $\widehat{F} = \widehat{f}|H^{\perp}$, and if $\widehat{f}|H^{\perp} \in L^1(H^{\perp})$ then (4.44) holds a.e. (See the remarks concerning formula (2.59), which generalizes (2.52), in §2.7.)

The classical **Poisson summation formula** is the case $G = \mathbb{R}$, $H = \mathbb{Z}$. In this case $H^{\perp} = \mathbb{Z}$ when we identify $\widehat{\mathbb{R}}$ with $\mathbb{R}$ via $\langle x, \xi \rangle = e^{2\pi i \xi x}$, and (4.44) becomes

$$\sum_{-\infty}^{\infty} f(x+n) = \sum_{-\infty}^{\infty} \widehat{f}(n) e^{2\pi i n x}.$$

## 4.4    Representations of Locally Compact Abelian Groups

In this section we show how to express an arbitrary unitary representation of a locally compact Abelian group $G$ in terms of irreducible representations, i.e., characters. The key to this result is the identification of $\widehat{G}$ with the spectrum of $L^1(G)$, and for our present purposes it will be more convenient to return to the original identification (4.1) of $\xi \in \widehat{G}$ with the functional $f \mapsto \xi(f) = \widehat{f}(\xi^{-1})$ rather than $f \mapsto \widehat{f}(\xi)$. The precise theorem is as follows.

**4.45 Theorem.** *Let $\pi$ be a unitary representation of the locally compact Abelian group $G$. There is a unique regular $\mathcal{H}_{\pi}$-projection-valued measure*

$P$ on $\widehat{G}$ such that

(4.46)
$$\pi(x) = \int \langle x, \xi \rangle \, dP(\xi) \quad for \ x \in G,$$
$$\pi(f) = \int \xi(f) \, dP(\xi) \quad for \ f \in L^1(G),$$

*where $\xi(f)$ is given by (4.1). Moreover, an operator $T \in \mathcal{L}(\mathcal{H}_\pi)$ belongs to $\mathcal{C}(\pi)$ if and only if $T$ commutes with $P(E)$ for every Borel set $E \subset \widehat{G}$.*

*Proof.* By Theorem 1.54 and Theorem 4.3, there is a unique regular projection-valued measure $P$ on $\widehat{G}$ such that $\pi(f)$ is given by (4.46) for all $f \in L^1(G)$. Moreover, the assertion about commuting operators follows from Theorems 1.54 and 3.12(b), so it remains only to show that the formula for $\pi(x)$ in (4.46) is valid.

The proof of Theorem 3.11 shows that $\pi(x)$ is the strong limit of $\pi(L_x \psi_U)$ where $\{\psi_U\}$ is an approximate identity. By a slight modification of (4.15),

$$\pi(L_x \psi_U) = \int \xi(L_x \psi_U) \, dP(\xi) = \int \langle x, \xi \rangle \xi(\psi_U) \, dP(\xi),$$

so it suffices to show that the integrals on the right converge weakly to $\int \langle x, \xi \rangle \, dP(\xi)$. If $G$ is first countable, so that the net $\{\psi_U\}$ can be taken to be a sequence, this is an easy consequence of the dominated convergence theorem since $|\xi(\psi_U)| \leq 1$ and $\xi(\psi_U) \to 1$ for each $\xi$. For the general case we need a slightly more involved argument.

**4.47 Lemma.** *If $\{\psi_U\}$ is an approximate identity, then $\widehat{\psi}_U \to 1$ uniformly on compact subsets of $\widehat{G}$ as $U \to \{1\}$.*

*Proof.* Suppose $K \subset \widehat{G}$ is compact. In view of Pontrjagin duality, the topology on $G$ is the topology of uniform convergence of the functions $\xi \to \langle x, \xi \rangle$ on compact subsets of $\widehat{G}$, so if $\epsilon > 0$, the set

$$V = \{x \in G : |\langle x, \xi \rangle - 1| < \epsilon \ for \ \xi \in K\}$$

is a neighborhood of 1 in $G$. But then, if $U \subset V$ and $\xi \in K$,

$$|\widehat{\psi}_U(\xi) - 1| = \left| \int_V (\overline{\langle x, \xi \rangle} - 1) \psi_U(x) \, dx \right| < \epsilon. \qquad \square$$

Now we can complete the proof of Theorem 4.45. If $u, v \in \mathcal{H}_\pi$ and $\epsilon > 0$, $\mu_{u,v}(E) = \langle P(E)u, v \rangle$ is a finite Radon measure on $\widehat{G}$, so there is a compact $K \subset \widehat{G}$ such that $|\mu_{u,v}|(\widehat{G} \setminus K) < \epsilon$. Moreover,

$$\left\langle \left[ \pi(L_x \psi_U) - \int \langle x, \xi \rangle \, dP(\xi) \right] u, \, v \right\rangle = \int_{\widehat{G}} \langle x, \xi \rangle (\widehat{\psi}_U(\xi^{-1}) - 1) \, d\mu_{u,v}(\xi).$$

We write the integral on the right as $\int_K + \int_{\widehat{G}\setminus K}$. By Lemma 4.47, the integral over $K$ is at most $\epsilon$ if $U$ is sufficiently small, and the integral over $\widehat{G}\setminus K$ is at most $2\epsilon$ since $|\widehat{\psi}_U - 1| \leq 2$. Hence

$$\langle \pi(x)u, v \rangle = \lim \langle \pi(L_x\psi_U)u, v \rangle = \int \langle x, \xi \rangle \, d\mu_{u,v}(\xi),$$

which means that $\pi(x) = \int \langle x, \xi \rangle \, dP(\xi)$.                         $\square$

Let us examine the meaning of Theorem 4.45 a bit. First suppose the measure $P$ is discrete, that is, $P(E) = \sum_{\xi \in E} P(\{\xi\})$ for any Borel $E \subset \widehat{G}$. Let $A = \{\xi : P(\{\xi\}) \neq 0\}$, and for $\xi \in A$ let $\mathcal{H}_\xi$ be the range of $P(\{\xi\})$. Then $\mathcal{H}_\pi = \bigoplus_{\xi \in A} \mathcal{H}_\xi$, each $\mathcal{H}_\xi$ is invariant under $\pi$, and the subrepresentation of $\pi$ on $\mathcal{H}_\xi$ is just $\xi$ times the identity, a direct sum of copies of the irreducible representation $\xi$. Thus the formula $\pi(x) = \sum_{\xi \in A} \langle x, \xi \rangle P(\{\xi\})$ of Theorem 4.45 exhibits $\pi$ as a direct sum of irreducible representations. If $P$ is not discrete, Theorem 4.45 can be thought of as exhibiting $\mathcal{H}_\pi$ as a direct integral of "infinitesimal pieces," the ranges of the projections $dP(\xi)$, on which $\pi$ acts as a sum of copies of the character $\xi$. We shall make the connection of this result with the general theory of direct integrals in §7.4.

*Example.* Let $\pi$ be the left regular representation of $G$ on $L^2(G)$, $\pi(x)f = L_x f$. Then

$$\mathcal{F}^{-1}(L_x f)(\xi) = \int \langle y, \xi \rangle f(x^{-1}y) \, dy = \int \langle xy, \xi \rangle f(y) \, dy$$
$$= \langle x, \xi \rangle \mathcal{F}^{-1} f(\xi).$$

Comparing this with (4.46), we see that $\mathcal{F}^{-1}[P(E)f]$ must be $\chi_E \mathcal{F}^{-1} f$, so that

$$P(E) = \mathcal{F}(\text{multiplication by } \chi_E)\mathcal{F}^{-1}.$$

When $G = \mathbb{R}$, Theorem 4.45 is equivalent to the classical theorem of Stone on the structure of one-parameter unitary groups (see Rudin [123]). Stone's theorem, in its usual formulation, says that if $\pi$ is a unitary representation of $\mathbb{R}$ on $\mathcal{H}$, there is a self-adjoint operator $A$ on $\mathcal{H}$ (usually unbounded) such that $\pi(x) = e^{2\pi i x A}$. If $P$ is the projection-valued measure on $\mathbb{R}$ associated to $A$, so that $A = \int \xi \, dP(\xi)$, this says that $\pi(x) = \int e^{2\pi i x \xi} \, dP(\xi)$, which is Theorem 4.45. Conversely, if $P$ is the projection-valued measure of Theorem 4.45 then $\pi(x) = e^{2\pi i x A}$ where $A = \int \xi \, dP(\xi)$.

## 4.5   Closed Ideals in $L^1(G)$

Suppose $\mathcal{A}$ is a commutative Banach algebra with spectrum $\sigma(\mathcal{A})$. The Gelfand transform provides a natural correspondence between closed subsets of $\sigma(\mathcal{A})$ and closed ideals in $\mathcal{A}$. Namely, if $\mathfrak{I} \subset \mathcal{A}$ is a closed ideal, let

$$\nu(\mathfrak{I}) = \left\{ h \in \sigma(\mathcal{A}) : \widehat{f}(h) = 0 \text{ for all } f \in \mathfrak{I} \right\}.$$

Then $\nu(\mathfrak{I})$ is a closed subset of $\sigma(\mathcal{A})$, called the **hull** of $\mathfrak{I}$. On the other hand, if $N \subset \sigma(\mathcal{A})$ is closed, let

$$\iota(N) = \left\{ f \in \mathcal{A} : \widehat{f} = 0 \text{ on } N \right\}.$$

Then $\iota(N)$ is a closed ideal in $\mathcal{A}$, called the **kernel** of $N$. We clearly have

$$(4.48) \qquad\qquad \iota(\nu(\mathfrak{I})) \supset \mathfrak{I}, \qquad \nu(\iota(N)) \supset N$$

for any $\mathfrak{I}$ and $N$, so the obvious question is: when are these inclusions equalities? In other words, to what extent are $\nu$ and $\iota$ inverses of each other?

The simplest case is that of a commutative C* algebra. In this case, by the Gelfand-Naimark theorem we may assume that $\mathcal{A} = C_0(X)$ and $\sigma(\mathcal{A}) = X$, where $X$ is a locally compact Hausdorff space, and the Gelfand transform is then the identity map. Here the situation is as nice as possible: the maps $\nu$ and $\iota$ are mutually inverse bijections.

**4.49 Theorem.** *Let $X$ be a locally compact Hausdorff space. Then $\iota(\nu(\mathfrak{I})) = \mathfrak{I}$ for every closed ideal $\mathfrak{I} \subset C_0(X)$, and $\nu(\iota(N)) = N$ for every closed set $N \subset X$.*

*Proof.* If $N \subset X$ is closed and $x \notin N$, by Urysohn's lemma there exists $f \in \iota(N)$ such that $f(x) = 1$; hence $\nu(\iota(N)) = N$. Moreover, if $\mathfrak{I}$ is a closed ideal in $C_0(X)$ and $x$ and $y$ are distinct points of $X \setminus \nu(\mathfrak{I})$, there exist $f \in \mathfrak{I}$ such that $f(x) \neq 0$ and $g \in C_0(X)$ such that $g(x) = 1$ and $g = 0$ on $\nu(\mathfrak{I}) \cup \{y\}$ (by Urysohn again). Then $fg \in \mathfrak{I}$ and $(fg)(x) \neq 0 = (fg)(y)$, so $\mathfrak{I}$ separates points on $X \setminus \nu(\mathfrak{I})$. But if $N$ is any closed subset of $X$, $X \setminus N$ is a locally compact Hausdorff space and (by a simple argument that we leave to the reader)

$$C_0(X \setminus N) = \left\{ f|(X \setminus N) : f \in \iota(N) \right\}.$$

Hence the Stone-Weierstrass theorem on $X \setminus N$, with $N = \nu(\mathfrak{I})$, implies that $\mathfrak{I} = \iota(\nu(\mathfrak{I}))$. $\qquad\qquad\qquad\qquad\qquad\qquad\qquad\qquad\qquad$ $\square$

The remainder of this section is devoted to the study of the correspondences $\nu$ and $\iota$ for the algebra $L^1(G)$ where $G$ is a locally compact Abelian group. We identify $\sigma(L^1)$ with $\widehat{G}$ as in §4.2 so that the Gelfand transform becomes the Fourier transform. The simple part of the theory is that $\nu$ is a left inverse for $\iota$.

**4.50 Lemma.** *If $K \subset \widehat{G}$ is compact and $W$ is a neighborhood of $K$, there is a neighborhood $U$ of 1 such that $UK \subset W$.*

*Proof.* For each $x \in K$ there is a neighborhood $U_x$ of 1 such that $U_x U_x x \subset W$. Since $K$ is compact there exist $x_1, \ldots, x_n$ such that the sets $U_{x_i} x_i$ cover $K$. Let $U = \bigcap_1^n U_{x_i}$. If $x \in K$ then $x \in U_{x_i} x_i$ for some $i$, and hence $Ux \subset UU_{x_i} x_i \subset W$. In short, $UK \subset W$.     □

**4.51 Lemma.** *If $K \subset \widehat{G}$ is compact and $W \subset \widehat{G}$ is a neighborhood of $K$, there exists $f \in L^1(G)$ such that $\widehat{f} = 1$ on $K$ and supp $\widehat{f} \subset W$.*

*Proof.* By Lemma 4.50 there is a compact symmetric neighborhood $U$ of 1 in $\widehat{G}$ such that $UUK \subset W$. Then $\chi_U, \chi_{UK} \in L^2(\widehat{G})$. Let $g, h$ be the inverse Fourier transforms of $\chi_U, \chi_{UK}$ (as given by the Plancherel theorem), and let $f = |U|^{-1} gh$. Then $f \in L^1$ and $\widehat{f} = \widehat{g} * \widehat{h}$ by Proposition 4.37; thus $\widehat{f}(\xi) = |U|^{-1} \int_U \chi_{UK}(\eta^{-1}\xi)\,d\eta$ has the desired properties.     □

**4.52 Theorem.** *If $N \subset \widehat{G}$ is closed, then $\nu(\iota(N)) = N$.*

*Proof.* If $\xi \notin N$, take $K = \{\xi\}$ and $W = \widehat{G} \setminus N$ in Lemma 4.51 to obtain $f \in \iota(N)$ such that $\widehat{f}(\xi) \neq 0$.     □

When $G$ is compact, the other half of the correspondence is easily analyzed. First, a simple lemma that will also be useful elsewhere.

**4.53 Lemma.** *If $f \in L^1(G)$ and $\xi \in \widehat{G}$ ($\subset L^\infty(G)$) then $f * \xi = \widehat{f}(\xi)\xi$.*

*Proof.* For any $x \in G$,

$$f * \xi(x) = \int f(y)\langle y^{-1}x, \xi\rangle\,dy = \langle x, \xi\rangle \int f(y)\overline{\langle y, \xi\rangle}\,dy = \widehat{f}(\xi)\langle x, \xi\rangle.$$

□

**4.54 Theorem.** *If $G$ is compact, then $\iota(\nu(\mathfrak{I})) = \mathfrak{I}$ for every closed ideal $\mathfrak{I} \subset L^1(G)$.*

*Proof.* Since $G$ is compact, we have $\widehat{G} \subset L^\infty \subset L^2 \subset L^1$. Suppose $f \in \iota(\nu(\mathfrak{I}))$. Then $f * \xi = \widehat{f}(\xi)\xi$ by Lemma 4.53, and either $\widehat{f}(\xi) = 0$ or $\xi \notin \nu(\mathfrak{I})$. In the first case, $f * \xi = 0$; in the second case, there exists $g \in \mathfrak{I}$ such that $\widehat{g}(\xi) = 1$, so that $\xi = g * \xi \in \mathfrak{I}$ by Lemma 4.53 again. In either case we have $f * \xi \in \mathcal{I}$, and hence $f * g \in \mathcal{I}$ for any $g$ in the linear span of $\widehat{G}$. The latter is dense in $L^2$ by Corollary 4.27, so $f * g \in \mathcal{I}$ for all $g \in L^2$ since $\mathcal{I}$ is closed. Finally, we can take $g$ to be an approximate identity to conclude that $f \in \mathcal{I}$. □

When $G$ is noncompact, the question of whether $\iota(\nu(\mathcal{I})) = \mathcal{I}$ is much more delicate. We now exhibit a simple example to show that the answer can be negative.

**4.55 Theorem.** *Let $G = \mathbb{R}^n$ with $n \geq 3$, and let $S$ be the unit sphere in $\mathbb{R}^n$. There is a closed ideal $\mathcal{I}$ in $L^1(\mathbb{R}^n)$ such that $\nu(\mathcal{I}) = S$ but $\mathcal{I} \neq \iota(S)$.*

*Proof.* First we observe that if $f$ and $x_1 f$ ($=$ the function whose value at $x$ is $x_1 f(x)$) are in $L^1(\mathbb{R}^n)$ then

$$
\begin{aligned}
-2\pi i (x_1 f)^\smallfrown(\xi) &= \int (-2\pi i x_1 e^{-2\pi i \xi \cdot x}) f(x) \, dx \\
&= \int \frac{\partial e^{-2\pi i \xi \cdot x}}{\partial \xi_1} f(x) \, dx = \frac{\partial \widehat{f}}{\partial \xi_1}(\xi).
\end{aligned}
$$

(4.56)

Hence $\partial \widehat{f} / \partial \xi_1$ exists and is continuous.

Let $I$ be the set of all $f \in L^1$ such that $x_1 f \in L^1$ and $\widehat{f}|S = (\partial \widehat{f}/\partial \xi_1)|S = 0$, and let $\mathcal{I}$ be the closure of $I$ in $L^1$. Since $(L_y f)^\smallfrown(\xi) = e^{-2\pi i y \xi} \widehat{f}(\xi)$, $I$ is translation-invariant, so $\mathcal{I}$ is a closed ideal by Theorem 2.45. Moreover, $\{\widehat{f} : f \in I\}$ contains all $\phi \in C_c^\infty(\mathbb{R}^n)$ such that $(\operatorname{supp}\phi) \cap S = \varnothing$, so $\nu(\mathcal{I}) = S$. To show that $\mathcal{I} \neq \iota(S)$, we shall exhibit a bounded linear functional on $L^1$ that annihilates $\mathcal{I}$ but not $\iota(S)$.

**4.57 Lemma.** *Let $\mu$ denote surface measure on $S$. Then $|\widehat{\mu}(x)| \leq C(1 + |x|)^{-1}$ for some $C > 0$.*

*Proof.* Given $x \neq 0 \in \mathbb{R}^n$, let $r = |x|$ and $x' = r^{-1} x$. If $\xi \in S$, let $\theta = \arccos(x' \cdot \xi)$ be the angle between $x$ and $\xi$. Then

$$
\widehat{\mu}(x) = \int e^{-2\pi i x \cdot \xi} d\mu(\xi) = c_n \int_0^\pi e^{-2\pi i r \cos\theta} \sin^{n-2}\theta \, d\theta,
$$

where $c_n$ is the area of the unit $(n-2)$-sphere. (We are writing the integral for $\widehat{\mu}(x)$ in spherical coordinates with $x'$ as the "north pole" and $\theta$ as the polar angle. The portion of $S$ where $\theta = \theta_0$ is an $(n-2)$-sphere of radius $\sin\theta_0$, and its area is $c_n \sin^{n-2}\theta_0$.) Integration by parts shows that $\widehat{\mu}(x)$ equals

$$
\frac{-c_n}{2\pi i r} \left[ e^{-2\pi i r \cos\theta} \sin^{n-3}\theta \Big|_0^\pi - (n-3) \int_0^\pi e^{-2\pi i r \cos\theta} \sin^{n-4}\theta \cos\theta \, d\theta \right].
$$

(Note that this works only when $n \geq 3$.) The quantity in square brackets remains bounded as $r = |x| \to \infty$, so the result follows. □

Returning to the proof of Theorem 4.55, let $\phi(x) = -2\pi i x_1 \widehat{\mu}(x)$. By Lemma 4.57, $\phi \in L^\infty$. If $f \in L^1$ and $x_1 f \in L^1$, by (4.56) we have

$$\int \phi f = -2\pi i \iint x_1 e^{-2\pi i x \cdot \xi} f(x) \, d\mu(\xi) \, dx$$

$$= -2\pi i \int (x_1 f)\widehat{\ }(\xi) \, d\mu(\xi) = \int \frac{\partial \widehat{f}}{\partial \xi_1}(\xi) \, d\mu(\xi).$$

From this we deduce two things. On the one hand, $\int \phi f = 0$ if $f \in I$ and hence if $f \in \mathcal{J}$. On the other hand, if $f$ is the inverse Fourier transform of a function in $C_c^\infty(\mathbb{R}^n)$ that equals $\xi_1(|\xi|^2 - 1)$ when $|\xi| \leq 2$, we have $\widehat{f} = 0$ on $S$ and $\partial \widehat{f} / \partial \xi_1 = |\xi|^2 - 1 + 2\xi_1^2 = 2\xi_1^2$ on $S$, so $\int \phi f = \int 2\xi_1^2 d\mu(\xi) > 0$. Hence $f \in \iota(S)$ but $f \notin \mathcal{J}$.                                                        □

We now aim toward some positive results in the noncompact case. Before proceeding to the main results, we need some approximation lemmas. The proof of the first one contains most of the hard technical work in this chain of arguments; once it is accomplished, the others are rather straightforward.

**4.58 Lemma.** *Suppose $f \in L^1(G)$, $\xi_0 \in \widehat{G}$, $\widehat{f}(\xi_0) = 0$, and $\epsilon > 0$. There exists $h \in L^1(G)$ such that $\widehat{h} = 1$ near $\xi_0$ and $\|f * h\|_1 < \epsilon$.*

*Proof.* Without loss of generality, we assume $\xi_0 = 1$. (If not, replace $f$ and $h$ by $\xi_0 f$ and $\xi_0 h$.) Let $\delta = \epsilon/(3 + 3\|f\|_1)$.

Choose (i) a compact $F \subset G$ such that $\int_{G \setminus F} |f| < \delta$, (ii) a compact symmetric neighborhood $V$ of $1$ in $\widehat{G}$ such that $|\langle x, \xi \rangle - 1| < \delta$ when $x \in F$ and $\xi \in VV$, and (iii) an open $W \supset V$ such that $|W \setminus V| < |V|$. By Lemma 4.50 there is a compact neighborhood $K$ of $1$ in $\widehat{G}$ such that $K \subset V$ and $KV \subset W$, so that $|KV| < 2|V|$. Let $\phi$ and $\psi$ be the inverse Fourier transforms (in the $L^2$ sense) of $\chi_V$ and $\chi_{KV}$, and let $h = |V|^{-1}\phi\psi$. We claim that $h$ does the job.

To begin with, by Hölder's inequality and the Plancherel theorem,

$$\|h\|_1 \leq |V|^{-1}\|\phi\|_2\|\psi\|_2 = |V|^{-1}|V|^{1/2}|KV|^{1/2} < \sqrt{2}.$$

Next, by Proposition 4.37, $\widehat{h}(\xi) = |V|^{-1}\chi_V * \chi_{KV}(\xi)$, which equals $1$ when $\xi \in K$ since $V$ is symmetric. Finally, we estimate $\|f * h\|_1$. Since $\int f = \widehat{f}(1) = 0$,

$$f * h(x) = \int f(y)[h(y^{-1}x) - h(x)] \, dy = \int f(y)[L_y h - h](x) \, dy,$$

so

$$\|f * h\|_1 \leq \int_F |f(y)| \, \|L_y h - h\|_1 \, dy + \int_{G \setminus F} |f(y)| \, \|L_y h - h\|_1 \, dy.$$

The second term on the right is bounded by $2\|h\|_1 \int_{G\setminus F} |f| < 3\delta$, and the first is bounded by $\|f\|_1 \sup_F \|L_y h - h\|_1$, so it suffices to show that $\|L_y h - h\|_1 < 3\delta$ for $y \in F$. But

$$|V|(L_y h - h) = (L_y\phi)(L_y\psi - \psi) + (L_y\phi - \phi)\psi.$$

If $y \in F$, the Plancherel theorem gives

$$\|L_y\phi - \phi\|_2^2 = \int |\langle y,\xi\rangle - 1|^2 |\widehat{\phi}(\xi)|^2 \, d\xi = \int_V |\langle y,\xi\rangle - 1|^2 \, d\xi < \delta^2 |V|.$$

Likewise, $\|L_y\psi - \psi\|_2^2 < \delta^2|KV|$, so by Hölder's inequality,

$$|V|\,\|L_y h - h\|_1 \le 2|V|^{1/2}|KV|^{1/2} < 3\delta|V|,$$

and we are done. $\qquad\square$

**4.59 Lemma.** *Suppose $f \in L^1(G)$, $\xi_0 \in \widehat{G}$, $\widehat{f}(\xi_0) = 0$, and $\epsilon > 0$. There exists $u \in L^1(G)$ such that $\widehat{u} = 0$ near $\xi_0$ and $\|f - f * u\|_1 < \epsilon$.*

*Proof.* By means of an approximate identity we can find $\psi \in L^1$ such that $\|f - f * \psi\|_1 < \frac{1}{2}\epsilon$. Then $(f * \psi)\widehat{\phantom{i}}(\xi_0) = \widehat{f}(\xi_0)\widehat{\psi}(\xi_0) = 0$, so we can apply Lemma 4.58 with $f$ and $\epsilon$ replaced by $f * \psi$ and $\frac{1}{2}\epsilon$ to obtain $h \in L^1$ such that $\widehat{h} = 1$ near $\xi_0$ and $\|f * \psi * h\|_1 < \frac{1}{2}\epsilon$. Then $u = \psi - \psi * h$ does the job. $\qquad\square$

**4.60 Lemma.** *Suppose $f \in L^1(G)$, $\xi_0 \in \widehat{G}$, and $\epsilon > 0$. There exists $v \in L^1(G)$ such that $\|v\|_1 < \epsilon$ and $\widehat{f}(\xi) + \widehat{v}(\xi) = \widehat{f}(\xi_0)$ in a neighborhood of $\xi_0$.*

*Proof.* By Lemma 4.51 there exists $g \in L^1$ such that $\widehat{g}(\xi) = \widehat{f}(\xi_0)$ in a neighborhood of $\xi_0$. Apply Lemma 4.58 to $g - f$ to obtain $h \in L^1$ such that $\widehat{h} = 1$ near $\xi_0$ and $\|(g - f) * h\|_1 < \epsilon$. Then $v = (g - f) * h$ does the job. $\qquad\square$

**4.61 Lemma.** *Suppose $f \in L^1(G)$ and $\epsilon > 0$. There exists $w \in L^1(G)$ such that $\operatorname{supp}\widehat{w}$ is compact and $\|f - f * w\|_1 < \epsilon$.*

*Proof.* By means of an approximate identity we can find $\psi \ge 0$ such that $\|\psi\|_1 = 1$ and $\|f - f * \psi\|_1 < \frac{1}{2}\epsilon$. Let $\phi = \sqrt{\psi}$; then $\|\widehat{\phi}\|_2 = \|\phi\|_2 = 1$. Multiplying $\phi$ by the characteristic function of a large compact set, we can obtain $\theta \in L^2$ such that $\operatorname{supp}\widehat{\theta}$ is compact, $\|\theta\|_2 \le 1$, and $\|\phi - \theta\|_2 < \epsilon/4\|f\|_1$. Let $w = \theta^2$. Then $\widehat{w} = \widehat{\theta} * \widehat{\theta}$ by Proposition 4.37, so $\operatorname{supp}\widehat{w} = (\operatorname{supp}\widehat{\theta})(\operatorname{supp}\widehat{\theta})$ is compact, and by Hölder's inequality,

$$\|f - f * w\|_1 \le \|f - f * \psi\|_1 + \|f\|_1\|\psi - w\|_1$$

$$< \tfrac{1}{2}\epsilon + \|f\|_1\big[\|\phi(\phi - \theta)\|_1 + \|(\phi - \theta)\theta\|_1\big] < \tfrac{1}{2}\epsilon + \|f\|_1\frac{2\epsilon}{4\|f\|_1} = \epsilon.$$

$\qquad\square$

Now suppose $\mathcal{I}$ is an ideal in $L^1(G)$. If $f \in L^1(G)$ and $\xi_0 \in \widehat{G}$, we say that $f$ is **locally in** $\mathcal{I}$ **at** $\xi_0$ if there exists $g \in \mathcal{I}$ such that $\widehat{f} = \widehat{g}$ in a neighborhood of $\xi_0$.

**4.62 Lemma.** *Suppose $\mathcal{I}$ is an ideal in $L^1(G)$ and $f \in L^1(G)$. If $\widehat{f}$ has compact support and $f$ is locally in $\mathcal{I}$ at every $\xi \in \widehat{G}$, then $f \in \mathcal{I}$.*

*Proof.* For each $\xi \in \widehat{G}$, choose $g_\xi \in \mathcal{I}$ such that $\widehat{f} = \widehat{g}$ on an open set $U_\xi$ containing $\xi$. By passing to a finite subcover of supp $\widehat{f}$, we obtain open sets $U_1, \ldots, U_n$ in $\widehat{G}$ and $g_1, \ldots, g_n$ in $\mathcal{I}$ such that $\widehat{f} = \widehat{g_j}$ on $U_j$ and supp $\widehat{f} \subset \bigcup_1^n U_j$. Next, each $\xi \in U_j \cap$ supp $\widehat{f}$ has a compact neighborhood contained in $U_j$; by passing to a finite subcover of supp $\widehat{f}$ again, we obtain compact sets $K_1, \ldots, K_n$ such that $K_j \subset U_j$ and supp $\widehat{f} \subset \bigcup_1^n K_j$. By Lemma 4.51, there exist $h_1, \ldots, h_n \in L^1(G)$ such that $\widehat{h_j} = 1$ on $K_j$ and supp $\widehat{h_j} \subset U_j$. Then $\prod_1^n (1 - \widehat{h_j}) = 0$ on supp $\widehat{f}$, so

$$\widehat{f} = \widehat{f}\Big[1 - \prod_1^n (1 - \widehat{h_j})\Big].$$

If we multiply out the product inside the square brackets, each term of the resulting sum (except the 1's, which cancel) is a product of $\widehat{h_j}$'s, i.e., the Fourier transform of a convolution of $h_j$'s. Collecting terms, we see that $f = \sum f * H_j$ where $H_j \in L^1$ and supp $\widehat{H_j} \subset U_j$. But then $(f * H_j)\widehat{\phantom{x}} = \widehat{f}\widehat{H_j} = \widehat{g_j}\widehat{H_j} = (g_j * H_j)\widehat{\phantom{x}}$, so $f * H_j = g_j * h_j \in \mathcal{I}$, and hence $f \in \mathcal{I}$.                                                                                 $\square$

**4.63 Lemma.** *If $\mathcal{I}$ is a closed ideal in $L^1(G)$ and $\xi_0 \notin \nu(\mathcal{I})$, then every $f \in L^1(G)$ is locally in $\mathcal{I}$ at $\xi_0$.*

*Proof.* Pick $g \in \mathcal{I}$ such that $\widehat{g}(\xi_0) = 1$. By Lemma 4.60, there exists $v \in L^1$ such that $\|v\|_1 < \frac{1}{2}$ and $\widehat{v} + \widehat{g} = 1$ in a neighborhood of $\xi_0$. Let $v_n = v * \cdots * v$ ($n$ factors); then $\|\widehat{v_n}\|_\infty \leq \|v_n\|_1 < 2^{-n}$. Hence, if $f \in L^1$, the series $f + \sum_1^\infty f * v_n$ converges in $L^1$ to a function $h$ such that

$$\widehat{h} = \sum_0^\infty \widehat{f v_n} = \sum_0^\infty \widehat{f}(\widehat{v})^n = \frac{\widehat{f}}{1 - \widehat{v}}.$$

But $1 - \widehat{v} = \widehat{g}$ near $\xi_0$, so $\widehat{f} = \widehat{h}(1 - \widehat{v}) = \widehat{h}\widehat{g} = (h * g)\widehat{\phantom{x}}$ near $\xi_0$. Since $h * g \in \mathcal{I}$, $f$ is locally in $\mathcal{I}$ at $\xi_0$.                                                        $\square$

We have now proved the most fundamental result of this subject:

**4.64 Theorem** (Wiener's Theorem). *If $\mathfrak{I}$ is a closed ideal in $L^1(G)$ and $\nu(\mathfrak{I}) = \varnothing$, then $\mathfrak{I} = L^1(G)$.*

*Proof.* $\{f \in L^1 : \operatorname{supp}\widehat{f} \text{ is compact}\}$ is dense in $L^1$ by Lemma 4.61 and is contained in $\mathfrak{I}$ by Lemmas 4.62 and 4.63. $\qquad\square$

We shall discuss Wiener's theorem in greater detail below. Right now, we take the extra step needed to obtain a substantially more general result. If $f \in L^1(G)$, we set

$$\nu(f) = \{\xi \in \widehat{G} : \widehat{f}(\xi) = 0\}.$$

The assertion that $\iota(\nu(\mathfrak{I})) = \mathfrak{I}$, for $\mathfrak{I}$ a closed ideal in $L^1(G)$, is then equivalent to the assertion that $f \in \mathfrak{I}$ whenever $\nu(f) \supset \nu(\mathfrak{I})$. We also recall that a **perfect** set in a topological space is a closed set with no isolated points, and we denote the topological boundary of a set $S$ by $\partial S$.

**4.65 Lemma.** *Suppose $\mathfrak{I}$ is a closed ideal in $L^1(G)$, $f \in L^1(G)$, and $\nu(f) \supset \nu(\mathfrak{I})$. Let $D(f)$ be the set of all $\xi \in \widehat{G}$ such that $f$ is not locally in $\mathfrak{I}$ at $\xi$. Then $D(f)$ is perfect.*

*Proof.* The complement of $D(f)$ is open by definition, so we must show that $D(f)$ has no isolated points. Suppose $\xi_0 \in D(f)$ has a compact neighborhood $W$ that contains no point of $D(f)$ except $\xi_0$ itself. By Lemma 4.51, there exists $h \in L^1(G)$ such that $\widehat{h} = 1$ near $\xi_0$ and $\operatorname{supp}\widehat{h} \subset W$. By Lemma 4.63, $\xi_0 \in \nu(\mathfrak{I}) \subset \nu(f)$, so $\widehat{f}(\xi_0) = 0$. By Lemma 4.59, there is a sequence $\{u_n\} \subset L^1$ such that $\widehat{u}_n = 0$ near $\xi_0$ and $\|f - f * u_n\|_1 \to 0$. Now, $h * f * u_n$ is locally in $\mathfrak{I}$ at every $\xi \in W \setminus \{\xi_0\}$ since $f$ is, and it is locally in $\mathfrak{I}$ at $\xi_0$ and at every $\xi \notin W$ since its Fourier transform vanishes in a neighborhood of these points. Hence $h * f * u_n$ is locally in $\mathfrak{I}$ at every $\xi \in \widehat{G}$, and $\operatorname{supp}(\widehat{h}\widehat{f}\widehat{u}_n) \subset W$ is compact, so $h * f * u_n \in \mathfrak{I}$ by Lemma 4.62. Since $\mathfrak{I}$ is closed, $h * f \in \mathfrak{I}$. But $\widehat{f} = \widehat{h}\widehat{f}$ near $\xi_0$, so $f$ is locally in $\mathfrak{I}$ at $\xi_0$, contrary to the assumption that $\xi_0 \in D(f)$. $\qquad\square$

**4.66 Lemma.** *If $E$ and $F$ are subsets of a topological space and $E \subset F$, then $E \cap \partial F = \partial E \cap \partial F$.*

*Proof.* The interior of $E$ is contained in the interior of $F$, hence does not intersect $\partial F$. $\qquad\square$

**4.67 Theorem.** *Suppose $\mathfrak{I}$ is a closed ideal in $L^1(G)$, $f \in L^1(G)$, and $\nu(f) \supset \nu(\mathfrak{I})$. If $\partial\nu(\mathfrak{I}) \cap \partial\nu(f)$ contains no nonempty perfect set, then $f \in \mathfrak{I}$.*

*Proof.* First suppose $\widehat{f}$ has compact support, and let $D(f)$ be as in Lemma 4.65. By Lemma 4.63, $D(f) \subset \nu(\mathfrak{I}) \subset \nu(f)$, and $D(f)$ contains no interior point of $\nu(f)$ since $\widehat{f}$ agrees locally with 0 near such a point. Thus, by Lemma 4.66,

$$D(f) \subset \nu(\mathfrak{I}) \cap \partial\nu(f) = \partial\nu(\mathfrak{I}) \cap \partial\nu(f).$$

By Lemma 4.65, $D(f)$ is empty, so by Lemma 4.62, $f \in \mathfrak{I}$.

If $\widehat{f}$ does not have compact support, by Lemma 4.61 there is a sequence $\{w_n\}$ in $L^1$ such that $\operatorname{supp} w_n$ is compact and $f * w_n \to f$ in $L^1$. Let $f_n = f * w_n$. Then $\nu(f_n) \supset \nu(f) \supset \nu(\mathfrak{I})$, so by Lemma 4.66 again,

$$\nu(\mathfrak{I}) \cap \partial\nu(f_n) = \nu(\mathfrak{I}) \cap \nu(f) \cap \partial\nu(f_n) = \nu(\mathfrak{I}) \cap \partial\nu(f) \cap \partial\nu(f_n)$$
$$= \nu(\mathfrak{I}) \cap \partial\nu(f) \cap \nu(f_n) = \nu(\mathfrak{I}) \cap \partial\nu(f) = \partial\nu(\mathfrak{I}) \cap \partial\nu(f).$$

The preceding argument then shows that $f_n \in \mathfrak{I}$, and hence $f = \lim f_n \in \mathfrak{I}$. $\qquad\square$

**4.68 Corollary.** *If $\partial\nu(\mathfrak{I})$ does not contain any nonempty perfect set, then $\iota(\nu(\mathfrak{I})) = \mathfrak{I}$.*

The obvious situation in which this condition holds is when $\nu(\mathfrak{I})$ is discrete. It also holds when $\nu(\mathfrak{I})$ is open, a common occurrence when $\widehat{G}$ is totally disconnected — for example, when $G = \mathbb{Q}_p$ or when $\widehat{G}$ is an infinite product of finite groups.

**4.69 Corollary.** *If $f \in L^1$ and $\widehat{f} = 0$ on a neighborhood of $\nu(\mathfrak{I})$ then $f \in \mathfrak{I}$.*

If $N$ is a closed subset of $\widehat{G}$, $\iota(N)$ is clearly the largest closed ideal $\mathfrak{I}$ such that $\nu(\mathfrak{I}) = N$. There is also a smallest such ideal, and in view of Corollary 4.69 it is

$$\iota_0(N) = \text{closure of } \left\{ f \in L^1 : \widehat{f} = 0 \text{ on a neighborhood of } N \right\}.$$

By Corollary 4.68, these two ideals coincide if $\partial N$ contains no nonempty perfect set.

We conclude this section by discussing some of the ramifications of Wiener's theorem 4.64. The first remark to be made is that Wiener's theorem is a deep result only for noncompact, nondiscrete groups. Indeed, for the compact case we gave a simple proof of a more general result in Theorem 4.54. On the other hand, if $G$ is discrete then $L^1(G)$ has a unit, so every proper ideal is contained in a maximal ideal, and the latter are all of the form $\{f : \widehat{f}(\xi_0) = 0\}$ for some $\xi_0 \in \widehat{G}$; this immediately implies Wiener's theorem. Conversely, Wiener's theorem can be restated as an analogue of this result for nondiscrete groups:

**4.70 Corollary.** *Every proper closed ideal in $L^1(G)$ is contained in one of the maximal ideals $\iota(\{\xi_0\})$, $\xi_0 \in \widehat{G}$.*

When $G$ is nondiscrete, this assertion loses its validity if the word "closed" is omitted. For example, the set of all $f \in L^1$ such that supp $\widehat{f}$ is compact is a proper ideal not contained in any $\iota(\{\xi_0\})$.

Wiener's theorem is frequently used in the following form:

**4.71 Corollary.** *Suppose $f \in L^1(G)$. The closed linear span of the translates of $f$ is $L^1(G)$ if and only if $\widehat{f}$ never vanishes.*

*Proof.* By Theorem 2.45, the closed linear span of the translates of $f$ is the closed ideal generated by $f$. If $\widehat{f}(\xi_0) = 0$, this ideal is contained in $\iota(\{\xi_0\})$; if $\widehat{f}(\xi) \neq 0$ for all $\xi$, it equals $L^1$ by Wiener's theorem. $\qquad\square$

There is a corresponding result for $L^2$, but its proof is almost trivial:

**4.72 Proposition.** *Suppose $f \in L^2(G)$. The closed linear span of the translates of $f$ is $L^2(G)$ if and only if $\widehat{f} \neq 0$ almost everywhere.*

*Proof.* Let $\mathcal{M}$ be the closed subspace of $L^2$ spanned by the translates of $f$. Then $g \perp \mathcal{M}$ if and only if $\int (L_x f)\overline{g} = 0$ for all $x \in G$. By the Plancherel theorem and (4.15), this happens precisely when

$$0 = \int (L_x f)\widehat{\,}(\xi)\overline{\widehat{g}(\xi)}\, d\xi = \int \overline{\langle x,\xi\rangle}\widehat{f}(\xi)\overline{\widehat{g}(\xi)}\, d\xi = (\widehat{f}\,\overline{\widehat{g}})\widehat{\,}(x)$$

for all $x$, and by the Fourier uniqueness theorem 4.34 this is equivalent to $\widehat{f}\,\overline{\widehat{g}} = 0$ a.e. Thus there is a nonzero $g \in \mathcal{M}^\perp$ if and only if $\widehat{f} = 0$ on a set of positive measure. $\qquad\square$

Our proof of Wiener's theorem is complicated to some degree by the fact that we are working on general locally compact Abelian groups, but even for $G = \mathbb{R}$ (the setting of Wiener's original result) the argument is not easy. However, Mark Kac [69] has pointed out that a simple and elegant derivation of Corollary 4.71 for $G = \mathbb{R}$ is available if one is willing to assume that $f$ has some extra decay at infinity. Here is Kac's result and proof, generalized to $\mathbb{R}^n$, where we identify $(\mathbb{R}^n)\widehat{\,}$ with $\mathbb{R}^n$ via the pairing $\langle \xi, x \rangle = e^{2\pi i \xi \cdot x}$:

**4.73 Proposition.** *Suppose $f \in L^1(\mathbb{R}^n)$, $\widehat{f}(\xi) \neq 0$ for all $\xi \in \mathbb{R}^n$, and $\int |x|^{n+1}|f(x)|\, dx < \infty$. Then the closed linear span of the translates of $f$ is $L^1(\mathbb{R}^n)$.*

*Proof.* By the Hahn-Banach theorem, the conclusion of the proposition is equivalent to the assertion that if $\phi \in L^\infty(\mathbb{R}^n)$ and $\int \phi(y)f(y-x)\, dy =$

0 for all $x \in \mathbb{R}^n$, then $\phi = 0$ (a.e.). To prove this, let $\phi$ be such a function, and let $h$ belong to the space $C_c^{n+1}$ of compactly supported functions on $\mathbb{R}^n$ that possess continuous partial derivatives up to order $n+1$. Let $H(x) = \int h(\xi)e^{2\pi i \xi \cdot x}\, d\xi$ be the inverse Fourier transform of $h$. Integration by parts shows that $x^\alpha H(x)$ is bounded for every monomial $x^\alpha$ of degree at most $n+1$, so $|H(x)| \le C(1+|x|)^{-n-1}$ and hence $H \in L^1$. Thus we have $H * f \in L^1$ and $(H * f)\widehat{\phantom{f}} = h\widehat{f} \in L^1$, so the Fourier inversion theorem holds for $H*f$; moreover, the function $(x,y) \mapsto H(x)f(y-x)\phi(y)$ is in $L^1(\mathbb{R}^{2n})$. Hence,

$$0 = \iint H(x)f(y-x)\phi(y)\, dy\, dx = \int \phi(y)(H*f)(y)\, dy$$

$$= \iint \phi(y)h(\xi)\widehat{f}(\xi)e^{2\pi i y \cdot \xi}\, d\xi\, dy.$$

This holds for all $h \in C_c^{n+1}$. But the hypothesis on $f$ implies that $\widehat{f}$ is of class $C^{n+1}$, and since it never vanishes, the map $h \mapsto h\widehat{f}$ is a bijection on $C_c^{n+1}$. We therefore conclude that

$$\iint \phi(y)h(\xi)e^{2\pi i y \cdot \xi}\, d\xi\, dy = 0$$

for all $h \in C_c^{n+1}$. But this space is closed under translations, so for any $\eta \in \mathbb{R}^n$,

$$0 = \iint \phi(y)h(\xi - \eta)e^{2\pi i y \cdot \xi}\, d\xi\, dy = \iint \phi(y)h(\xi)e^{2\pi i y \cdot (\xi + \eta)}\, d\xi\, dy$$

$$= \int \phi(y)H(y)e^{2\pi i \eta \cdot y}\, dy.$$

By Fourier uniqueness, $\phi H = 0$ a.e. By taking $h(\xi) = h_0(R\xi)$ where $h_0 \ge 0$ and letting $R \to \infty$, we easily obtain $H$'s that are nonzero on any given ball about the origin. Thus $\phi = 0$ a.e. on such balls and hence on $\mathbb{R}^n$, as claimed.    $\square$

Wiener's principal motivation for proving Theorem 4.64 was to obtain part (a) of the following theorem (part (b) was added by Pitt). Some terminology: suppose $\phi$ is a bounded function on $G$. We say that $\phi(x) \to a$ as $x \to \infty$ if for every $\epsilon > 0$ there is a compact set $K \subset G$ such that $|\phi(x) - a| < \epsilon$ for $x \notin K$. We say that $\phi$ is **slowly oscillating** if for every $\epsilon > 0$ there is a compact $K \subset G$ and a neighborhood $V$ of 1 in $G$ such that $|\phi(x) - \phi(y^{-1}x)| < \epsilon$ whenever $y \in V$ and $x \notin K$. (For example, if $\phi$ is uniformly continuous on $G$ then $\phi$ is slowly oscillating, and one can take $K = \varnothing$.)

**4.74 Theorem** (The Wiener-Pitt Tauberian Theorem). *Suppose* $\phi \in L^\infty(G)$, $f \in L^1(G)$, $\int f = 1$, $\widehat{f}$ *never vanishes, and* $\phi * f(x) \to a$ *as* $x \to \infty$. *Then:*

a. $\phi * g(x) \to a \int g$ *as* $x \to \infty$ *for every* $g \in L^1$.
b. *If* $\phi$ *is slowly oscillating,* $\phi(x) \to a$ *as* $x \to \infty$.

*Proof.* The set $\mathcal{L}$ of all $g \in L^1$ satisfying $\phi * g(x) \to a \int g$ as $x \to \infty$ is clearly a linear subspace of $L^1$. $\mathcal{L}$ is closed because if $g_n \to g$ in $L^1$ then $\phi * g_n \to \phi * g$ uniformly and $\int g_n \to \int g$. $\mathcal{L}$ is translation-invariant because $\phi * (L_y g) = L_y(\phi * g)$, and $\lim_{x \to \infty} L_y(\phi * g)(x) = a \int g = a \int L_y g$ whenever $\lim_{x \to \infty} \phi * g(x) = a \int g$. $\mathcal{L}$ contains $f$ by assumption, and hence $\mathcal{L} = L^1$ by Corollary 4.71. Thus (a) is proved.

If $\phi$ is slowly oscillating and $\epsilon > 0$, choose $K$ and $V$ so that $|\phi(x) - \phi(y^{-1}x)| < \epsilon$ when $x \notin K$ and $y \in V$. Assuming, as we may, that $V$ has finite measure, let $g = |V|^{-1} \chi_V$. If $x \in K$ then

$$\phi(x) - \phi * g(x) = \frac{1}{|V|} \int_V [\phi(x) - \phi(y^{-1}x)] \, dy,$$

so $|\phi(x) - \phi * g(x)| < \epsilon$ for $x \notin K$. Since $\phi * g(x) \to a$ as $x \to \infty$ and $\epsilon$ is arbitrary, it follows that $\phi(x) \to a$ as $x \to \infty$. $\qquad \square$

When $G = \mathbb{R}$, the condition "$x \to \infty$" can be replaced by "$x \to +\infty$" (or "$x \to -\infty$") in the Wiener-Pitt theorem. Moreover, the condition of slow oscillation in part (b) can be replaced by slow oscillation at $+\infty$ (or $-\infty$), viz., that for every $\epsilon > 0$ there exist $A > 0$ and $\delta > 0$ such that $|\phi(x) - \phi(x - y)| < \epsilon$ whenever $|x - y| < \delta$ and $x > A$ (or $x < -A$).

---

## 4.6   Spectral Synthesis

The theory of closed ideals in $L^1$ (alias closed translation-invariant subspaces, in view of Theorem 2.45) has a dual formulation as a theory of weak* closed translation-invariant subspaces of $L^\infty$. We recall that if $\mathcal{M}$ is a subspace of a Banach space $\mathcal{X}$, its annihilator

$$\mathcal{M}^\perp = \{\xi \in \mathcal{X}^* : \xi = 0 \text{ on } \mathcal{M}\}$$

is a weak* closed subspace of $\mathcal{X}^*$; reciprocally, if $\mathcal{N}$ is a subspace of $\mathcal{X}^*$, its annihilator

$$^\perp \mathcal{N} = \{x \in \mathcal{X} : \xi(x) = 0 \text{ for } \xi \in \mathcal{N}\}$$

is a norm-closed subspace of $\mathfrak{X}$. It is an easy consequence of the Hahn-Banach theorem that $^\perp(\mathcal{M}^\perp)$ is the (norm) closure of $\mathcal{M}$, while $(^\perp\mathcal{N})^\perp$ is the weak* closure of $\mathcal{N}$. Hence the correspondences $\mathcal{M} \mapsto \mathcal{M}^\perp$ and $\mathcal{N} \mapsto {}^\perp\mathcal{N}$ are mutually inverse bijections between norm-closed subspaces of $\mathfrak{X}$ and weak* closed subspaces of $\mathfrak{X}^*$.

Now, when $\mathfrak{X} = L^1(G)$, if $\mathcal{M} \subset L^1$ is translation-invariant then so is $\mathcal{M}^\perp$, and if $\mathcal{N} \subset L^\infty$ is translation-invariant then so is $^\perp\mathcal{N}$, because

$$\int (L_y f)\phi = \int f(y^{-1}x)\phi(x)\,dx = \int f(x)\phi(yx)\,dx = \int f(L_{y^{-1}}\phi).$$

Moreover, these integrals are equal to $\check{f} * \phi(y)$ or $f * \check{\phi}(y^{-1})$ where $\check{f}(x) = f(x^{-1})$. For this reason it will be convenient to modify the correspondence $\mathcal{M} \mapsto \mathcal{M}^\perp$ by composing it with the map $f \mapsto \check{f}$. Thus, if $\mathfrak{I}$ and $\mathfrak{J}$ are translation-invariant subspaces of $L^1$ and $L^\infty$, respectively, we define

$$\mathfrak{I}_\perp = \{\phi \in L^\infty : \check{\phi} \in \mathfrak{I}^\perp\} = \{\phi \in L^\infty : f * \phi = 0 \text{ for } f \in \mathfrak{I}\},$$
$$_\perp\mathfrak{J} = \{f \in L^1 : \check{f} \in {}^\perp\mathfrak{J}\} = \{f \in L^1 : f * \phi = 0 \text{ for } \phi \in \mathfrak{J}\}.$$

We then have $_\perp(\mathfrak{I}_\perp) = \mathfrak{I}$ when $\mathfrak{I}$ is norm-closed, and $(_\perp\mathfrak{J})_\perp = \mathfrak{J}$ when $\mathfrak{J}$ is weak* closed, so $\mathfrak{I} \mapsto \mathfrak{I}_\perp$ and $\mathfrak{J} \mapsto {}_\perp\mathfrak{J}$ are mutually inverse correspondences between closed ideals in $L^1$ and weak* closed translation-invariant subspaces of $L^\infty$. We shall call the latter spaces **co-ideals** for short.

If $\mathcal{M}$ is any subspace of $L^\infty$, we define the **spectrum** of $\mathcal{M}$ to be the set of characters in $\mathcal{M}$:

$$\sigma(\mathcal{M}) = \mathcal{M} \cap \widehat{G}.$$

**4.75 Proposition.** *If $\mathfrak{I}$ is a closed ideal in $L^1$, then $\sigma(\mathfrak{I}_\perp) = \nu(\mathfrak{I})$.*

*Proof.* If $\xi \in \widehat{G}$, $\xi \in \nu(\mathfrak{I}_\perp)$ if and only if $\xi * f = 0$ for all $f \in \mathfrak{I}$. But by Lemma 4.53, $\xi * f = 0$ if and only if $\widehat{f}(\xi) = 0$. $\qquad\square$

On the other hand, if $N$ is a closed subset of $\widehat{G}$, let

$$\tau(N) = \text{the weak* closed linear span of } N \text{ in } L^\infty.$$

$\tau(N)$ is a co-ideal since each translate of a character is a scalar multiple of that character, and we have:

**4.76 Proposition.** *If $N$ is a closed set in $\widehat{G}$, then $_\perp(\tau(N)) = \iota(N)$.*

*Proof.* As in the preceding proof, we have $f \in {}_\perp(\tau(N))$ precisely when $\widehat{f}(\xi) = 0$ for all $\xi \in N$. $\qquad\square$

A co-ideal $\mathcal{J}$ in $L^\infty$ is said to admit **spectral synthesis** if $\mathcal{J} = \tau(\sigma(\mathcal{J}))$. By Propositions 4.75 and 4.76, this happens if and only if $\mathcal{I} = \iota(\nu(\mathcal{I}))$ where $\mathcal{I} = \bot\mathcal{J}$. Theorem 4.55 shows that not all co-ideals admit spectral synthesis, but Theorem 4.67 immediately yields the following positive results.

**4.77 Proposition.** *Let $\mathcal{J}$ be a co-ideal in $L^\infty$.*
  a. *If $\partial\sigma(\mathcal{J})$ contains no nonempty perfect set, then $\mathcal{J}$ admits spectral synthesis.*
  b. *If $\sigma(\mathcal{J}) = \{\xi_1, \ldots, \xi_n\}$, then $\mathcal{J}$ is the linear span of $\xi_1, \ldots, \xi_n$.*
  c. *If $\mathcal{J} \neq \{0\}$ then $\sigma(\mathcal{J}) \neq \varnothing$.*

*Proof.* (a) is a reformulation of Corollary 4.68. (b) is a consequence of (a), and (c) is a reformulation of Wiener's theorem. ☐

The motivation for the study of spectral synthesis is the desire to develop a theory of Fourier analysis for $L^\infty$ functions. When $G = \mathbb{R}^n$, this desire is largely satisfied by the theory of distributions, which allows one to define the Fourier transform of an $L^\infty$ function as a tempered distribution from which one can recover the function by a generalized form of the Fourier inversion formula, but spectral synthesis offers a somewhat different point of view that does not depend on a differentiable structure. Its aim is to "synthesize" $L^\infty$ functions as weak* limits of linear combinations of characters. Specifically, if $\phi \in L^\infty$, let $\mathcal{J}_\phi$ be the co-ideal generated by $\phi$, that is, the weak* closed linear span of the translates of $\phi$. Then $\mathcal{J}_\phi$ admits spectral synthesis if and only if $\phi$ can be synthesized from the characters in $\mathcal{J}_\phi$. We define the **spectrum** of $\phi$ to be the set of these characters:

$$\sigma(\phi) = \sigma(\mathcal{J}_\phi) = \mathcal{J}_\phi \cap \widehat{G}.$$

$\sigma(\phi)$ may be regarded as the "support of the Fourier transform of $\phi$" even though the Fourier transform is not defined on all of $L^\infty$, as the following propositions show.

**4.78 Proposition.** *If $\phi \in L^\infty \cap L^1$, then $\sigma(\phi) = \operatorname{supp} \widehat{\phi}$.*

*Proof.* First we observe that $f \in \bot\mathcal{J}_\phi$ if and only if $f * \phi = 0$. By Proposition 4.75, then, $\xi \in \sigma(\phi)$ if and only if $\widehat{f}(\xi) = 0$ for all $f \in L^1$ such that $f * \phi = 0$. But $(f * \phi)\widehat{\phantom{x}} = \widehat{f}\widehat{\phi}$, so by the Fourier uniqueness theorem 4.34, $f * \phi = 0$ precisely when $\widehat{f} = 0$ on $\operatorname{supp} \widehat{\phi}$. It is then immediate that $\operatorname{supp} \widehat{\phi} \subset \sigma(\phi)$, and the reverse inclusion also holds in view of Lemma 4.51. ☐

**4.79 Proposition.** *Suppose $\phi \in \mathcal{B}(G)$, so that $\phi(x) = \int \langle x, \xi \rangle d\mu(\xi)$ for some $\mu \in M(\widehat{G})$. Then $\sigma(\phi) = \operatorname{supp} \mu$, and $\mathcal{J}_\phi$ admits spectral synthesis.*

*Proof.* We have

$$f * \phi(x) = \iint f(y)\langle y^{-1}x, \xi\rangle d\mu(\xi)\, dy = \int \langle x, \xi\rangle \widehat{f}(\xi)\, d\mu(\xi),$$

so by Proposition 4.18, $f * \phi = 0$ precisely when $\widehat{f} = 0$ on supp $\mu$. As in the preceding proof, $\xi \in \sigma(\phi)$ if and only if $\widehat{f}(\xi) = 0$ for all $f$ such that $f * \phi = 0$, and it follows that $\sigma(\phi) = $ supp $\mu$. This proves the first assertion and shows that $f \in {}_\perp\mathcal{I}_\phi$ if and only if $\widehat{f} = 0$ on $\sigma(\phi)$, so ${}_\perp\mathcal{I}_\phi = \iota(\sigma(\phi))$. But also ${}_\perp\tau(\sigma(\phi)) = \iota(\sigma(\phi))$ by Proposition 4.76, so $\mathcal{I}_\phi = \tau(\sigma(\phi)) = \tau(\sigma(\mathcal{I}_\phi))$. $\qquad\square$

## 4.7    The Bohr Compactification

In this section $G$ will denote a *noncompact* locally compact Abelian group. $\widehat{G}$ is then nondiscrete by Proposition 4.36, and we denote by $\widehat{G}_d$ the group $\widehat{G}$ equipped with the discrete topology. By Proposition 4.5, the dual group of $\widehat{G}_d$ is a compact group, called the **Bohr compactification** of $G$ and denoted by $bG$.

By Pontrjagin duality, $G$ can be regarded as the set of group homomorphisms from $\widehat{G}$ to $\mathbb{T}$ that are continuous in the usual topology on $\widehat{G}$, while $bG$ is the set of *all* group homomorphisms from $\widehat{G}$ to $\mathbb{T}$. Thus $G$ is naturally embedded as a subgroup of $bG$. As such, $G$ is dense in $bG$, for if $\overline{G}$ is the closure of $G$ in $bG$, $\overline{G}^{\perp} \subset \widehat{G}_d$ is the set of characters on $G$ that are trivial on $G$, namely $\{1\}$; hence $\overline{G} = bG$ by Theorem 4.40 and Pontrjagin duality.

The embedding of $G$ into $bG$ is continuous, for the topology on $G$ is the topology of compact convergence on $\widehat{G}$ while the topology on $bG$ is the topology of pointwise convergence on $\widehat{G}$, which is weaker. The embedding is *not* a homeomorphism onto its range, however: if it were, $G$ would be closed in $bG$ by Lemma 4.31 and hence compact. (Thus $bG$ is not a "compactification" of $G$ in the usual sense of the word.)

The correspondence $G \mapsto bG$ is a functor from the category of locally compact Abelian groups to the category of compact Abelian groups, and it has the following universality property.

**4.80 Proposition.** *If $K$ is a compact group and $\rho : G \to K$ is a continuous homomorphism, then $\rho$ extends to a continuous homomorphism from $bG$ to $K$.*

*Proof.* The closure of $\rho(G)$ in $K$ is Abelian since $\rho(G)$ is, so we may assume that $K = \overline{\rho(G)}$ is Abelian. $\rho$ induces a continuous homomorphism

$\rho^* : \widehat{K} \to \widehat{G}$ by $\rho^*(\eta) = \eta \circ \rho$. Since $K$ is compact, $\widehat{K}$ is discrete, so $\rho^*$ is actually continuous from $\widehat{K}$ to $\widehat{G}_d$. Dualizing again gives a continuous homomorphism $\overline{\rho} : bG \to \widehat{\widehat{K}} = K$ whose restriction to $G$ is easily seen to be $\rho$. $\qquad\square$

The Bohr compactification is so named because of its connection with the theory of almost periodic functions, which was developed by Harald Bohr. A complete discussion of this matter is beyond our present scope, but we shall explain the principal link.

A bounded continuous function $f$ on $G$ is called **uniformly almost periodic** if the set of translates of $f$, $\{R_x f : x \in G\}$, is totally bounded in the uniform metric. The justification for the name is as follows. The almost periodicity of $f$ means that for any $\epsilon > 0$ there exist $x_1, \ldots, x_n \in G$ with the following property: for each $x \in G$ there is an $x_j$ such that $\|R_x f - R_{x_j} f\|_u < \epsilon$, and hence $\|R_{x_j^{-1}x} f - f\|_u < \epsilon$. If $K$ is any compact set in $G$ we can choose $x$ such that $x_j^{-1}x \notin K$ for $1 \le j \le n$; hence there exist "arbitrarily large" elements $z$ of $G$ (namely $z = x_j^{-1}x$ for suitable $x$ and $j$) that are "almost periods" of $f$ in the sense that $\|R_z f - f\|_u < \epsilon$.

**4.81 Theorem.** *If $f$ is a bounded continuous function on $G$, the following are equivalent:*
  *a. $f$ is the restriction to $G$ of a continuous function on $bG$.*
  *b. $f$ is the uniform limit of linear combinations of characters on $G$.*
  *c. $f$ is uniformly almost periodic.*

*Proof.* (i) implies (ii) because the linear combinations of characters on $bG$ are uniformly dense in $C(bG)$, by the Stone-Weierstrass theorem. Conversely, every character on $G$ extends continuously to $bG$ by Proposition 4.80, and it follows that (ii) implies (i).

To see that (i) implies (iii), suppose $f = \phi|G$ where $\phi \in C(bG)$. Since $x \mapsto R_x\phi$ is continuous from $bG$ to $C(bG)$, the set $\{R_x\phi : x \in bG\}$ is compact in $C(bG)$. Moreover, the restriction map $\phi \mapsto \phi|G$ is an isometry in the uniform metric since $G$ is dense in $bG$. It follows that $\{(R_x\phi)|G : x \in bG\}$ is a compact set containing $\{R_x f : x \in G\}$, so the latter is totally bounded.

To complete the proof, we need a lemma.

**4.82 Lemma.** *Suppose $K$ is a compact metric space. Then the group* $\mathrm{Iso}(K)$ *of isometric bijections of $K$ is compact in the uniform metric.*

*Proof.* The uniform limit of isometries is clearly an isometry (in particular, an injection), and the uniform limit of surjections is a surjection because its range must be both dense and compact. Hence $\mathrm{Iso}(K)$ is a closed subset of the metric space $C(K, K)$ of continuous maps of $K$ to itself. Moreover, Iso($K$) is an equicontinuous family simply because its members are isometries. That Iso($K$) is compact therefore follows from the Arzelà-Ascoli theorem. (See Folland [45]; the statement and proof given there pertain to complex-valued functions, but they work equally well for functions with values in a compact metric space.) $\qquad\Box$

Returning to the proof of Theorem 4.81, suppose $f$ is uniformly almost periodic, and let $K$ be the uniform closure of $\{R_x f : x \in G\}$; thus $K$ is a compact metric space. We claim that $f$ is uniformly continuous. If not, there is a net $\{x_\alpha\}$ in $G$ converging to 1 such that $R_{x_\alpha} f$ does not converge to $f$ uniformly. Since $K$ is compact, by passing to a subnet we can assume that $R_{x_\alpha} f \to g$ uniformly, where $g \neq f$. But this is impossible since $R_{x_\alpha} f \to f$ pointwise.

Now, Iso($K$) is a compact group by Lemma 4.82. The map $x \mapsto R_x$ is clearly a group homomorphism from $G$ to Iso($K$), and the uniform continuity of $f$ easily implies that it is continuous. By Proposition 4.80, it extends to a continuous homomorphism (still denoted $x \mapsto R_x$) from $bG$ to Iso($K$). But then the continuous extension of $f$ to $bG$ is given by $\phi(x) = R_x f(1)$, $x \in bG$. $\qquad\Box$

## 4.8   Notes and References

The theory of the Fourier transform on locally compact Abelian groups is a generalization of classical Fourier analysis on $\mathbb{R}$ and $\mathbb{T}$, which of course has a very long history. The abstract theory was first worked out by Weil [146]. §§4.1–3 are largely based on the elegant development of the theory by Cartan and Godement [24].

The **Fourier algebra** $A(G)$ is the set of continuous functions $f$ on $G$ possessing a Fourier expansion $f(x) = \int_{\widehat{G}} \langle x, \xi \rangle \phi(\xi)\, d\xi$ that is absolutely convergent for one, and hence every, value of $x$; in other words, $A(G) = \mathcal{F}^{-1}[L^1(\widehat{G})]$. By Proposition 4.14 and Pontrjagin duality, $A(G)$ is a dense subalgebra of $C_0(G)$. It has been the subject of intensive investigations in the classical case $G = \mathbb{T}$ (see Kahane [70] and Zygmund [152]), and some parts of this theory generalize to other locally compact groups (see Rudin [122] and Reiter [117]).

In particular, $A(G)$ is equal to $C_0(G)$ only in the trivial case where $G$ is finite. To see this, one can begin with the observation that if $A(G) = C_0(G)$, the Fourier transform is a Banach space isomorphism from $L^1(\widehat{G})$ to $C_0(G)$. One consequence of this is an estimate $\|\phi\|_1 \leq C \|\widehat{\phi}\|_{\sup}$ for $\phi \in L^1(\widehat{G})$, so one can prove that $A(G) \neq C_0(G)$ by displaying $\phi$'s with

$\|\phi\|_1$ arbitrarily large but $\|\widehat{\phi}\|_{\sup} \leq 1$. The Dirichlet kernel and related functions provide examples in the classical cases $G = \mathbb{T}$ and $\mathbb{R}$; the general case can be found in in Hewitt and Ross [67, §37.19]. On the other hand, another consequence is that by duality, the Fourier transform also gives an isomorphism from $M(G)$ to $L^\infty(\widehat{G})$. This immediately proves that $A(G) \neq C_0(G)$ whenever $L^\infty(\widehat{G})$ is not essentially identical to the space of bounded continuous functions on $\widehat{G}$, and it is the basis of a different proof of the result for general $G$; see Reiter [117, §5.4].

Bochner's theorem was first proved for $G = \mathbb{T}$ by Herglotz [65], and then for $G = \mathbb{R}$ by Bochner [15] in a paper that established the importance of functions of positive type (or positive definite functions) in harmonic analysis. The general case is due to Weil [146]. The Pontrjagin duality theorem was proved by Pontrjagin [113] in the case of compact second countable groups and by van Kampen [137] in the general case.

More extensive treatments of analysis on general locally compact Abelian groups can be found in Rudin [122], Reiter [117], and Hewitt and Ross [66], [67]. We shall also give a few references for the deeper and more detailed theory that is available on specific groups: Dym and McKean [35], Folland [44], Kahane [70], Körner [81], Stein and Weiss [131], and Zygmund [152] for the classical theory on $\mathbb{R}^n$ and $\mathbb{T}^n$; Golubov, Efimov, and Skvartsov [55] for Walsh functions; Taibleson [132] for Fourier analysis on the $p$-adic numbers and other local fields.

One topic we have not covered is the structure theory of locally compact Abelian groups, which may be found in Hewitt and Ross [66] and Rudin [122]. We mention only the main result:

**4.83 Theorem.** *Every locally compact Abelian group has an open subgroup of the form $\mathbb{R}^n \times G$ where $G$ is a compact group.*

Another topic we have omitted is the Fourier analysis of the measure algebra $M(G)$. When $G$ is discrete, of course, $M(G) = L^1(G)$. In other cases, the full spectrum of $M(G)$ is rather mysterious. In particular, $\widehat{G}$ is not dense in $\sigma(M(G))$, and the formula $(\mu^*)\widehat{\ }(\xi) = \overline{\widehat{\mu}(\xi)}$, which is easily seen to be true for $\xi \in \widehat{G}$, is *not* valid on all of $\sigma(M(G))$. In other words, $M(G)$ is not a symmetric Banach algebra. See Rudin [122] for this and other results concerning $M(G)$.

Theorem 4.45 was proved independently by Naimark [108], Ambrose [1], and Godement [53].

The correspondence between closed ideals in a commutative Banach algebra $\mathcal{A}$ and closed subsets of $\sigma(\mathcal{A})$ is an analogue of the correspondence between ideals in a polynomial ring and affine algebraic varieties that lies at the heart of algebraic geometry. Namely, each ideal in the ring $k[X_1, \ldots, X_n]$ ($k$ a field) determines the variety $V \subset k^n$ on which the elements of the ideal vanish, and each such variety determines the ideal of polynomials that vanish on it. In this context, the fundamental theorem that elucidates how these relations fit together is the Hilbert Nullstellensatz.

Theorem 4.55 is due to Schwartz [124]. More generally, Malliavin [101] has proved that for every noncompact locally compact Abelian group $G$ there is a closed ideal $\mathfrak{I} \subset L^1(G)$ such that $\iota(\nu(\mathfrak{I})) \neq \mathfrak{I}$. (The proof can also be found in Rudin [122] and Hewitt and Ross [67].)

Theorem 4.64 was proved by Wiener [148] for the case $G = \mathbb{R}$. Its generalization to arbitrary locally compact Abelian groups and to ideals with nonempty hulls (Theorem 4.67) is the result of the combined efforts of several people; detailed references may be found in Reiter [117] and Rudin [122]. These books, as well as Hewitt and Ross [67], contain much additional information on spectral synthesis, ideals in convolution algebras, and related topics.

When $G = \mathbb{R}$, there is a rather amazing result that interpolates between Corollary 4.71 and Proposition 4.72, due to Beurling [11]:

**4.84 Theorem.** *Suppose $f \in L^1(\mathbb{R}) \cap L^2(\mathbb{R})$. If $\nu(f)$ has Hausdorff dimension $\alpha$, where $0 < \alpha < 1$, then the linear span of the translates of $f$ is dense in $L^p(\mathbb{R})$ for $p > 2/(2 - \alpha)$.*

"Tauberian theorem" is a generic name for a theorem in which one assumes conditions on certain averages of a function and deduces corresponding conditions on other averages of the function or on the function itself. In the Wiener-Pitt theorem, $\phi * f$ is a weighted average of translates of $\phi$, and one is concerned with the behavior of such averages at infinity. (The original theorem of Tauber is a criterion for convergence of a numerical series $\sum_1^\infty c_j$: if the averages $a_n = n^{-1}(s_1 + \cdots + s_n)$ of the partial sums $s_k = \sum_1^k c_j$ converge to $s$ as $n \to \infty$ and if $j|c_j| \to 0$ as $j \to \infty$, then $\sum c_j$ converges to $s$.) For more about Tauberian theorems and their applications, see Wiener [148], [149], and Rudin [123].

A nice treatment of the classical theory of almost periodic functions on $\mathbb{R}$ can be found in Besicovitch [10]. The theory can be extended to non-Abelian groups, and there is an analogue of the Bohr compactification for an arbitrary locally compact group $G$ — namely, a compact group $bG$ with the universal mapping property of Proposition 4.80 — and an analogue of Theorem 4.81. (However, in general the canonical map $G \mapsto bG$ is not an injection, and indeed $bG$ may be trivial.) This theory was initiated by von Neumann [143]; its elements are outlined in Weil [146] and Loomis [84], and an exhaustive treatment can be found in Maak [86].

# 5

## Analysis on Compact Groups

In this chapter we present the basic theory of representations of compact groups and Fourier analysis on such groups, and we present as concrete examples four closely related non-Abelian connected compact Lie groups: $SU(2)$, $SO(3)$, $SO(4)$, and $U(2)$.

Throughout this chapter, $G$ will denote a compact group. Haar measure on $G$ is both left and right invariant (Corollary 2.28), and we always normalize it so that $|G| = 1$.

### 5.1 Representations of Compact Groups

We begin by establishing some basic facts about unitary representations of compact groups, the key to which is the following lemma. We recall from §3.1 that $\mathcal{C}(\pi, \rho)$ is the space of intertwining operators of the representations $\pi$ and $\rho$, and that $\mathcal{C}(\pi) = \mathcal{C}(\pi, \pi)$.

**5.1 Lemma.** *Suppose $\pi$ is a unitary representation of the compact group $G$. Fix a unit vector $u \in \mathcal{H}_\pi$, and define the operator $T$ on $\mathcal{H}_\pi$ by*

$$Tv = \int \langle v, \pi(x)u \rangle \pi(x)u \, dx.$$

*Then $T$ is positive, nonzero, and compact, and $T \in \mathcal{C}(\pi)$.*

(Observe that $v \mapsto \langle v, \pi(x)u \rangle \pi(x)u$ is the orthogonal projection of $v$ onto the line through $\pi(x)u$; $T$ is the average over $G$ of all these projections.)

*Proof.* For any $v \in \mathcal{H}_\pi$ we have

$$\langle Tv, v \rangle = \int \langle v, \pi(x)u \rangle \langle \pi(x)u, v \rangle \, dx = \int |\langle v, \pi(x)u \rangle|^2 \, dx \geq 0,$$

so $T$ is positive. Moreover, if we take $v = u$, $|\langle u, \pi(x)u \rangle|^2$ is strictly positive on a neighborhood of 1, so $\langle Tu, u \rangle > 0$ and hence $T \neq 0$.

Since $G$ is compact, $x \mapsto \pi(x)u$ is uniformly continuous. Hence, given $\epsilon > 0$, we can find a partition of $G$ into disjoint sets $E_1, \ldots, E_n$ and points $x_j \in E_j$ such that $\|\pi(x)u - \pi(x_j)u\| < \frac{1}{2}\epsilon$ for $x \in E_j$. Then

$$\|\langle v, \pi(x)u\rangle\pi(x)u - \langle v, \pi(x_j)u\rangle\pi(x_j)u\|$$
$$\leq \|\langle v, [\pi(x) - \pi(x_j)]u\rangle\pi(x)u\| + \|\langle v, \pi(x_j)u\rangle[\pi(x) - \pi(x_j)]u\| < \epsilon\|v\|$$

for $x \in E_j$, so if we set

$$T_\epsilon v = \sum_1^n |E_j|\langle v, \pi(x_j)u\rangle\pi(x_j)u = \sum_1^n \int_{E_j} \langle v, \pi(x_j)u\rangle\pi(x_j)u \, dx,$$

we have $\|Tv - T_\epsilon v\| < \epsilon\|v\|$ for all $v$. But the range of $T_\epsilon$ is the linear span of $\{\pi(x_j)u\}_1^n$, so $T_\epsilon$ has finite rank. $T$ is therefore compact, being the norm limit of operators of finite rank.

Finally, $T \in \mathcal{C}(\pi)$ because

$$\pi(y)Tv = \int \langle v, \pi(x)u\rangle\pi(yx)u \, dx = \int \langle v, \pi(y^{-1}x)u\rangle\pi(x)u \, dx$$

$$= \int \langle \pi(y)v, \pi(x)u\rangle\pi(x)u \, dx = T\pi(y)v. \quad \square$$

**5.2 Theorem.** *If $G$ is compact, then every irreducible representation of $G$ is finite-dimensional, and every unitary representation of $G$ is a direct sum of irreducible representations.*

*Proof.* Suppose $\pi$ is irreducible, and let $T$ be as in Lemma 5.1. By Schur's lemma, $T = cI$ with $c \neq 0$. So the identity operator on $\mathcal{H}_\pi$ is compact, and hence $\dim \mathcal{H}_\pi < \infty$.

Now let $\pi$ be an arbitrary unitary representation of $G$, and again let $T$ be as in Lemma 5.1. Since $T$ is compact, nonzero, and self-adjoint, by Theorem 1.52 it has a nonzero eigenvalue $\lambda$ whose eigenspace $\mathcal{M}$ is necessarily finite-dimensional. Since $T \in \mathcal{C}(\pi)$, $\mathcal{M}$ is invariant under $\pi$; hence $\pi$ has a finite-dimensional subrepresentation. But an easy inductive argument using Proposition 3.1 shows that every finite-dimensional representation is a direct sum of irreducible representations, so $\pi$ has an irreducible subrepresentation.

By Zorn's lemma there is a maximal family $\{\mathcal{M}_\alpha\}$ of mutually orthogonal irreducible invariant subspaces for $\pi$. If $\mathcal{N}$ is the orthogonal complement of $\bigoplus \mathcal{M}_\alpha$, then $\mathcal{N}$ is invariant, and $\pi^{\mathcal{N}}$ has an irreducible subspace by the above argument, contradicting maximality unless $\mathcal{N} = \{0\}$. Thus $\mathcal{H}_\pi = \bigoplus \mathcal{M}_\alpha$. $\quad \square$

We denote by $\widehat{G}$ the set of unitary equivalence classes of irreducible unitary representations of $G$, which is sometimes called the **dual space** of $G$. We denote the equivalence class of $\pi$ by $[\pi]$. Thus, "$[\pi] \in \widehat{G}$" will be a convenient shorthand for the statement "$\pi$ is an irreducible unitary representation of $G$." This definition of $\widehat{G}$ is in essential agreement with our earlier one in the case when $G$ is Abelian, since a character of an Abelian group is nothing but a representation of the group on $\mathbb{C}$.

The decomposition of a unitary representation $\rho$ of $G$ into irreducible subrepresentations is in general not unique. (For example, let $\rho$ be the trivial representation of $G$ on a Hilbert space $\mathcal{H}$ of dimension $> 1$. Then any orthonormal basis of $\mathcal{H}$ gives a decomposition of $\mathcal{H}$ into irreducible [one-dimensional] invariant subspaces.) However, the decomposition of $\rho$ into subspaces corresponding to different irreducible equivalence classes is uniquely determined. Namely, for each $[\pi] \in \widehat{G}$, let $\mathcal{M}_\pi$ be the closed linear span of all irreducible subspaces of $\mathcal{H}_\rho$ on which $\rho$ is equivalent to $\pi$. Then $\mathcal{M}_\pi$ is invariant, and we have:

**5.3 Proposition.** $\mathcal{M}_\pi \perp \mathcal{M}_{\pi'}$ *if* $[\pi] \neq [\pi']$. *If* $\mathcal{N}$ *is any irreducible subspace of* $\mathcal{M}_\pi$ *then* $\rho^{\mathcal{N}}$ *is equivalent to* $\pi$.

*Proof.* Suppose $\mathcal{L}_\pi$ and $\mathcal{L}_{\pi'}$ are irreducible subspaces on which $\rho$ is equivalent to $\pi$ and $\pi'$ respectively, and let $P$ be the orthogonal projection onto $\mathcal{L}_\pi$. Then $P|\mathcal{L}_{\pi'} \in \mathcal{C}(\rho^{\mathcal{L}_{\pi'}}, \rho^{\mathcal{L}_\pi})$, so $P|\mathcal{L}_{\pi'} = 0$ by Schur's lemma 3.6. It follows that $\mathcal{L}_\pi \perp \mathcal{L}_{\pi'}$ and hence that $\mathcal{M}_\pi \perp \mathcal{M}_{\pi'}$.

If $\mathcal{N}$ is an irreducible subspace of $\mathcal{M}_\pi$, by definition of $\mathcal{M}_\pi$ there is an irreducible space $\mathcal{L} \subset \mathcal{M}_\pi$ such that $\rho^{\mathcal{L}}$ is equivalent to $\pi$ and $P(\mathcal{N}) \neq \{0\}$, where $P$ is the orthogonal projection onto $\mathcal{L}$. Again, $P|\mathcal{N} \in \mathcal{C}(\rho^{\mathcal{N}}, \rho^{\mathcal{L}})$, so $\rho^{\mathcal{N}}$ and $\rho^{\mathcal{L}}$ are equivalent by Schur's lemma since $P|\mathcal{N} \neq 0$. Hence $\rho^{\mathcal{N}} \cong \rho^{\mathcal{L}} \cong \pi$. $\qquad\square$

In view of Theorem 5.2, it follows that $\mathcal{H}_\rho = \bigoplus_{[\pi] \in \widehat{G}} \mathcal{M}_\pi$. It also follows that $\mathcal{M}_\pi$ can be decomposed as $\bigoplus_{\alpha \in A} \mathcal{L}_\alpha$ where $\rho^{\mathcal{L}_\alpha}$ is equivalent to $\pi$ for each $\alpha$. The latter decomposition is not unique if $\mathcal{M}_\pi$ is not irreducible, but the following proposition shows that the cardinality of $A$ is the same for all such decompositions (a result that is also an easy consequence of the finite dimensionality of $\pi$). This cardinality is called the **multiplicity** of $[\pi]$ in $\rho$ and is denoted by $\text{mult}(\pi, \rho)$.

**5.4 Proposition.** $\text{mult}(\pi, \rho) = \dim \mathcal{C}(\pi, \rho)$.

(The proof will make clear what this means when the multiplicity is infinite.)

*Proof.* Let $\mathcal{M}_\pi = \bigoplus_{\alpha \in A} \mathcal{L}_\alpha$ as above. For each $\alpha$ let $T_\alpha : \mathcal{H}_\pi \to \mathcal{L}_\alpha$ be a unitary equivalence. Fix a unit vector $u \in \mathcal{H}_\pi$ and let $v_\alpha = T_\alpha u$. Then $\{v_\alpha\}_{\alpha \in A}$ is an orthonormal set in $\mathcal{M}_\pi$; let $\mathcal{V}$ be the closed subspace it spans. We shall prove the proposition by showing that the map $T \mapsto Tu$ is a linear isomorphism from $\mathcal{C}(\pi, \rho)$ to $\mathcal{V}$.

Every nonzero $T \in \mathcal{C}(\pi, \rho)$ is an injection because its nullspace is invariant under $\pi$ (if $Tv = 0$ then $T\pi(x)v = \rho(x)Tv = 0$), so the map $T \mapsto Tu$ is injective. Moreover, the range of $T$ is a subspace of $\mathcal{H}_\rho$ on which $\rho$ is equivalent to $\pi$, and hence is contained in $\mathcal{M}_\pi$. Thus we may write $Tu = \sum u_\alpha$ with $u_\alpha \in \mathcal{L}_\alpha$. If $P_\alpha$ is the orthogonal projection onto $\mathcal{L}_\alpha$, then $P_\alpha T \in \mathcal{C}(\pi, \rho^{\mathcal{L}_\alpha})$ and so $P_\alpha T = c_\alpha T_\alpha$ by Schur's lemma. It follows that $u_\alpha = P_\alpha Tu = c_\alpha v_\alpha$, so that $Tu \in \mathcal{V}$. Finally, if $v = \sum c_\alpha v_\alpha \in \mathcal{V}$, the sum $T = \sum c_\alpha T_\alpha$ converges strongly in $\mathcal{C}(\pi, \rho)$ since $\sum |c_\alpha|^2 < \infty$ and the ranges of the $T_\alpha$'s are mutually orthogonal, and we have $v = Tu$. $\qquad\square$

We remark that the definition of $\mathcal{M}_\pi$ and Propositions 5.3 and 5.4 do not depend on the compactness of $G$. However, if $G$ is noncompact it is usually false that $\mathcal{H}_\rho = \bigoplus \mathcal{M}_\pi$.

It is worth noting that if $\rho$ is a possibly nonunitary representation of the compact group $G$ on a finite-dimensional space $\mathcal{V}$, that is, any continuous homomorphism from $G$ to the group of invertible operators on $\mathcal{V}$, then there is an inner product on $\mathcal{V}$ with respect to which $\rho$ is unitary. Namely, start with any inner product $\langle \ , \ \rangle_0$ on $\mathcal{V}$ and define a new inner product by

$$\langle u, v \rangle = \int \langle \rho(x)u, \rho(x)v \rangle_0 \, dx.$$

Then $\langle \ , \ \rangle$ is a $\rho$-invariant inner product, for

$$\langle \rho(y)u, \rho(y)v \rangle = \int \langle \rho(xy)u, \rho(xy)v \rangle_0 \, dx = \int \langle \rho(x)u, \rho(x)v \rangle_0 \, dx = \langle u, v \rangle.$$

Hence the theory of unitary representations of $G$ essentially includes the theory of all finite-dimensional representations of $G$, unitary or not.

## 5.2    The Peter-Weyl Theorem

When $G$ is Abelian, $\widehat{G}$ is a set of continuous functions on $G$. The corresponding set of functions in the non-Abelian case is the set of matrix elements of the irreducible representations of $G$. If $\pi$ is any unitary representation of $G$, the functions

$$\phi_{u,v}(x) = \langle \pi(x)u, v \rangle \qquad (u, v \in \mathcal{H}_\pi)$$

are called **matrix elements** of $\pi$. If $u$ and $v$ are members of an orthonormal basis $\{e_j\}$ for $\mathcal{H}_\pi$, $\phi_{u,v}(x)$ is indeed one of the entries of the matrix for $\pi(x)$ with respect to that basis, namely

(5.5) $$\pi_{ij}(x) = \phi_{e_j, e_i}(x) = \langle \pi(x) e_j, e_i \rangle.$$

We denote the linear span of the matrix elements of $\pi$ by $\mathcal{E}_\pi$. $\mathcal{E}_\pi$ is a subspace of $C(G)$, and hence also of $L^p(G)$ for all $p$.

**5.6 Proposition.** *$\mathcal{E}_\pi$ depends only on the unitary equivalence class of $\pi$. It is invariant under left and right translations and is a two-sided ideal in $L^1(G)$. If $\dim \mathcal{H}_\pi = n < \infty$ then $\dim \mathcal{E}_\pi \leq n^2$.*

*Proof.* If $T$ is a unitary equivalence of $\pi$ and $\pi'$, so that $\pi'(x) = T\pi(x)T^{-1}$, then $\langle \pi(x)u, v \rangle = \langle \pi'(x)Tu, Tv \rangle$; this proves the first assertion. We have

$$\phi_{u,v}(y^{-1}x) = \langle \pi(y^{-1}x)u, v \rangle = \langle \pi(x)u, \pi(y)v \rangle = \phi_{u,\pi(y)v}(x),$$

and likewise $\phi_{u,v}(xy) = \phi_{\pi(y)u,v}(x)$. That $\mathcal{E}_\pi$ is a two-sided ideal now follows from Theorem 2.45; alternatively, it is easy to check directly that $f * \phi_{u,v} = \phi_{u,\pi(\tilde{f})v}$ and $\phi_{u,v} * f = \phi_{\pi(\check{f})u,v}$ where $\check{f}(x) = f(x^{-1})$. Finally, if $\dim \mathcal{H}_\pi = n$, $\mathcal{E}_\pi$ is clearly spanned by the $n^2$ functions $\pi_{ij}$ given by (5.5). $\qquad \square$

**5.7 Proposition.** *If $\pi = \pi_1 \oplus \cdots \oplus \pi_n$ then $\mathcal{E}_\pi = \sum_1^n \mathcal{E}_{\pi_j}$. (The sum need not be direct.)*

*Proof.* Clearly $\mathcal{E}_{\pi_j} \subset \mathcal{E}_\pi$ for all $j$ (take $u, v \in \mathcal{H}_{\pi_j} \subset \mathcal{H}_\pi$). On the other hand, if $u = \sum u_j$ and $v = \sum v_j$ with $u_j, v_j \in \mathcal{H}_{\pi_j}$, then $\langle \pi(x)u_j, v_k \rangle = 0$ for $j \neq k$ and hence $\phi_{u,v} = \sum \phi_{u_j, v_j} \in \sum \mathcal{E}_{\pi_j}$. $\qquad \square$

The matrix elements of irreducible representations can be used to make an orthonormal basis for $L^2(G)$. The first main step in proving this is the following. Here and in the sequel we shall set

$$d_\pi = \dim \mathcal{H}_\pi,$$

and we denote the trace of a matrix $A$ by $\operatorname{tr} A$.

**5.8 Theorem** (The Schur Orthogonality Relations). *Let $\pi$ and $\pi'$ be irreducible unitary representations of $G$, and consider $\mathcal{E}_\pi$ and $\mathcal{E}_{\pi'}$ as subspaces of $L^2(G)$.*
  a. *If $[\pi] \neq [\pi']$ then $\mathcal{E}_\pi \perp \mathcal{E}_{\pi'}$.*
  b. *If $\{e_j\}$ is any orthonormal basis for $\mathcal{H}_\pi$ and $\pi_{ij}$ is given by (5.5), then $\{\sqrt{d_\pi}\, \pi_{ij} : i, j = 1, \ldots, d_\pi\}$ is an orthonormal basis for $\mathcal{E}_\pi$.*

*Proof.* If $A$ is any linear map from $\mathcal{H}_\pi$ to $\mathcal{H}_{\pi'}$, let

$$\widetilde{A} = \int \pi'(x^{-1})A\pi(x)\,dx.$$

Then

$$\widetilde{A}\pi(y) = \int \pi'(x^{-1})A\pi(xy)\,dx = \int \pi'(yx^{-1})A\pi(x)\,dx = \pi'(y)\widetilde{A},$$

so $\widetilde{A} \in \mathcal{C}(\pi, \pi')$. Given $v \in \mathcal{H}_\pi$ and $v' \in \mathcal{H}_{\pi'}$, let us define $A$ by $Au = \langle u, v \rangle v'$. Then for any $u \in \mathcal{H}_\pi$ and $u' \in \mathcal{H}_{\pi'}$,

$$\langle \widetilde{A}u, u' \rangle = \int \langle A\pi(x)u, \pi'(x)u' \rangle\,dx = \int \langle \pi(x)u, v \rangle \langle v', \pi(x)u' \rangle\,dx$$

$$= \int \phi_{u,v}(x)\overline{\phi_{u',v'}(x)}\,dx.$$

We now apply Schur's lemma. If $[\pi] \neq [\pi']$ then $\widetilde{A} = 0$, so $\mathcal{E}_\pi \perp \mathcal{E}_{\pi'}$ and (a) is proved. If $\pi' = \pi$ then $\widetilde{A} = cI$, so if we take $u = e_i$, $u' = e_{i'}$, $v = e_j$, and $v' = e_{j'}$ we get

$$\int \pi_{ij}(x)\overline{\pi_{i'j'}(x)}\,dx = c\langle e_i, e_{i'} \rangle = c\delta_{ii'}.$$

But

$$cd_\pi = \operatorname{tr} \widetilde{A} = \int \operatorname{tr}[\pi(x^{-1})A\pi(x)]\,dx = \operatorname{tr} A,$$

and since $Au = \langle u, e_j \rangle e_{j'}$ we have $\operatorname{tr} A = \delta_{jj'}$ Hence

$$\int \pi_{ij}(x)\overline{\pi_{i'j'}(x)}\,dx = \frac{1}{d_\pi}\delta_{ii'}\delta_{jj'},$$

so $\{\sqrt{d_\pi}\,\pi_{ij}\}$ is an orthonormal set. Since $\dim \mathcal{E}_\pi \leq d_\pi^2$ by Proposition 5.6, it is a basis.                                                                    □

We observed in Proposition 5.6 that $\mathcal{E}_\pi$ is invariant under the left and right regular representations $L$ and $R$. (We retain our usual notation $L_x$ and $R_x$ for left and right translation by $x$.) The question then arises: what are the irreducible subrepresentations of $L$ and $R$ in $\mathcal{E}_\pi$? The answer is simple and elegant. We recall that $\overline{\pi}$ denotes the contragredient of the representation $\pi$, whose matrix elements $\overline{\pi}_{ij}$ are the complex conjugates of those of $\pi$.

**5.9 Theorem.** *Suppose $\pi$ is irreducible, and let $\pi_{ij}$ be given by (5.5). For $i = 1, \ldots, d_\pi$ let $\mathcal{R}_i$ be the linear span of $\pi_{i1}, \ldots, \pi_{id_\pi}$ (the ith row* of the matrix $(\pi_{ij})$) and let $\mathcal{C}_i$ be the linear span of $\pi_{1i}, \ldots, \pi_{d_\pi i}$ (the *i*th column). Then $\mathcal{R}_i$ (resp. $\mathcal{C}_i$) is invariant under the right (resp. left) regular representation, and $R^{\mathcal{R}_i}$ (resp. $L^{\mathcal{C}_i}$) is equivalent to $\pi$ (resp. $\bar{\pi}$). The equivalence is given by

$$\sum c_j e_j \mapsto \sum c_j \pi_{ij} \qquad \left( resp. \ \sum c_j e_j \mapsto \sum c_j \pi_{ji} \right).$$

*Proof.* In terms of the basis $\{e_j\}$ for $\mathcal{H}_\pi$, $\pi$ is given by

$$\pi(x)\left(\sum_j c_j e_j\right) = \sum_{kj} \pi_{kj}(x) c_j e_k.$$

Moreover, $\pi(yx) = \pi(y)\pi(x)$, so $\pi_{ij}(yx) = \sum_k \pi_{ik}(y)\pi_{kj}(x)$. In other words, $R_x \pi_{ij} = \sum_k \pi_{kj}(x)\pi_{ik}$, so

$$R_x\left(\sum_j c_j \pi_{ij}\right) = \sum_{jk} \pi_{kj}(x) c_j \pi_{ik}.$$

Comparison of the two displayed lines proves the desired result for right translations. In the same way, for left translations we see that

$$L_x\left(\sum_j c_j \pi_{ji}\right) = \sum_{jk} \pi_{jk}(x^{-1}) c_j \pi_{ki},$$

and since $\pi$ is unitary, $\pi_{jk}(x^{-1}) = \bar{\pi}_{kj}(x)$; the result follows. $\qquad \square$

Now let
$$\mathcal{E} = \text{the linear span of} \bigcup_{[\pi] \in \widehat{G}} \mathcal{E}_\pi.$$

That is, $\mathcal{E}$ consists of *finite* linear combinations of matrix elements of irreducible representations. By Proposition 5.7, $\mathcal{E}$ is also the linear span of the $\mathcal{E}_\pi$ as $\pi$ ranges over all finite-dimensional representations of $G$. $\mathcal{E}$ can be considered as the space of "trigonometric polynomials" on $G$.

**5.10 Proposition.** $\mathcal{E}$ *is an algebra.*

*Proof.* It suffices to show that if $[\pi], [\pi'] \in \widehat{G}$ and $\pi_{ij}, \pi'_{kl}$ are as in (5.5), then $\pi_{ij}\pi'_{kl}$ is a matrix element of some finite-dimensional representation of $G$. The appropriate representation is the (inner) tensor product of $\pi$ and $\pi'$. Rather than refer to the general theory of tensor products in Appendix 3 and §7.3, we present the construction *ad hoc*. Via the choice of bases for $\mathcal{H}_\pi$ and $\mathcal{H}_{\pi'}$ implicit in the definition of $\pi_{ij}$ and $\pi'_{kl}$, we identify $\mathcal{H}_\pi$ and $\mathcal{H}'_\pi$ with $\mathbb{C}^n$ and $\mathbb{C}^{n'}$ where $n = d_\pi$, $n' = d_{\pi'}$. We consider $\mathbb{C}^{nn'}$ as the space of $n \times n'$ matrices over $\mathbb{C}$, and we define the representation $\pi \otimes \pi'$ on $\mathbb{C}^{nn'}$ by

$$(\pi \otimes \pi')(x)T = \pi(x)T\overline{\pi}'(x^{-1}).$$

The matrices $e_{jk} \in \mathbb{C}^{nn'}$ whose $(j,k)$th entry is 1 and whose other entries are 0 are an orthonormal basis of $\mathbb{C}^{nn'}$, and a simple calculation (using the fact that $\overline{\pi}'_{kl}(x^{-1}) = \pi'_{lk}(x)$) gives

$$\langle (\pi \otimes \pi')(x)e_{jl}, e_{ik} \rangle = \pi_{ij}(x)\pi'_{kl}(x). \qquad \square$$

We come now to the main technical lemma in the Peter-Weyl theory of representations of compact groups. This lemma is an immediate consequence of the preceding results together with the Gelfand-Raikov and Stone-Weierstrass theorems. However, it antedates these theorems by about fifteen years, and the original argument of Peter and Weyl is interesting and elegant. It also gives an independent proof of the Gelfand-Raikov theorem for the case of compact groups. Accordingly, we shall give both proofs.

**5.11 Theorem.** $\mathcal{E}$ *is dense in* $C(G)$ *in the uniform norm, and dense in* $L^p(G)$ *in the* $L^p$ *norm for* $p < \infty$.

*Proof.* It is enough to show that $\mathcal{E}$ is dense in $C(G)$. But $\mathcal{E}$ is an algebra (Proposition 5.10) that separates points (by the Gelfand-Raikov theorem), is closed under complex conjugation (because each representation has a contragredient), and contains constants (because of the trivial representation of $G$ on $\mathbb{C}$). The result therefore follows from the Stone-Weierstrass theorem.

Now, here is the original proof of Peter and Weyl. Given a function $\psi \in C(G)$ that is real and symmetric ($\psi(x^{-1}) = \psi(x)$), let $T_\psi f = \psi * f$ for $f \in L^2$. The conditions on $\psi$ imply that $T_\psi$ is self-adjoint on $L^2$. Moreover, by Proposition 2.41, $T$ maps $L^2$ into $C(G)$, $\|T_\psi f\|_{\sup} \le \|f\|_2 \|\psi\|_2$, and

$$\|L_x(T_\psi f) - T_\psi f\|_{\sup} = \|(L_x\psi - \psi) * f\|_{\sup} \le \|f\|_2 \|L_x\psi - \psi\|_2.$$

Thus, if $B$ is a bounded set in $L^2$, $\{T_\psi f : f \in B\}$ is uniformly bounded and equicontinuous, so it follows from the Arzelà-Ascoli theorem that $T_\psi$ is compact as a map from $L^2$ to $C(G)$ and *a fortiori* as an operator on $L^2$. Hence, by Theorem 1.52,

$$L^2(G) = \bigoplus_{\alpha \in \sigma(T_\psi)} \mathcal{M}_\alpha, \qquad \mathcal{M}_\alpha = \{f : T_\psi f = \alpha f\}.$$

Each eigenspace $\mathcal{M}_\alpha$ is invariant under right translations since

$R_x(\psi * f) = \psi * R_x f$. Moreover, if $\alpha \neq 0$, $\dim \mathcal{M}_\alpha < \infty$. Let $f_1, \ldots, f_n$ be an orthonormal basis for $\mathcal{M}_\alpha$ and let $\rho_{jk}(x) = \langle R_x f_k, f_j \rangle$. Then $f_k(yx) = \sum_j \rho_{jk}(x) f_j(y)$, so $f_k(x) = \sum_j f_j(1) \rho_{jk}(x)$. But this means that $\mathcal{M}_\alpha \subset \mathcal{E}_\rho$ where $\rho$ is the right regular representation on $\mathcal{M}_\alpha$. In short, $\mathcal{M}_\alpha \subset \mathcal{E}$ for each $\alpha \neq 0$.

If $f \in L^2$ then $f = \sum_\alpha f_\alpha$ with $f_\alpha \in \mathcal{M}_\alpha$, the series converging in $L^2$. Then $T_\psi f = \sum_{\alpha \neq 0} \alpha f_\alpha$, the series converging uniformly since $T$ is bounded from $L^2$ to $C(G)$. By the result just proved, it follows that $\mathcal{E} \cap \mathrm{Range}(T_\psi)$ is uniformly dense in $\mathrm{Range}(T_\psi)$. But the union of the ranges of $T_\psi$ as $\psi$ runs through an approximate identity is dense in $C(G)$, so the proof is complete. $\qquad\square$

If we combine Theorem 5.11 with the Schur orthogonality relations, we see that $L^2(G)$ is the orthogonal direct sum of the spaces $\mathcal{E}_\pi$ as $[\pi]$ ranges over $\widehat{G}$, and that we obtain an orthonormal basis for $L^2(G)$ by fixing an element $\pi$ of each irreducible equivalence class $[\pi]$ and taking the matrix elements corresponding to an orthonormal basis of $\mathcal{H}_\pi$. The necessity of choosing a particular $\pi \in [\pi]$ is a minor annoyance; it serves mainly to avoid certain abuses of language in the results that follow. *We henceforth assume that such a choice has been made once and for all.*

We can now summarize all our main results in a single theorem.

**5.12 Theorem** (The Peter-Weyl Theorem). *Let $G$ be a compact group. Then $\mathcal{E}$ is uniformly dense in $C(G)$, $L^2(G) = \bigoplus_{[\pi] \in \widehat{G}} \mathcal{E}_\pi$, and if $\pi_{ij}$ is given by (5.5),*

$$\left\{ \sqrt{d_\pi}\, \pi_{ij} : i, j = 1, \ldots, d_\pi,\ [\pi] \in \widehat{G} \right\}$$

*is an orthonormal basis for $L^2(G)$. Each $[\pi] \in \widehat{G}$ occurs in the right and left regular representations of $G$ with multiplicity $d_\pi$. More precisely, for $i = 1, \ldots, d_\pi$ the subspace of $\mathcal{E}_\pi$ (resp. $\mathcal{E}_{\overline{\pi}}$) spanned by the $i$th row (resp. the $i$th column) of the matrix $(\pi_{ij})$ (resp. $(\overline{\pi}_{ij})$) is invariant under the right (resp. left) regular representation, and the latter representation is equivalent to $\pi$ there.*

As an application, we give a representation-theoretic characterization of compact Lie groups. The proof naturally requires some Lie theory, for which we refer to Helgason [64].

**5.13 Theorem.** *Let $G$ be a compact group. Then $G$ is a Lie group if and only if $G$ has a faithful finite-dimensional representation.*

*Proof.* If $\pi$ is a faithful representation of $G$ on $\mathbb{C}^n$, then $G$ is isomorphic to the compact subgroup $\pi(G)$ of the Lie group $U(n)$. But any closed subgroup of a Lie group is a Lie group.

On the other hand, if $G$ is a Lie group, there is an open neighborhood $U$ of 1 in $G$ that contains no subgroups except $\{1\}$. (Take $U = \exp(\frac{1}{2}V)$, where $V$ is any bounded open neighborhood of 0 in the Lie algebra on which the exponential map is a diffeomorphism. If $x \neq 1 \in U$ then $x^n \in \exp(V \setminus \frac{1}{2}V) \subset G \setminus U$ for some positive integer $n$, so no subgroup containing $x$ is contained in $U$.) If $[\pi] \in \widehat{G}$, let $K_\pi$ be the kernel of $\pi$. Then $\bigcap_{[\pi] \in \widehat{G}} K_\pi = \{1\}$ by the Gelfand-Raikov theorem or Theorem 5.11, so $\bigcap_{[\pi] \in \widehat{G}}(K_\pi \setminus U) = \varnothing$. By compactness of $G \setminus U$, there exist $\pi_1, \ldots, \pi_n$ such that $\bigcap_1^n (K_{\pi_j} \setminus U) = \varnothing$, or $\bigcap_1^n K_{\pi_j} \subset U$. By the choice of $U$, $\bigcap_1^n K_{\pi_j} = \{1\}$, which means that $\pi_1 \oplus \cdots \oplus \pi_n$ is faithful.   $\square$

## 5.3   Fourier Analysis on Compact Groups

According to the Peter-Weyl theorem, if $f \in L^2(G)$ we have

$$(5.14) \qquad f = \sum_{[\pi] \in \widehat{G}} \sum_{i,j=1}^{d_\pi} c_{ij}^\pi \pi_{ij}, \qquad c_{ij}^\pi = d_\pi \int f(x) \overline{\pi_{ij}(x)} \, dx.$$

The drawback to this decomposition of $L^2$ is that it depends on choosing an orthonormal basis for each $\mathcal{H}_\pi$. However, it is possible to reformulate (5.14) in a way that does not involve such choices. Namely, if $f \in L^1(G)$, $[\pi] \in \widehat{G}$, and $\pi$ is the chosen representative of the class $[\pi]$ (see the remarks preceding Theorem 5.12), we define the **Fourier transform** of $f$ at $\pi$ to be the operator

$$\widehat{f}(\pi) = \int f(x) \pi(x^{-1}) \, dx = \int f(x) \pi(x)^* \, dx$$

on $\mathcal{H}_\pi$. This agrees with our earlier definition in the Abelian case, where $\mathcal{H}_\pi = \mathbb{C}$ for all $[\pi] \in \widehat{G}$. If we choose an orthonormal basis for $\mathcal{H}_\pi$ so that $\pi(x)$ is represented by the matrix $(\pi_{ij}(x))$, then $\widehat{f}(\pi)$ is given by the matrix

$$\widehat{f}(\pi)_{ij} = \int f(x) \overline{\pi_{ji}(x)} \, dx = \frac{1}{d_\pi} c_{ji}^\pi$$

where $c_{ji}^\pi$ is as in (5.14). But then

$$\sum_{i,j} c_{ij}^\pi \pi_{ij}(x) = d_\pi \sum_{i,j} \widehat{f}(\pi)_{ji} \pi_{ij}(x) = d_\pi \operatorname{tr}[\widehat{f}(\pi)\pi(x)]$$

(where tr denotes the trace of a matrix), so (5.14) becomes a Fourier inversion formula,

$$(5.15) \qquad f(x) = \sum_{[\pi] \in \widehat{G}} d_\pi \operatorname{tr}[\widehat{f}(\pi)\pi(x)].$$

(This must be taken with a grain of salt, as the series usually does not converge pointwise; it converges in the $L^2$ norm if $f \in L^2$. We shall say more about this below.) Moreover, the Parseval equation

$$\|f\|_2^2 = \sum_{[\pi] \in \widehat{G}} \sum_{i,j=1}^{d_\pi} \frac{1}{d_\pi} |c_{ij}^\pi|^2$$

becomes

$$(5.16) \qquad \|f\|_2^2 = \sum_{[\pi] \in \widehat{G}} d_\pi \operatorname{tr}[\widehat{f}(\pi)^* \widehat{f}(\pi)].$$

The Fourier transform as defined above is more cumbersome to use than the Abelian Fourier transform because its values are operators on a whole family of Hilbert spaces. Nonetheless, it has a number of properties analogous to those of the Abelian Fourier transform. We observe that $\widehat{f}(\pi) = \pi(\overline{f})^*$; it is then easy to verify the following variant of Theorem 3.9:

$$(5.17) \qquad (f * g)\widehat{\phantom{x}}(\pi) = \widehat{g}(\pi)\widehat{f}(\pi), \quad (f^*)\widehat{\phantom{x}}(\pi) = \widehat{f}(\pi)^*,$$

$$(5.18) \qquad (L_x f)\widehat{\phantom{x}}(\pi) = \widehat{f}(\pi)\pi(x^{-1}), \quad (R_x f)\widehat{\phantom{x}}(\pi) = \pi(x)\widehat{f}(\pi).$$

We can express (5.14) in yet another way. If $\pi$ is a finite-dimensional unitary representation of $G$, the **character** $\chi_\pi$ of $\pi$ is the function

$$\chi_\pi(x) = \operatorname{tr} \pi(x).$$

Note that this depends only on the equivalence class of $\pi$, since similar matrices have the same trace; hence we could (and perhaps should) write $\chi_{[\pi]}$ instead of $\chi_\pi$. We have

$$\operatorname{tr}[\widehat{f}(\pi)\pi(x)] = \int f(y) \operatorname{tr}[\pi(y^{-1})\pi(x)]\, dy = \int f(y) \operatorname{tr} \pi(y^{-1}x)\, dy$$
$$= f * \chi_\pi(x),$$

so (5.15) becomes

$$(5.19) \qquad f = \sum_{[\pi] \in \widehat{G}} d_\pi f * \chi_\pi.$$

In particular, $d_\pi f * \chi_\pi$ is the orthogonal projection of $f$ onto $\mathcal{E}_\pi$.

If we apply (5.19) and the remark following it to the case where $f$ is itself an irreducible character, we obtain the following useful convolution formulas: if $[\pi], [\pi'] \in \widehat{G}$, then

$$(5.20) \qquad \chi_\pi * \chi_{\pi'} = \begin{cases} d_\pi^{-1}\chi_\pi & \text{if } [\pi'] = [\pi], \\ 0 & \text{if } [\pi'] \neq [\pi]. \end{cases}$$

A function $f$ on $G$ is called a **central function** or **class function** if $f$ is constant on conjugacy classes, that is, if $f(yxy^{-1}) = f(x)$ for all $x, y \in G$, or equivalently (by substituting $xy$ for $x$) $f(xy) = f(yx)$ for all $x, y \in G$. (In the case of $L^p$ functions, these relations are to hold almost everywhere.) For example, the character of any finite-dimensional representation is central, because $\text{tr}[\pi(x)\pi(y)] = \text{tr}[\pi(y)\pi(x)]$. We denote the set of central functions in a function space by prefixing the letter $Z$ to the name of the space:

$$ZL^p(G) = \text{the set of central functions in } L^p(G),$$
$$ZC(G) = \text{the set of central functions in } C(G).$$

**5.21 Proposition.** *The spaces $L^p(G)$ and $C(G)$ are Banach algebras under convolution, and $ZL^p(G)$ and $ZC(G)$ are their respective centers.*

*Proof.* We have $\|f * g\|_p \leq \|f\|_1 \|g\|_p \leq \|f\|_p \|g\|_p$ by Proposition 2.40(a) and the fact that $|G| = 1$, and likewise with the $L^p$ norm replaced by the uniform norm. This shows that $L^p(G)$ and $C(G)$ are Banach algebras. Next, suppose $f \in L^p$. Then $f * g = g * f$ if and only if

$$\int f(xy)g(y^{-1})\, dy = \int g(y)f(y^{-1}x)\, dy = \int f(yx)g(y^{-1})\, dy$$

for almost every $x$. This holds for every $g \in L^p$ precisely when $f(xy) = f(yx)$ for almost every $x$ and $y$. Similarly for $f \in C(G)$. $\qquad \square$

Central functions play a crucial role in analysis on compact groups. Here are a few of the basic results concerning them.

**5.22 Lemma.** *If $f \in ZL^1(G)$ and $[\pi] \in \widehat{G}$, then $d_\pi f * \chi_\pi = (\int f\overline{\chi}_\pi)\chi_\pi$.*

*Proof.* We have

$$\widehat{f}(\pi)\pi(x) = \int f(y)\pi(y^{-1}x)\, dy = \int f(xy^{-1})\pi(y)\, dy$$
$$= \int f(y^{-1}x)\pi(y)\, dy = \int f(y)\pi(xy^{-1})\, dy = \pi(x)\widehat{f}(\pi).$$

Hence, by Schur's lemma, $\widehat{f}(\pi) = c_\pi I$. Taking the trace, we obtain

$$d_\pi c_\pi = \operatorname{tr}\widehat{f}(\pi) = \int f(y)\operatorname{tr}\pi(y)^*\,dy = \int f(y)\overline{\chi_\pi(y)}\,dy,$$

so by the calculation leading to (5.19),

$$d_\pi f * \chi_\pi = d_\pi \operatorname{tr}[\widehat{f}(\pi)\pi] = d_\pi c_\pi \operatorname{tr}\pi = \left[\int f\overline{\chi_\pi}\right]\chi_\pi. \qquad \square$$

**5.23 Proposition.** $\{\chi_\pi : [\pi] \in \widehat{G}\}$ *is an orthonormal basis for* $ZL^2(G)$.

*Proof.* We have already observed that the $\chi_\pi$'s are central. They are also orthonormal by the Schur orthogonality relations, since $\chi_\pi = \sum \pi_{ii}$. They are a basis because if $f \in ZL^2(G)$, by (5.19) and Lemma 5.22 we have $f = \sum\langle f, \chi_\pi\rangle\chi_\pi$. $\qquad \square$

If $f \in L^2(G)$, the "Fourier series" (5.15) or (5.19) for $f$ converges to $f$ in the $L^2$ norm. As we know from the classical theory of Fourier series (the case $G = \mathbb{T}$; see Zygmund [152]), it usually does not converge uniformly when $f$ is continuous, or in the $L^1$ norm when $f \in L^1$, or unconditionally in the $L^p$ norm when $f \in L^p$ and $p \neq 2$. However, as in the classical case, there are "summability methods" that enable one to recover $f$ from its Fourier series in these topologies. We now describe one such, for which we need a lemma.

**5.24 Lemma.** *The neighborhoods of* 1 *in* $G$ *that are invariant under conjugations constitute a neighborhood base at* 1.

*Proof.* Let $U$ be any neighborhood of 1, and let $V$ be a symmetric neighborhood of 1 such that $VVV \subset U$. Choose $x_1, \ldots, x_n \in G$ such that $G = \bigcup_1^n Vx_j$, and let $W = \bigcap_1^n x_j^{-1}Vx_j$. If $x \in G$ then $x \in Vx_j$ for some $j$, so $xWx^{-1} \subset Vx_jWx_j^{-1}V \subset VVV \subset U$. Hence $\bigcup_{x \in G} xWx^{-1}$ is a neighborhood of 1 that is invariant under conjugations and contained in $U$. $\qquad \square$

If $f \in L^1$ and $g \in L^2$, then $\sum d_\pi(f * g) * \chi_\pi$ converges in $L^2$ to $f * g$ (Proposition 2.40(a)), and $\sum d_\pi g * \chi_\pi$ converges in $L^2$ to $g$. Now, if $\Phi$ is any finite subset of $\widehat{G}$, we have

$$\left[\sum_{[\pi]\in\Phi} d_\pi(f * g) * \chi_\pi\right] * \left[\sum_{[\pi]\in\Phi} d_\pi g * \chi_\pi\right]$$

$$= \sum_{[\pi],[\pi']\in\Phi} d_\pi d_{\pi'}(f * g * g) * (\chi_\pi * \chi_{\pi'}) = \sum_{[\pi]\in\Phi} d_\pi(f * g * g) * \chi_\pi,$$

where we have used the fact that characters are central, Proposition 5.21, and (5.20). By Proposition 2.41, it follows that the series $\sum d_\pi (f * g * g) * \chi_\pi$ converges *uniformly* to $f * g * g$.

This is the key to the summability method. For each neighborhood $U$ of 1 in $G$, let us take $g = |V|^{-1}\chi_V$ where $V$ is a central neighborhood of 1 such that $VV \subset U$ — such a $V$ exists by Lemma 5.24 — and let $\psi_U = g * g$. Then $\{\psi_U\}$ is an approximate identity consisting of central functions. Also, by Lemma 5.22, $\psi_U * \chi_\pi = c_U(\pi)\chi_\pi$ for some scalar $c_U(\pi)$. Therefore

$$f * \psi_U = \sum_{[\pi] \in \widehat{G}} c_U(\pi)d_\pi f * \chi_\pi,$$

where the series on the right converges uniformly by the result of the preceding paragraph, and $f$ is the limit of $f * \psi_U$ as $U \to \{1\}$ in the $L^p$ or uniform norm whenever $f \in L^p(G)$ ($p < \infty$) or $f \in C(G)$.

We have $\sum_{[\pi] \in \widehat{G}} |c_U(\pi)|^2 = \|\psi_U\|_2^2 < \infty$, and $c_U(\pi) \to 1$ as $U \to \{1\}$ since $\psi_U * \chi_\pi \to \chi_\pi$. The constants $c_U(\pi)$ can thus be regarded as "summability factors" whose introduction into the Fourier series of $f$ causes the series to converge uniformly; $f$ is then obtained as the limit in the appropriate norm as these factors tend to 1 in a suitable way. As an immediate application of this construction, we have the following result.

**5.25 Proposition.** *The linear span of* $\{\chi_\pi : [\pi] \in \widehat{G}\}$ *is dense in* $ZC(G)$ *and also in* $ZL^p(G)$ *for* $1 \le p < \infty$.

*Proof.* If $f$ is central, the partial sums of the series $\sum c_U(\pi)d_\pi f * \chi_\pi$ are linear combinations of characters by Lemma 5.22. They converge uniformly and hence in $L^p$ ($p < \infty$) to $f * \psi_U$, and $f * \psi_U \to f$ in the appropriate norm as $U \to \{1\}$. $\qquad\square$

The spaces $ZL^p(G)$ and $ZC(G)$ are commutative Banach algebras under convolution, so one naturally wishes to know their spectra. In fact, we have the following analogue of Theorem 4.3.

**5.26 Theorem.** *Let* $\mathcal{A}$ *be any of the convolution algebras* $ZL^p(G)$ *($1 \le p < \infty$) or* $ZC(G)$. *For* $f \in \mathcal{A}$ *and* $[\pi] \in \widehat{G}$, *let* $h_\pi(f) = d_\pi \int f \overline{\chi_\pi}$. *Then the map* $[\pi] \mapsto h_\pi$ *is a bijection from* $\widehat{G}$ *to* $\sigma(\mathcal{A})$ *which is a homeomorphism if* $\widehat{G}$ *is given the discrete topology.*

*Proof.* Each $h_\pi$ is clearly a bounded linear functional on $\mathcal{A}$. Moreover, by (5.20), Proposition 5.21, and Lemma 5.22,

$$h_\pi(f * g)\chi_\pi = d_\pi^2 f * g * \chi_\pi = d_\pi^3 f * g * \chi_\pi * \chi_\pi$$
$$= d_\pi^3 f * \chi_\pi * g * \chi_\pi = d_\pi h_\pi(f)h_\pi(g)\chi_\pi * \chi_\pi = h_\pi(f)h_\pi(g)\chi_\pi,$$

so $h_\pi$ is multiplicative, i.e., $h_\pi \in \sigma(\mathcal{A})$. Next, if $h \in \sigma(\mathcal{A})$, 5.20 shows that

$$h(\chi_\pi)h(\chi_{\pi'}) = \begin{cases} d_\pi^{-1}h(\chi_\pi) & \text{if } [\pi'] = [\pi], \\ 0 & \text{if } [\pi'] \neq [\pi]. \end{cases}$$

It follows that $h(\chi_\pi)$ must be either $d_\pi^{-1}$ or 0, and that there can only be one $[\pi]$, say $[\pi_0]$, for which $h(\chi_\pi) \neq 0$. But then $h$ agrees with $h_{\pi_0}$ on the linear span of the characters, so $h = h_{\pi_0}$ by Proposition 5.25. Finally, the topology on $\sigma(\mathcal{A})$ is discrete because $\{h_\pi\} = \{h \in \sigma(\mathcal{A}) : h(\chi_\pi) \neq 0\}$ is open for every $\pi$. $\qquad\square$

We conclude this section with an amusing application of our general theory to the theory of finite groups.

**5.27 Proposition.** *Let $G$ be a finite group. Then the cardinality of $\widehat{G}$ is the number of conjugacy classes in $G$, and $\sum_{[\pi]\in\widehat{G}} d_\pi^2$ is the order of $G$.*

*Proof.* In view of Proposition 5.23, $\{\chi_\pi : [\pi] \in \widehat{G}\}$ and the set of characteristic functions of the conjugacy classes in $G$ are both bases for the space of central functions on $G$; this proves the first assertion. The second one is true since $\sum_{[\pi]\in\widehat{G}} d_\pi^2$ is the dimension of $L^2(G)$ by the Peter-Weyl theorem. $\qquad\square$

---

## 5.4   Examples

In this section we describe the irreducible representations and the decomposition of $L^2$ for the group $SU(2)$, then use the results to obtain the irreducible representations of the closely related groups $SO(3)$ and $U(2)$.

   **1. SU(2).**   We recall that $U(n)$ is the group of unitary transformations of $\mathbb{C}^n$, alias the group of $n \times n$ complex matrices $T$ such that $T^*T = I$, and that $SU(n)$ is the subgroup of $U(n)$ consisting of the matrices whose determinant is 1. We have $T \in U(n)$ if and only if $TT^* = I$, which says precisely that the rows of $T$ form an orthonormal set.
   When $n = 2$, we therefore have

$$\begin{pmatrix} a & b \\ c & d \end{pmatrix} \in U(2) \quad \Longleftrightarrow \quad |a|^2 + |b|^2 = |c|^2 + |d|^2 = 1 \text{ and } a\bar{c} + b\bar{d} = 0.$$

Thus $(a,b)$ must be a unit vector and $(c,d) = e^{i\theta}(-\bar{b}, \bar{a})$ for some $\theta \in \mathbb{R}$.

But then $\det T = e^{i\theta}(|a|^2 + |b|^2) = e^{i\theta}$, so $T \in SU(2)$ precisely when $e^{i\theta} = 1$. In short, if we set

$$(5.28) \qquad U_{a,b} = \begin{pmatrix} a & b \\ -\bar{b} & \bar{a} \end{pmatrix},$$

we have

$$SU(2) = \{U_{a,b} : a, b \in \mathbb{C}, \ |a|^2 + |b|^2 = 1\}.$$

We observe that

$$U_{a,b}^{-1} = U_{a,b}^* = U_{\bar{a},-b},$$

and that the action of $U_{a,b}$ on $\mathbb{C}^2$ is given by

$$\begin{pmatrix} a & b \\ -\bar{b} & \bar{a} \end{pmatrix} \begin{pmatrix} z \\ w \end{pmatrix} = \begin{pmatrix} az + bw \\ -\bar{b}z + \bar{a}w \end{pmatrix}, \quad \text{or} \quad U_{a,b}(z, w) = (az + bw, -\bar{b}z + \bar{a}w).$$

The correspondence $U_{a,b} \leftrightarrow (a, b) = U_{a,b}(1, 0)$ identifies $SU(2)$ as a set with the unit sphere $S^3 \subset \mathbb{C}^2$ in such a way that the identity element is identified with the "north pole" $(1, 0)$.

We single out the following three one-parameter subgroups of $SU(2)$ for special attention:

$$(5.29)$$
$$F(\theta) = \begin{pmatrix} e^{i\theta} & 0 \\ 0 & e^{-i\theta} \end{pmatrix},$$
$$G(\phi) = \begin{pmatrix} \cos\phi & \sin\phi \\ -\sin\phi & \cos\phi \end{pmatrix}, \qquad H(\psi) = \begin{pmatrix} \cos\psi & i\sin\psi \\ i\sin\psi & \cos\psi \end{pmatrix}.$$

Geometrically, these are three mutually orthogonal great circles in the sphere that intersect at $\pm \left( \begin{smallmatrix} 1 & 0 \\ 0 & 1 \end{smallmatrix} \right)$.

**5.30 Proposition.** *Each $T \in SU(2)$ is conjugate to precisely one matrix $F(\theta)$ as in (5.29) with $0 \leq \theta \leq \pi$.*

*Proof.* Every unitary matrix $T$ is normal, so by the spectral theorem there exists $V \in U(2)$ such that $VTV^{-1} = \left( \begin{smallmatrix} \alpha & 0 \\ 0 & \beta \end{smallmatrix} \right)$ where $\alpha$ and $\beta$ are the eigenvalues of $T$. If $T \in SU(2)$ then $\beta = \bar{\alpha}$, so $VTV^{-1} = F(\theta)$ for some $\theta \in [-\pi, \pi]$. By replacing $V$ with $(\det V)^{-1/2}V$ we can assume that $V \in SU(2)$, and since $F(-\theta) = H(\frac{1}{2}\pi)F(\theta)H(-\frac{1}{2}\pi)$ we can obtain $\theta \in [0, \pi]$. Finally, if $\theta_1, \theta_2 \in [0, \pi]$ then $F(\theta_1)$ and $F(\theta_2)$ have different eigenvalues, and hence are not conjugate, unless $\theta_1 = \theta_2$. $\qquad\square$

**5.31 Corollary.** *If $g$ is a continuous function on $SU(2)$, let $g^0(\theta) = g(F(\theta))$. Then $g \mapsto g^0$ is an isomorphism from the algebra of continuous central functions on $SU(2)$ (with pointwise multiplication) to $C([0, \pi])$.*

Proposition 5.30 has the following geometrical interpretation. The angle between the vectors $(a, b)$ and $(1, 0)$ in $\mathbb{C}^2$ is $\arccos(\mathrm{Re}\,a)$. It is an easy exercise, which we leave to the reader, to show that if $U_{c,d}U_{a,b}U_{c,d}^{-1} = U_{\alpha,\beta}$ then $\mathrm{Re}\,a = \mathrm{Re}\,\alpha$. Thus the set $\{F(\theta) : 0 \le \theta \le \pi\}$ is a meridian joining the north and south poles, and the conjugacy classes are the surfaces of constant latitude.

Next, we describe a family of unitary representations of $SU(2)$. Let $\mathcal{P}$ be the space of all polynomials $P(z, w) = \sum c_{jk} z^j w^k$ in two complex variables, and let $\mathcal{P}_m \subset \mathcal{P}$ be the space of homogeneous polynomials of degree $m$:

$$\mathcal{P}_m = \left\{ P : P(z, w) = \sum_0^m c_j z^j w^{m-j}, \; c_0, \ldots, c_m \in \mathbb{C} \right\}.$$

We define an inner product on $\mathcal{P}$ by regarding it as a subset of $L^2(\sigma)$, where $\sigma$ is surface measure on the unit sphere $S^3$ normalized so that $\sigma(S^3) = 1$:

$$\langle P, Q \rangle = \int_{S^3} P\overline{Q}\,d\sigma.$$

($\mathcal{P}$ is not complete with respect to this inner product, but each $\mathcal{P}_m$ is, being finite-dimensional.)

We shall show that the monomials $z^j w^k$ form an orthogonal set in $\mathcal{P}$. In the following calculations, we employ the polar coordinate notation

$$(z, w) = Z = rZ', \text{ where } r = |Z| = \left[ |z|^2 + |w|^2 \right]^{1/2} \text{ and } Z' \in S^3.$$

We also denote Lebesgue measure on $\mathbb{C}^2$ by $d^4 Z$ and Lebesgue measure on $\mathbb{C}$ by $d^2 z$ or $d^2 w$. Thus,

$$d^4 Z = d^2 z\, d^2 w = c r^3\, dr\, d\sigma(Z').$$

Here $c$ is the Euclidean surface measure of $S^3$; as the following argument shows, it equals $2\pi^2$.

**5.32 Lemma.** *If $f : \mathbb{C}^2 \to \mathbb{C}$ satisfies $f(aZ) = a^m f(Z)$ for $a > 0$, then*

$$\int_{S^3} f(Z')\,d\sigma(Z') = \frac{1}{\pi^2 \Gamma(\frac{1}{2}m + 2)} \int_{\mathbb{C}^2} f(Z)e^{-|Z|^2}\,d^4 Z.$$

*Proof.* We integrate in polar coordinates:

$$\int_{\mathbb{C}^2} f(Z)e^{-|Z|^2}\,d^4 Z = c \int_0^\infty \int_{S^3} f(rZ')e^{-r^2} r^3\,d\sigma(Z')\,dr$$

$$= c \int_0^\infty r^{m+3} e^{-r^2}\,dr \int_{S^3} f(Z')\,d\sigma(Z')$$

$$= \frac{c}{2} \Gamma(\tfrac{1}{2}m + 2) \int_{S^3} f(Z')\,d\sigma(Z').$$

If we take $f = 1$, we see that

$$\frac{c}{2} = \int_{\mathbb{C}^2} e^{-|Z|^2} \, d^4 Z = \left[ \int_{-\infty}^{\infty} e^{-t^2} \, dt \right]^4 = \pi^2,$$

which completes the proof.                                                  □

**5.33 Lemma.** *If $p, q, r, s$ are nonnegative integers,*

$$\int z^p \overline{z}^q w^r \overline{w}^s \, d\sigma(z,w) = \begin{cases} 0 & \text{if } q \neq p \text{ or } s \neq r, \\ p! r! / (p + r + 1)! & \text{if } q = p \text{ and } s = r. \end{cases}$$

*Proof.* By Lemma 5.32,

$$\int z^p \overline{z}^q w^r \overline{w}^s \, d\sigma(z,w)$$

$$= \frac{1}{\pi^2 \Gamma(\frac{1}{2}(p+q+r+s) + 2)} \int z^p \overline{z}^q e^{-|z|^2} \, d^2 z \int w^r \overline{w}^s e^{-|w|^2} \, d^2 w.$$

We do the latter integrals in polar coordinates too:

$$\int z^p \overline{z}^q e^{-|z|^2} \, d^2 z = \int_0^{\infty} \int_0^{2\pi} e^{i(p-q)\theta} r^{p+q+1} e^{-r^2} \, dr \, d\theta$$

$$= \begin{cases} 0 & \text{if } p \neq q, \\ 2\pi \cdot \frac{1}{2}\Gamma(p+1) = \pi p! & \text{if } q = p, \end{cases}$$

and similarly for the integral in $w$. The result follows.            □

**5.34 Proposition.** *The spaces $\mathcal{P}_m$ are mutually orthogonal in $L^2(\sigma)$, and*

$$\left\{ \sqrt{\frac{(m+1)!}{j!(m-j)!}} z^j w^{m-j} : 0 \leq j \leq m \right\}$$

*is an orthonormal basis for $\mathcal{P}_m$.*

*Proof.* This is an immediate consequence of Lemma 5.33.           □

$SU(2)$ acts on $\mathcal{P}$ via its natural action on $\mathbb{C}^2$, yielding a representation $\pi$:

(5.35)    $[\pi(U_{a,b})P](z,w) = P(U_{a,b}^{-1}(z,w)) = P(\overline{a}z - bw, \overline{b}z + aw).$

Clearly $\mathcal{P}_m$ is invariant under $\pi$; we denote the subrepresentation of $\pi$ on $\mathcal{P}_m$ by $\pi_m$. Then $\pi_m$ is a unitary representation of $SU(2)$ on $\mathcal{P}_m$ with respect to the inner product in $L^2(\sigma)$, since $\sigma$ is rotation-invariant. (In the physics literature, $\pi_m$ is normally labeled by $\frac{1}{2}m$ instead of $m$, because of its connection with particles of spin $\frac{1}{2}m$.) We shall show that $\pi_m$ is irreducible.

**5.36 Lemma.** *Suppose* $\mathcal{M}$ *is a* $\pi$-*invariant subspace of* $\mathcal{P}_m$. *If* $P \in \mathcal{M}$ *then* $z(\partial P/\partial w)$ *and* $w(\partial P/\partial z)$ *are in* $\mathcal{M}$.

*Proof.* If $G(\phi)$ is as in (5.29), then $\phi^{-1}\big[\pi(G(\phi))P - P\big]$ lies in $\mathcal{M}$ for all $\phi \neq 0$. As $\phi \to 0$, its coefficients approach those of $(d/d\phi)\pi(G(\phi))P\big|_{\phi=0}$; hence the latter polynomial also belongs to $\mathcal{M}$ because $\mathcal{P}_m$ has finite dimension and so $\mathcal{M}$ is closed in it. But

$$\frac{d}{d\phi}\pi(G(\phi))P\big|_{\phi=0} = \frac{d}{d\phi}P(z\cos\phi - w\sin\phi,\, z\sin\phi + w\cos\phi)\big|_{\phi=0}$$
$$= z\frac{\partial P}{\partial w} - w\frac{\partial P}{\partial z}.$$

Similarly, with $H(\psi)$ as in (5.29),

$$z\frac{\partial P}{\partial w} + w\frac{\partial P}{\partial z} = \frac{1}{i}\frac{d}{d\psi}\pi(H(\psi))P\big|_{\psi=0} \in \mathcal{M}.$$

Adding and subtracting these, we obtain the desired result. $\qquad\square$

**5.37 Theorem.** $\pi_m$ *is irreducible for each* $m \geq 0$.

*Proof.* Suppose $\mathcal{M}$ is an invariant subspace of $\mathcal{P}_m$ and $P \neq 0 \in \mathcal{M}$. Let $P(z,w) = \sum_0^m c_j z^j w^{m-j}$ and let $J$ be the largest value of $j$ such that $c_j \neq 0$. Then

$$\left(w\frac{\partial}{\partial z}\right)^J P(z,w) = c_J J! w^m.$$

By Lemma 5.36 we see that $w^m \in \mathcal{M}$, and then applying $z(\partial/\partial w)$ successively we obtain $zw^{m-1} \in \mathcal{M}$, $z^2 w^{m-2} \in \mathcal{M}$, ..., $z^m \in \mathcal{M}$. Hence $\mathcal{M} = \mathcal{P}_m$. $\qquad\square$

Let $\chi_m$ be the character of $\pi_m$, and let $\chi_m^0(\theta) = \chi_m(F(\theta))$ as in Corollary 5.31. Since the orthogonal basis vectors $z^j w^{m-j}$ for $\mathcal{P}_m$ are eigenvectors for $\pi_m(F(\theta))$,

$$\pi_m(F(\theta))(z^j w^{m-j}) = e^{i(2j-m)\theta} z^j w^{m-j},$$

we have

$$
\begin{aligned}
\text{(5.38)} \qquad \chi_m^0(\theta) &= \sum_0^m e^{i(2j-m)\theta} = \frac{e^{i(m+2)\theta} - e^{-im\theta}}{e^{2i\theta} - 1}\\
&= \frac{e^{i(m+1)\theta} - e^{-i(m+1)\theta}}{e^{i\theta} - e^{-i\theta}} = \frac{\sin(m+1)\theta}{\sin\theta}.
\end{aligned}
$$

From this we can show that the $\pi_m$'s form a complete list of irreducible representations of $SU(2)$.

**5.39 Theorem.** $[SU(2)]\hat{\ } = \{[\pi_m] : m \geq 0\}.$

*Proof.* First, the $\pi_m$'s are all inequivalent because they have different dimensions (and different characters). Next, we observe that by (5.38),

$$\chi_0^0(\theta) = 1, \qquad \chi_1^0(\theta) = 2\cos\theta,$$

$$\chi_m^0(\theta) - \chi_{m-2}^0(\theta) = \frac{\sin(m+1)\theta - \sin(m-1)\theta}{\sin\theta} = 2\cos m\theta \text{ for } m \geq 2.$$

Hence the linear span of $\{\chi_m^0\}_0^\infty$ is the linear span of $\{\cos m\theta\}_0^\infty$, which is uniformly dense in $C([0,\pi])$. By Corollary 5.31, the linear span of $\{\chi_m\}_0^\infty$ is uniformly dense in the space of continuous central functions on $SU(2)$; in particular, the only such function that is orthogonal to all $\chi_m$ is the zero function. By Proposition 5.23, then, the $\chi_m$'s exhaust all the irreducible characters. $\square$

We remark that by (5.38) and Corollary 5.31, the characters $\chi_m$ are all real-valued. It follows that $\pi_m$ is equivalent to its contragredient $\bar{\pi}_m$ for every $m$, and in view of Theorem 5.2, the same is true of every unitary representation of $SU(2)$.

Let us compute the matrix elements of $\pi_m$ with respect to the basis given in Proposition 5.32, namely

$$e_j(z,w) = \sqrt{\frac{(m+1)!}{j!(m-j)!}} z^j w^{m-j}.$$

To improve the aesthetics of the final result, we reparametrize $SU(2)$ by replacing $b$ by $\bar{b}$. That is, we set

$$\pi_m(a,b) = \pi_m(U_{a,\bar{b}}), \qquad \pi_m^{jk}(a,b) = \langle \pi_m(a,b)e_k, e_j \rangle,$$

where $U_{a,\bar{b}}$ is given by (5.28) with $\bar{b}$ substituted for $b$. We then have

$$\sqrt{\frac{(m+1)!}{k!(m-k)!}} (\bar{a}z - \bar{b}w)^k (bz + aw)^{m-k} = [\pi_m(a,b)e_k](z,w)$$

$$= \sum_j \pi_m^{jk}(a,b)e_j(z,w) = \sum_j \sqrt{\frac{(m+1)!}{j!(m-j)!}} \pi_m^{jk}(a,b)z^j w^{m-j}.$$

Thus,

$$(5.40) \quad \sum_j \sqrt{\frac{k!(m-k)!}{j!(m-j)!}} \pi_m^{jk}(a,b)z^j w^{m-j} = (\bar{a}z - \bar{b}w)^k (bz + aw)^{m-k}.$$

To solve this for $\pi_m^{jk}(a,b)$ one can simply multiply out the right side and equate the coefficients of $z^j w^{m-j}$ on the two sides. Another method is to set $z = e^{2\pi it}$ and $w = 1$; then the sum on the left is a Fourier series, and one can compute $\pi_m^{jk}(a,b)$ by the usual formula for Fourier coefficients:

$$\pi_m^{jk}(a,b) = \sqrt{\frac{j!(m-j)!}{k!(m-k)!}} \int_0^1 (\bar{a}e^{2\pi it} - \bar{b})^k (be^{2\pi it} + a)^{m-k} e^{-2\pi ijt} \, dt.$$

When $k = 0$ we have

$$\sum_j \sqrt{\frac{m!}{j!(m-j)!}} \pi_m^{j0}(a,b) z^j w^{m-j} = (bz + aw)^m$$

$$= \sum_j \frac{m!}{j!(m-j)!} b^j a^{m-j} z^j w^{m-j},$$

so

$$\pi_m^{j0}(a,b) = \sqrt{\frac{m!}{j!(m-j)!}} b^j a^{m-j} = \frac{1}{\sqrt{m+1}} e_j(b,a).$$

Thus the functions $\{\pi_m^{j0} : 0 \le j \le m\}$ span $\mathcal{P}_m$, the space that gave us $\pi_m$ originally. (The Peter-Weyl theorem predicts that we should get $\overline{\pi}_m$ instead of $\pi_m$, but we have observed above that these representations are equivalent. The factor $\sqrt{m+1}$ is the factor $\sqrt{d_{\pi_m}}$ needed to normalize the matrix elements according to the Schur orthogonality relations.) Similarly, the functions $\pi_m^{jm}$, $0 \le j \le m$, span the complex conjugate space $\overline{\mathcal{P}}_m$.

We can also say something interesting about the linear span of $\{\pi_m^{jk} : 0 \le j \le m\}$ for general $k$. First, if we use (5.40) to define $\pi_m^{jk}(a,b)$ for all $a, b \in \mathbb{C}^2$, it is clear that $\pi_m^{jk}(a,b)$ is a polynomial in the variables $a, b, \bar{a}, \bar{b}$ that is homogeneous of degree $m - k$ in $(a,b)$ and homogenous of degree $k$ in $(\bar{a}, \bar{b})$; we say that it has **bidegree** $(m - k, k)$. Second, as a function on $\mathbb{C}^2$, $\pi_m^{jk}$ is *harmonic*: it satisfies Laplace's equation $\sum_{n=1}^4 (\partial^2 \pi_m^{jk}/\partial x_n^2) = 0$, where $a = x_1 + ix_2$ and $b = x_3 + ix_4$. Indeed, it suffices to observe that the right side of (5.40) is harmonic as a function of $(a,b)$ for all $(z,w)$, since then the coefficient of each $z^j w^{m-j}$ must also be harmonic. This is easy to check as it stands, and it becomes even more so if one writes the Laplacian in complex coordinates:

$$\frac{\partial^2}{\partial x_1^2} + \cdots + \frac{\partial^2}{\partial x_4^2} = 4\frac{\partial^2}{\partial a \partial \bar{a}} + 4\frac{\partial^2}{\partial b \partial \bar{b}}.$$

In short, when we identify $SU(2)$ with the unit sphere in $\mathbb{C}^3$ via $U_{a,b} \leftrightarrow (a, \bar{b})$, the Peter-Weyl decomposition $L^2 = \bigoplus_0^\infty \mathcal{E}_{\pi_m}$ is the usual decomposition of functions on the sphere into spherical harmonics (see, e.g., Stein and Weiss [131]), and the further decomposition

$$\mathcal{E}_{\pi_m} = \bigoplus_{p+q=m} \mathcal{H}_{p,q}, \qquad \mathcal{H}_{p,q} = \text{linear span of } \{\pi_{p+q}^{jq} : 0 \le j \le p+q\},$$

yields a refinement in which the spherical harmonics are grouped according to their bidegree.

**2. SO(3).** $SU(2)$ is intimately related to both $SO(3)$ and $SO(4)$, the groups of rotations of $\mathbb{R}^3$ and $\mathbb{R}^4$. The most elegant way to explain these connections is via the algebra of **quaternions**, denoted by $\mathbb{H}$ in honor of their discoverer Hamilton. There are several ways of describing $\mathbb{H}$; we shall begin as follows. As a real vector space, $\mathbb{H}$ is $\mathbb{R}^4 \cong \mathbb{R} \times \mathbb{R}^3$; we write elements of $\mathbb{H}$ as $\xi = (a, \mathbf{x})$ with $a \in \mathbb{R}$ and $\mathbf{x} \in \mathbb{R}^3$. The multiplication in $\mathbb{H}$ is given by

$$(5.41) \qquad (a, \mathbf{x})(b, \mathbf{y}) = (ab - \mathbf{x} \cdot \mathbf{y}, \, b\mathbf{x} + a\mathbf{y} + \mathbf{x} \times \mathbf{y}),$$

$\mathbf{x} \cdot \mathbf{y}$ and $\mathbf{x} \times \mathbf{y}$ being the usual dot and cross products. It is routine to check (although not obvious at first glance) that this multiplication is associative and that the Euclidean norm on $\mathbb{R}^4$ respects it: $|\xi\eta| = |\xi| \, |\eta|$. The subspace $\mathbb{R} \times \{\mathbf{0}\}$ is the center of $\mathbb{H}$. We identify it simply with $\mathbb{R}$ and think of it as the "real axis" in $\mathbb{H}$, and we likewise identify $\{0\} \times \mathbb{R}^3$ with $\mathbb{R}^3$; thus, instead of writing an element of $\mathbb{H}$ as $(a, \mathbf{x})$ we may write it as $a + \mathbf{x}$. Moreover, we denote the canonical basis vectors for $\mathbb{R}^3$ by $i$, $j$, and $k$, so that

$$(5.42) \qquad (a, \mathbf{x}) = a + \mathbf{x} = a + x_1 i + x_2 j + x_3 k.$$

The multiplication law (5.41) is completely specified by giving the products of these basis vectors, which are as follows:

$$i^2 = j^2 = k^2 = -1,$$
$$ij = -ji = k, \quad jk = -kj = i, \quad ki = -ik = j.$$

The **conjugate** of the quaternion $\xi = (a, \mathbf{x})$ is $\bar{\xi} = (a, -\mathbf{x})$. Conjugation is an anti-automorphism of $\mathbb{H}$: $\overline{\xi\eta} = \bar{\eta}\bar{\xi}$. Also, just as in the complex numbers, if $\xi = (a, \mathbf{x})$ we have $\xi\bar{\xi} = \bar{\xi}\xi = a^2 + |\mathbf{x}|^2 = |\xi|^2$. It follows that $\mathbb{H}$ is a division algebra: every nonzero $\xi \in \mathbb{H}$ is invertible, with $\xi^{-1} = \bar{\xi}/|\xi|^2$. In particular, if $|\xi| = 1$ then $\xi^{-1} = \bar{\xi}$, so the set of elements of $\mathbb{H}$ of norm 1 form a group, which — for the moment — we denote by $U(\mathbb{H})$:

$$U(\mathbb{H}) = \{\xi \in \mathbb{H} : |\xi| = 1\}.$$

If $\xi \in U(\mathbb{H})$, the operation $\eta \mapsto \xi\eta\xi^{-1}$ is an isometric linear transformation of $\mathbb{H}$ that leaves the center $\mathbb{R}$ pointwise fixed and hence also leaves the orthogonal subspace $\mathbb{R}^3$ invariant. The restriction of this map to $\mathbb{R}^3$ is therefore an element of $SO(3)$ that we denote by $\kappa(\xi)$:

$$(5.43) \qquad \kappa(\xi)\mathbf{x} = \xi\mathbf{x}\xi^{-1}.$$

($\kappa(\xi)$ belongs to $SO(3)$ rather than just $O(3)$ because $U(\mathbb{H})$, the unit sphere in $\mathbb{H}$, is connected, and $\kappa : U(\mathbb{H}) \to SO(3)$ is clearly continuous.) We note that $\kappa(\xi\eta) = \kappa(\xi)\kappa(\eta)$, that is, $\kappa$ is a homomorphism from $U(\mathbb{H})$ to $SO(3)$.

We can describe the rotation $\kappa(\xi)$ very explicitly. Each $\xi \in U(\mathbb{H})$ can be written in the form

$$\xi = \cos\theta + (\sin\theta)\mathbf{u}$$

for some $\theta \in [0, \pi]$ and some unit vector $\mathbf{u} \in \mathbb{R}^3$; $\theta$ is uniquely determined, and so is $\mathbf{u}$ except when $\sin\theta = 0$, i.e., when $\xi = \pm 1$. Clearly $\kappa(\pm 1) = I$, so we may assume that $0 < \theta < \pi$. Pick a unit vector $\mathbf{v} \in \mathbb{R}^3$ that is orthogonal to $\mathbf{u}$, and let $\mathbf{w} = \mathbf{u} \times \mathbf{v}$. Thus $\mathbf{u}$, $\mathbf{v}$, and $\mathbf{w}$ form an orthonormal basis for $\mathbb{R}^3$, and by (5.41) their quaternion products are given by $\mathbf{uv} = -\mathbf{vu} = \mathbf{w}$, $\mathbf{wu} = -\mathbf{uw} = \mathbf{v}$. A simple calculation then shows that

$$\xi(a\mathbf{u} + b\mathbf{v} + c\mathbf{w})\xi^{-1} = a\mathbf{u} + (b\cos 2\theta - c\sin 2\theta)\mathbf{v} + (c\cos 2\theta + b\sin 2\theta)\mathbf{w}.$$

In other words, $\kappa(\xi)$ *is a rotation through the angle* $2\theta$ *about the* $\mathbf{u}$-*axis.* In particular, $\kappa(\xi) = I$ only when $\xi = \pm 1$.

Moreover, *every* rotation of $\mathbb{R}^3$ is of this form. Indeed, if $T \in SO(3)$, the eigenvalues of $T$ have absolute value 1, their product is 1, and the nonreal ones form a conjugate pair. It follows easily that 1 is an eigenvalue, of multiplicity one unless $T = I$, so $T$ is a rotation about the $\mathbf{u}$-axis where $\mathbf{u}$ is the corresponding unit eigenvector. To sum up, we have proved the following:

**5.44 Theorem.** *The map $\kappa$ defined by (5.43) is a 2-to-1 homomorphism from $U(\mathbb{H})$ onto $SO(3)$, and hence $SO(3) \cong U(\mathbb{H})/\{\pm 1\}$.*

Now let us look at $\mathbb{H}$ from a slightly different angle. Using the representation (5.42), we observe that

$$a + bi + cj + dk = (a + bi) + (c + di)j.$$

Also, the algebra structure on $\mathbb{H}$ restricted to elements of the form $a + bi$ coincides with the algebra structure on $\mathbb{C}$, and $j(a + bi) = (a - bi)j$. We can therefore identify $\mathbb{H}$ with $\mathbb{C}^2$, with the multiplication now given by

$$(5.45) \quad (z + wj)(u + vj) = (zu - w\bar{v}) + (zv + w\bar{u})j \qquad (z, w, u, v \in \mathbb{C}).$$

Now, one further twist: the reader will recall that the correspondence $a + bi \leftrightarrow \left( \begin{smallmatrix} a & b \\ -b & a \end{smallmatrix} \right)$ is an isomorphism from $\mathbb{C}$ to the algebra of $2 \times 2$ real matrices of the stated form. Likewise, (5.45) shows that the correspondence

$$(5.46) \qquad z + wj \leftrightarrow \begin{pmatrix} z & w \\ -\overline{w} & \overline{z} \end{pmatrix}$$

is an isomorphism from $\mathbb{H}$ to the algebra of $2 \times 2$ complex matrices of the stated form. In particular, if we compare this with (5.28), we see that *the correspondence $z + wj \leftrightarrow U_{z,w}$ identifies the groups $U(\mathbb{H})$ and $SU(2)$.*

Hence, Theorem 5.44 can be restated thus:

**5.47 Corollary.** $SO(3) \cong SU(2)/\{\pm I\}$.

Another way to describe this relation between $SU(2)$ and $SO(3)$ is as follows. The vector subspace $\mathbb{R}^3 \subset \mathbb{H}$ is the set of $z + wj$ with $z$ purely imaginary. The correspondence (5.46), restricted to this subspace, turns $\mathbb{R}^3$ into the space of trace-zero skew-Hermitian $2 \times 2$ complex matrices, which (equipped with the Lie product $[A, B] = AB - BA$) is the Lie algebra $\mathfrak{su}(2)$ of $SU(2)$. With this identification, the action $\kappa(\xi)\mathbf{x} = \xi\mathbf{x}\xi^{-1}$ of $U(\mathbb{H})$ on $\mathbb{R}^3$ turns into the adjoint representation $\mathrm{Ad}(U)A = UAU^{-1}$ of $SU(2)$ on $\mathfrak{su}(2)$.

From Corollary 5.47 we immediately see that the representations of $SO(3)$ are just the representations of $SU(2)$ that are trivial on $\pm I$. Since the irreducible representation $\pi_m$ of $SU(2)$ on $\mathcal{P}_m$ satisfies $\pi_m(-I) = (-1)^m I$, we have:

**5.48 Corollary.** $[SO(3)]\hat{} = \{[\rho_k] : k = 0, 1, 2, \ldots\}$, where $\rho_k(\mathrm{Ad}(U)) = \pi_{2k}(U)$.

In §6.2 we shall identify the irreducible decompositions of the natural representations of $SO(3)$ on functions and differential forms on the unit sphere $S^2 \subset \mathbb{R}^3$.

**3. SO(4).** In this section we shall identify $\mathbb{R}^4$ with $\mathbb{H}$ and $SU(2)$ with $U(\mathbb{H})$.

In the preceding section we exploited the action of $SU(2)$ on $\mathbb{H}$ by conjugation. We shall now exploit two other actions of $SU(2)$ on $\mathbb{H}$, by left and right translation:

$$L(\xi)\zeta = \xi\zeta, \qquad R(\xi)\zeta = \zeta\xi^{-1}.$$

Since the Euclidean norm on $\mathbb{H}$ is multiplicative, the images of $SU(2)$ under $L$ and $R$ are both closed subgroups of $SO(4)$. These subgroups commute with each other (this is just the associative law, $\xi(\zeta\eta) = (\xi\zeta)\eta$), and they intersect only in $\pm 1$. Indeed, if $L(\xi) = R(\eta)$ then $\xi = L(\xi)1 = R(\eta)1 = \eta^{-1}$, so $L(\xi) = R(\xi^{-1})$. But this says that $\xi$ belongs to the center of $\mathbb{H}$, namely $\mathbb{R}$; hence $\xi = \pm 1$ and also $\eta = \xi^{-1} = \pm 1$.

**5.49 Theorem.** *If $T \in SO(4)$, there exist $\xi, \eta \in SU(2)$, unique up to a common factor of $\pm 1$, such that $T = L(\xi)R(\eta)$. Hence, $SO(4) \cong [SU(2) \times SU(2)]/\{\pm(1,1)\}$.*

*Proof.* Let $\zeta = T(1)$, and let $S = L(\zeta)^{-1}T$. Then $S$ is an element of $SO(4)$ that satisfies $S(1) = 1$ and hence leaves the real axis pointwise fixed, so it is in essence a rotation of the orthogonal subspace $\mathbb{R}^3$. But we saw in Theorem 5.44 that every such $S$ is conjugation by some $\eta$, i.e., $S = L(\eta)R(\eta)$. Hence $T = L(\zeta)S = L(\xi)R(\eta)$ where $\xi = \zeta\eta$. The uniqueness and the fact that the map $(\xi, \eta) \mapsto L(\xi)R(\eta)$ is a group homomorphism follow from the preceding remarks. $\qquad\square$

*Remark.* The fact that $SO(4)$ has two rather large normal subgroups of which it is almost the direct product is a peculiarity of dimension 4. If $n$ is even and $> 4$, $SO(n)$ has no nontrivial closed normal subgroups except $\{\pm I\}$, and if $n$ is odd, it has no such subgroups at all.

We can use Theorem 5.49 to describe the irreducible representations of $SO(4)$ in terms of those of $SU(2) \times SU(2)$, and the latter are given in terms of the irreducible representations of $SU(2)$ by results that we shall prove in the next chapter. Indeed, recall the representations $\pi_n$ of $SU(2)$ on $\mathcal{P}_n$ ($n \geq 0$) from §5.4.1. For $m, n \geq 0$, define the representation $\pi_{mn}$ of $SU(2) \times SU(2)$ on $\mathcal{P}_m \otimes \mathcal{P}_n$ (but retaining the quaternion notation for elements of $SU(2)$) by

$$\pi_{mn}(\xi, \eta) = \pi_m(\xi) \otimes \pi_n(\eta).$$

(See Appendix 3 for more about tensor products.) Then, by Theorems 7.12 and 7.17 and Corollary 7.14, the $\pi_{mn}$'s are irreducible and inequivalent, and every irreducible representation of $SU(2) \times SU(2)$ is equivalent to some $\pi_{mn}$. Moreover, by Theorem 5.49, the irreducible representations of $SO(4)$ are essentially the irreducible representations of $SU(2) \times SU(2)$ that are trivial on $\pm(1,1)$. Since $\pi_m(-1) = (-1)^m I$, these are precisely the ones where $m$ and $n$ have the same parity. We have therefore proved:

**5.50 Corollary.** $[SO(4)]\hat{} = \{[\rho_{mn}] : m, n \geq 0, \ m \equiv n \bmod 2\}$, *where* $\rho_{mn}(L(\xi)R(\eta)) = \pi_{mn}(\xi, \eta)$.

**4. U(2).** In this section we revert to $2 \times 2$ matrix notation for elements of $SU(2)$. The center $Z$ of $U(2)$ consists of the scalar multiples of the identity,

$$Z = \{e^{i\theta}I : \theta \in \mathbb{R}\},$$

since $U(2)$ acts irreducibly on $\mathbb{C}^2$. $U(2)$ is almost the product of the subgroups $Z$ and $SU(2)$, via the map

$$(e^{i\theta}I,\, U) \mapsto e^{i\theta}U.$$

This map is onto, for if $V \in U(2)$ we choose $\theta$ so that $\det V = e^{2i\theta}$ and take $U = e^{-i\theta}V$. Its kernel is $\pm(I, I)$ since $Z \cap SU(2) = \{\pm I\}$. Hence,

$$U(2) \cong [Z \times SU(2)]/(\pm I) \cong [\mathbb{T} \times SU(2)]/(\pm I).$$

From this we can easily determine the irreducible representations of $U(2)$. Suppose $\rho$ is such a representation. On $Z$, $\rho$ must act as scalar multiples of $I$ by Schur's lemma, and hence (since $\widehat{\mathbb{T}} \cong \mathbb{Z}$) $\rho(e^{i\theta}I) = e^{in\theta}I$ for some $n \in \mathbb{Z}$. By the same token, if $\mathcal{M}$ is a subspace of $\mathcal{H}_\rho$ that is invariant under $\rho|SU(2)$, it is invariant under all of $\rho$. Hence $\rho|SU(2)$ is an irreducible representation of $SU(2)$, so it is equivalent to $\pi_m$ for some $m \geq 0$. In short, if $T \in SU(2)$,

$$(5.51) \qquad \rho(e^{i\theta}T) = e^{in\theta}\pi_m(T).$$

Moreover, since $-I$ can be expressed either as $e^{i\pi}I$ or as $e^{i0}(-I)$, we must have $(-1)^n I = \pi_m(-I) = (-1)^m I$, so $m$ and $n$ must have the same parity. Conversely, if this is the case, (5.51) gives a well-defined representation of $U(2)$. Different $m$ or $n$ give inequivalent representations, since their restrictions to $Z$ or $SU(2)$ are inequivalent. In summary:

**5.52 Theorem.** $[U(2)]^\wedge = \{[\rho_{m,n}] : m \geq 0,\ n \in \mathbb{Z},\ m \equiv n \pmod{2}\}$, *where*

$$\rho_{m,n}(e^{i\theta}T) = e^{in\theta}\pi_m(T) \qquad (\theta \in \mathbb{R},\ T \in SU(2)).$$

Another description of $\rho_{m,n}$ is as follows. If we define

$$\rho_{2k}(e^{i\theta}T) = \rho_{2k,0}(e^{i\theta}T) = \pi_{2k}(T),$$

$$\rho_{2k+1}(e^{i\theta}T) = \rho_{2k+1,1}(e^{i\theta}T) = e^{i\theta}\pi_{2k+1}(T),$$

then

$$\rho_{2k,2j}(V) = (\det V)^j \rho_{2k}(V), \qquad \rho_{2k+1,2j+1}(V) = (\det V)^j \rho_{2k+1}(V).$$

We conclude by giving the decomposition of the natural representation $\lambda$ of $U(2)$ on $L^2$ of the unit sphere in $\mathbb{C}^2$, namely $[\lambda(V)f](z,w) = f(V^{-1}(z,w))$. In fact, this is nothing but the decomposition $L^2(S^3) = \bigoplus_{p,q=0}^\infty \mathcal{H}_{p,q}$ that we obtained at the end of §5.4.1. We simply observe that if $P$ is any homogeneous polynomial of bidegree $(p,q)$ then

$$[\lambda(e^{i\theta}I)P](z,w) = P(e^{-i\theta}z, e^{-i\theta}w) = e^{i(q-p)\theta}P(z,w).$$

It then follows that $\mathcal{H}_{p,q}$ is $U(2)$-invariant and that $\lambda^{\mathcal{H}_{p,q}}$ is equivalent to $\rho_{q+p,q-p}$. Thus $\lambda$ is equivalent to $\bigoplus_{p,q=0}^\infty \rho_{q+p,q-p}$, each summand occurring with multiplicity one.

## 5.5  Notes and References

The theory in §5.2 was first developed in the classic paper of Peter and Weyl [112]. That paper deals with compact Lie groups, but everything in it applies without change to general compact groups once one has the existence of Haar measure. An extensive treatment of various topics in Fourier analysis on general compact groups can be found in Hewitt and Ross [67].

There is an analogue of the Pontrjagin duality theorem for non-Abelian compact groups, which we now describe. Suppose $G$ is a compact group. Let $\mathcal{V}$ be the smallest set of Hilbert spaces containing $\mathbb{C}^n$ for all positive integers $n$ and closed under formation of finite direct sums, tensor products (see Appendix 3), and duals, and let $\mathcal{R}$ be the set of all unitary representations of $G$ on spaces in $\mathcal{V}$. (We do not identify isomorphic spaces in $\mathcal{V}$, nor do we identify equivalent representations in $\mathcal{R}$. $\mathcal{R}$ is, in essence, the set of all finite-dimensional representations of $G$; the purpose of $\mathcal{V}$ is to make this a genuine set.) Let $\Gamma$ be the set of all maps $\gamma$ that assign to each $\pi \in \mathcal{R}$ a unitary operator $\gamma(\pi)$ on $\mathcal{H}_\pi$, such that the following properties hold for all $\pi, \pi_1, \pi_2 \in \mathcal{R}$:

   i.  If $U : \mathcal{H}_{\pi_1} \to \mathcal{H}_{\pi_2}$ is a unitary equivalence of $\pi_1$ and $\pi_2$, then $\gamma(\pi_2) = U\gamma(\pi_1)U^{-1}$.

   ii.  $\gamma(\pi_1 \oplus \pi_2) = \gamma(\pi_1) \oplus \gamma(\pi_2)$.

   iii.  $\gamma(\pi_1 \otimes \pi_2) = \gamma(\pi_1) \otimes \gamma(\pi_2)$, where $\pi_1 \otimes \pi_2$ is the inner tensor product of $\pi_1$ and $\pi_2$ (see §7.3).

   iv.  $\gamma(\overline{\pi})$ is the inverse transpose of $\gamma(\pi)$, acting on $\mathcal{H}_\pi^*$.

Each $x \in G$ defines a $\gamma_x \in \Gamma$ by $\gamma_x(\pi) = \pi(x)$. Define a multiplication on $\Gamma$ by $(\gamma_1\gamma_2)(\pi) = \gamma_1(\pi)\gamma_2(\pi)$, and impose on $\Gamma$ the weakest topology that makes the maps $\gamma \mapsto \langle \gamma(\pi)u, v \rangle$ ($\pi \in \mathcal{R}$, $u, v \in \mathcal{H}_\pi$) continuous. Then:

**5.53 Theorem** (The Tannaka Duality Theorem). *$\Gamma$ is a compact group, and the map $x \mapsto \gamma_x$ is an isomorphism from $G$ to $\Gamma$.*

For the proof, as well as another formulation of the theorem due to Krein, we refer the reader to Hewitt and Ross [67].

There is much more one can say about the representation theory and Fourier analysis of compact connected Lie groups, because these groups and their irreducible representations are known quite explicitly. To begin with, the simply connected compact Lie groups are the finite products of groups from the following list:

- the double covers of the rotation groups $SO(n)$ ($n \geq 3$), usually called Spin($n$);

- the special unitary groups $SU(n)$ $(n \geq 2)$;

- the quaternionic unitary groups $Sp(n)$ $(n \geq 1)$;

- the five exceptional simply connected compact groups $E_6$, $E_7$, $E_8$, $F_4$, and $G_2$.

The most general connected compact Lie group is then isomorphic to $(G \times \mathbb{T}^n)/Z$, where $G$ is simply connected, $n \geq 0$ (in particular, either factor $G$ or $\mathbb{T}^n$ may be missing), and $Z$ is a finite subgroup of the center of $G \times \mathbb{T}^n$. Since the centers of all the groups in the above list are all known explicitly (and are all of order $\leq 4$ except for $SU(n)$), one has a rather complete picture of the connected compact Lie groups. See Helgason [64] and Simon [129]. Moreover, the theory of highest weights provides a concrete description of $\widehat{G}$ when $G$ is a compact connected Lie group. See Bröcker and tom Dieck [20], Knapp [79], and Simon [129]. See also Boerner [16] and Weyl [147] for more information about the representations of the classical matrix groups, including their connections with Young diagrams and the representations of permutation groups.

The theory of Fourier series on compact connected Lie groups has many connections with classical Fourier analysis and special functions. See Stanton and Tomas [130] and the references given there for studies of convergence and summability of Fourier series on compact Lie groups, and Coifman and Weiss [25] for an extensive discussion of analysis on $SU(2)$.

# 6

## Induced Representations

The inducing construction is a way of manufacturing a unitary representation of a locally compact group $G$ out of a unitary representation of a closed subgroup $H$. Geometrically speaking, these induced representations are the unitary representations of $G$ arising from the action of $G$ on functions or sections of homogeneous vector bundles on the homogeneous space $G/H$.

Unless $G/H$ is a finite set, the representations of $G$ induced from $H$ are always infinite-dimensional. In particular, when $G$ is compact or Abelian, they are highly reducible. On the other hand, induced representations are the single most important source of *irreducible* representations of noncompact, non-Abelian groups. Indeed, for many such groups, the inducing construction (with certain generalizations and modifications) suffices to provide a list of irreducible representations that is complete up to unitary equivalence. One way in which this happens is via the "Mackey machine," a body of techniques for analyzing representations of a group $G$ in terms of the representations of a normal subgroup $N$ and the representations of various subgroups of $G/N$. However, induced representations are also important for groups that have few if any normal subgroups, such as the semisimple Lie groups.

In §6.1 we explain the construction of induced representations. §6.2 is devoted to the Frobenius reciprocity theorem for compact groups, which provides a powerful tool for finding the irreducible decomposition of an induced representation of a compact group. In §§6.3–5 we develop the notion of pseudomeasures of positive type (a generalization of functions of positive type) and use it to prove the theorem on induction in stages and the imprimitivity theorem, which is the deepest result of the chapter and forms the basis for the Mackey machine. In §6.6 we work out the simplest case of the Mackey machine, that of regular semi-direct products, and in §§6.7–8 we present some concrete examples.

## 6.1   The Inducing Construction

Let $G$ be a locally compact group, $H$ a closed subgroup, $q : G \to G/H$ the canonical quotient map, and $\sigma$ a unitary representation of $H$ on $\mathcal{H}_\sigma$. We denote the norm and inner product on $\mathcal{H}_\sigma$ by $\|u\|_\sigma$ and $\langle u, v \rangle_\sigma$, and we denote by $C(G, \mathcal{H}_\sigma)$ the space of continuous functions from $G$ to $\mathcal{H}_\sigma$. If $f \in C(G, \mathcal{H}_\sigma)$, we shall frequently wish to apply the operators $\sigma(\xi)$ to the values $f(x)$, and to avoid clutter we shall usually write $\sigma(\xi)f(x)$ instead of the more precise $\sigma(\xi)[f(x)]$.

The main ingredient in the inducing construction is the following space of vector-valued functions:

$$\mathcal{F}_0 = \{ f \in C(G, \mathcal{H}_\sigma) :$$
$$q(\operatorname{supp} f) \text{ is compact and } f(x\xi) = \sigma(\xi^{-1})f(x) \text{ for } x \in G, \ \xi \in H \}.$$

Here is how to produce functions in $\mathcal{F}_0$:

**6.1 Proposition.** *If $\alpha : G \to \mathcal{H}_\sigma$ is continuous with compact support, then the function*

$$f_\alpha(x) = \int_H \sigma(\eta)\alpha(x\eta) \, d\eta$$

*belongs to $\mathcal{F}_0$ and is left uniformly continuous on $G$. Moreover, every element of $\mathcal{F}_0$ is of the form $f_\alpha$ for some $\alpha \in C_c(G, \mathcal{H}_\sigma)$.*

*Proof.* Clearly $q(\operatorname{supp} f_\alpha) \subset q(\operatorname{supp} \alpha)$, and

$$f_\alpha(x\xi) = \int_H \sigma(\eta)\alpha(x\xi\eta) \, d\eta = \int_H \sigma(\xi^{-1}\eta)\alpha(x\eta) \, d\eta = \sigma(\xi^{-1})f_\alpha(x).$$

Hence, to prove the first assertion it remains to show that $f_\alpha$ is left uniformly continuous. Fix a compact neighborhood $N$ of 1 in $G$, let $K$ be a compact subset of $G$ such that $q(K) = q(\operatorname{supp} \alpha)$ (Lemma 2.48), and let $J = K^{-1}N^{-1}(\operatorname{supp} \alpha) \cap H$, a compact subset of $H$. Given $\epsilon > 0$, by Proposition 2.6 there is a neighborhood $N_\epsilon \subset N$ of 1 such that $\|\alpha(yx) - \alpha(x)\|_\sigma < \epsilon$ for $x \in G$ and $y \in N_\epsilon$. Then for $x \in K$ and $y \in N_\epsilon$,

$$\|f_\alpha(yx) - f_\alpha(x)\|_\sigma = \left\| \int_J \sigma(\eta)[\alpha(yx\eta) - \alpha(x\eta)] \, d\eta \right\|_\sigma \leq \epsilon |J|.$$

This remains true for $x \in KH$ because $f_\alpha(yx\xi) - f_\alpha(x\xi) = \sigma(\xi^{-1})[f_\alpha(yx) - f_\alpha(x)]$ for $\xi \in H$. Since $f_\alpha = 0$ outside $KH$, $f_\alpha$ is left uniformly continuous on $G$.

On the other hand, if $f \in \mathcal{F}_0$, by Lemma 2.49 there exists $\psi \in C_c(G)$ such that $\int_H \psi(x\eta) \, d\eta = 1$ for $x \in \operatorname{supp} f$. Let $\alpha = \psi f$; then

$$f_\alpha(x) = \int_H \psi(x\eta)\sigma(\eta)f(x\eta) \, d\eta = \int_H \psi(x\eta)f(x) \, d\eta = f(x),$$

so $f = f_\alpha$. $\qquad\square$

$G$ acts on $\mathcal{F}_0$ by left translation, $f \mapsto L_x f$, so we obtain a unitary representation of $G$ if we can impose an inner product on $\mathcal{F}_0$ with respect to which these translations are isometries. When $G/H$ admits an invariant measure $\mu$ (necessarily unique up to scalar multiples, by Theorem 2.51), this is easy. If $f, g \in \mathcal{F}_0$, $\langle f(x), g(x) \rangle_\sigma$ depends only on the coset $q(x)$ of $x$ since $\sigma$ is unitary, so it defines a function in $C_c(G/H)$ that can be integrated with respect to $\mu$, and we set

$$\langle f, g \rangle = \int_{G/H} \langle f(x), g(x) \rangle_\sigma \, d\mu(xH).$$

This is an inner product on $\mathcal{F}_0$ (it is positive definite by Proposition 2.60), and it is preserved by left translations since $\mu$ is invariant. Hence, if we denote by $\mathcal{F}$ the Hilbert space completion of $\mathcal{F}_0$, the translation operators $L_x$ extend to unitary operators on $\mathcal{F}$. It follows easily from Proposition 6.1 that $x \mapsto L_x f$ is continuous from $G$ to $\mathcal{F}$ for each $f \in \mathcal{F}_0$; and then since the operators $L_x$ are uniformly bounded, they are strongly continuous on all of $\mathcal{F}$. Hence they define a unitary representation of $G$, called the **representation induced by** $\sigma$ and denoted by $\operatorname{ind}_H^G(\sigma)$.

For example, let $\sigma$ be the trivial representation of $H$ on $\mathbb{C}$. Then $\mathcal{F}_0$ consists of functions on $G$ that are constant on cosets of $H$, so $\mathcal{F}_0$ can be naturally identified with $C_c(G/H)$. The same identification makes $\mathcal{F}$ into $L^2(G/H)$, and $\operatorname{ind}_H^G(\sigma)$ is then just the natural representation of $G$ on $L^2(G/H)$ by left translations. In the general case, a similar interpretation of the elements of $\mathcal{F}$ as objects living on $G/H$ rather than $G$ is available; we shall explain it at the end of this section.

When $G/H$ has no $G$-invariant measure, there are two ways one can modify the preceding construction to obtain a unitary representation of $G$: one can replace the invariant measure $\mu$ above by a quasi-invariant measure in defining the Hilbert space $\mathcal{F}$ and modify the action of $G$ to make it unitary, or one can modify the construction of $\mathcal{F}$ by adding a "twist" to $\mathcal{F}_0$. We shall present both of these ways, as they each have advantages. The first is a little more straightforward, but the second has the pleasant feature of not relying on the choice of a quasi-invariant measure, and it is the one we shall use in the rest of the chapter.

Here is the first construction. Let $\mu$ be a strongly quasi-invariant measure on $G/H$. For $f, g \in \mathcal{F}_0$, we define

$$(6.2) \qquad \langle f, g \rangle_\mu = \int_{G/H} \langle f(x), g(x) \rangle_\sigma \, d\mu(xH).$$

As above, this is an inner product on $\mathcal{F}_0$, so by completing $\mathcal{F}_0$ with respect to it we obtain a Hilbert space $\mathcal{F}_\mu$. Let $\rho$ be the rho-function associated to $\mu$, and let

$$\lambda(x, yH) = \frac{\rho(xy)}{\rho(y)} = \frac{d\mu_x}{d\mu}(yH) \qquad (x, y \in G)$$

(see Theorems 2.58 and 2.61). We define operators $\Pi_\mu(x)$ on $\mathcal{F}_0$ for $x \in G$ by

$$[\Pi_\mu(x)f](y) = \sqrt{\lambda(x^{-1}, yH)} \, f(x^{-1}y).$$

Then

$$\langle \Pi_\mu(x)f, \Pi_\mu(x)g \rangle_\mu = \int \lambda(x^{-1}, yH) \langle f(x^{-1}y), g(x^{-1}y) \rangle_\sigma \, d\mu(yH)$$

$$= \int \langle f(y), g(y) \rangle_\sigma \, d\mu(yH) = \langle f, g \rangle_\mu,$$

so $\Pi_\mu(x)$ extends to an isometry of $\mathcal{F}_\mu$ which is actually unitary since $\Pi_\mu(\mathcal{F}_0) = \mathcal{F}_0$. It follows from Proposition 6.1, as in the case of an invariant measure, that $\Pi_\mu(x)$ is strongly continuous in $x$ on $\mathcal{F}_0$ and hence on $\mathcal{F}_\mu$, so $\Pi_\mu$ is a unitary representation of $G$.

   $\Pi_\mu$ depends on the choice of $\mu$, but its unitary equivalence class does not. If $\mu'$ is another strongly quasi-invariant measure, and $\rho$ and $\rho'$ are the rho-functions associated to $\mu$ and $\mu'$, then $d\mu' = \phi \, d\mu$ where $\phi(xH) = \rho'(x)/\rho(x)$, by Theorem 2.61 and its proof. The map $f \mapsto \sqrt{\rho'/\rho} \, f$ is a bijection of $\mathcal{F}_0$ (here we are using the fact that $\rho'$ and $\rho$ are continuous!); it extends to a unitary isomorphism from $\mathcal{F}_\nu$ to $\mathcal{F}_\mu$ since $\int \phi |f|^2 \, d\mu = \int |f|^2 \, d\nu$, and we have

$$[\Pi_\mu(x)(\sqrt{\rho'/\rho} \, f)](y) = \sqrt{\frac{\rho'(x^{-1}y)}{\rho(y)}} \, f(x^{-1}y) = [\sqrt{\rho'/\rho} \, \Pi_{\mu'}(x)f](y).$$

Hence $f \mapsto \sqrt{\rho'/\rho} \, f$ is a unitary equivalence of $\Pi_{\mu'}$ and $\Pi_\mu$. Any of the representations $\Pi_\mu$ may be called the representation of $G$ **induced** by the representation $\sigma$ of $H$ and denoted by $\mathrm{ind}_H^G(\sigma)$.

   *Remark 1.* Our construction of $\mathcal{F}_\mu$ as the completion of a space of continuous functions finesses most technical problems associated with the study of measurable vector-valued functions, but it is occasionally important to note that $\mathcal{F}_\mu$ can be identified with a space of $\mathcal{H}_\sigma$-valued functions on $G$, in which two functions are identified if they are equal locally a.e. Namely, if $f \in \mathcal{F}_\mu$, pick a sequence $\{f_n\}_0^\infty$ in $\mathcal{F}_0$ such that $\|f_n - f\| < C2^{-\epsilon n}$ for some $C, \epsilon > 0$. Then $\sum_1^\infty \|f_n - f_{n-1}\| < 2C \sum_1^\infty 2^{-\epsilon n} < \infty$, hence $\sum_1^\infty \|f_n(x) - f_{n-1}(x)\|_\sigma < \infty$ for every $x$ except those in a set of the form $q^{-1}(E)$ where $\mu(E) = 0$. But then the series $f_0(x) + \sum_1^\infty [f_n(x) - f_{n-1}(x)]$ converges in $\mathcal{H}_\sigma$ to an element $f(x)$ for all $x \notin q^{-1}(E)$. (If it converges for some $x_0$, it does so for all $x$ in the coset $x_0 H$.) Moreover, $f(x)$ is independent of the choice of sequence $\{f_n\}$, since any two such sequences can be combined to yield a third one; and $q^{-1}(E)$ is locally null by Theorem 2.66. Hence we can identify $f$ with the locally a.e.-defined function $x \mapsto f(x)$, and the formula (6.2) for the inner product continues to hold for such functions. Moreover, if $\phi \in C_c(G)$, the integral $\int \phi(x) f(x)\, dx$ makes sense: it can be defined as $\lim \int \phi(x) f_n(x)\, dx$ with $f_n$ as above.

*Remark 2.* Suppose there is a Borel cross-section for $G/H$ in $G$, i.e., a Borel set $M \subset G$ that meets each coset of $H$ in exactly one point. (This is always the case if $G$ is second countable; see §2.7.) Then each $x \in G$ can be written uniquely as $x = x_M x_H$ with $x_M \in M$ and $x_H \in H$; each $f \in \mathcal{F}_0$ is completely determined by its restriction $\tilde{f} = f|M$; and a quasi-invariant measure $\mu$ on $G/H$ yields a measure $\tilde\mu$ on $M$ by $\tilde\mu(E) = \mu(q(E))$. The map $f \mapsto \tilde{f}$ gives a unitary indentification of $\mathcal{F}_\mu$ with the space of $\mathcal{H}_\sigma$-valued functions on $M$ that are square-integrable with respect to $\tilde\mu$, and under this identification the representation $\Pi_\mu$ turns into

$$[\tilde\Pi_\mu(x)]\tilde{f}(y) = \sqrt{\lambda(x^{-1}, yH)}\, \sigma((x^{-1}y)_H)^{-1}\tilde{f}((x^{-1}y)_M).$$

This gives another realization of induced representations that is often useful.

We now give the second, "intrinsic" construction of the induced representation, which starts with a modification of the space $\mathcal{F}_0$. Namely, let $\mathcal{F}^0$ be the space of continuous functions $f : G \to \mathcal{H}_\sigma$ such that $q(\operatorname{supp} f)$ is compact and

$$(6.3) \qquad f(x\xi) = \sqrt{\frac{\Delta_H(\xi)}{\Delta_G(\xi)}}\, \sigma(\xi^{-1}) f(x) \qquad (x \in G,\ \xi \in H),$$

where $\Delta_G$ and $\Delta_H$ are the modular functions of $G$ and $H$.

**6.4 Proposition.** *If $\alpha : G \to \mathcal{H}_\sigma$ is continuous with compact support, then the function*

$$(6.5) \qquad f_\alpha(x) = \int_H \sqrt{\frac{\Delta_G(\eta)}{\Delta_H(\eta)}}\, \sigma(\eta)\alpha(x\eta)\, d\eta$$

*belongs to $\mathcal{F}_0$ and is left uniformly continuous on $G$. Moreover, every element of $\mathcal{F}_0$ is of the form $f_\alpha$ for some $\alpha \in C_c(G, \mathcal{H}_\sigma)$.*

*Proof.* Essentially identical to the proof of Proposition 6.1.     □

We recall the canonical map $P : C_c(G) \to C_c(G/H)$ defined in §2.6:

$$P\phi(xH) = \int_H \phi(x\xi)\,d\xi.$$

If $f \in \mathcal{F}^0$, then $x \mapsto \|f(x)\|_\sigma^2$ satisfies all the requirements for a rho-function on $G$ except that it may not be strictly positive. Positivity is not needed, however, in the proof of Lemma 2.57, so that lemma together with Proposition 2.50 implies that

$$P\phi \mapsto \int_G \phi(x)\|f(x)\|_\sigma^2\,dx \qquad (\phi \in C_c(G))$$

is a well-defined positive linear functional on $C_c(G/H)$. Hence there is a Radon measure $\mu_f$ on $G/H$ such that $\int P\phi\,d\mu_f = \int \phi\|f\|_\sigma^2$ for all $\phi \in C_c(G)$. Since $\int \phi\|f\|_\sigma^2 = 0$ if $\operatorname{supp}\phi \cap \operatorname{supp} f = \varnothing$, $\operatorname{supp}\mu_f$ is contained in $q(\operatorname{supp} f)$ and hence is compact. In particular, $\mu_f(G/H) < \infty$.

It now follows by polarization (see Appendix 1) that if $f, g \in \mathcal{F}^0$, there is a complex Radon measure $\mu_{f,g}$ on $G/H$ such that

$$\int_{G/H} P\phi\,d\mu_{f,g} = \int_G \phi(x)\langle f(x), g(x)\rangle_\sigma\,dx \qquad (\phi \in C_c(G)),$$

namely $\mu_{f,g} = \frac{1}{4}(\mu_{f+g} - \mu_{f-g} + i\mu_{f+ig} - i\mu_{f-ig})$. We define

$$\langle f, g\rangle = \mu_{f,g}(G/H).$$

It is an easy exercise to verify that $(f, g) \mapsto \langle f, g\rangle$ is an inner product on $\mathcal{F}^0$; we denote by $\mathcal{F}$ the Hilbert space completion of $\mathcal{F}^0$. For future reference we note that the norm of $f \in \mathcal{F}^0$ may be computed as follows: by Lemma 2.49 there exists $\phi \in C_c(G)$ such that $P\phi = 1$ on $q(\operatorname{supp} f)$; then

$$(6.6) \qquad \|f\|^2 = \int_G \phi(x)\|f(x)\|_\sigma^2\,dx.$$

For $x \in G$ we define the operator $\Pi(x)$ on $\mathcal{F}^0$ by

$$[\Pi(x)f](y) = f(x^{-1}y).$$

Then $\Pi(x)$ is bijective on $\mathcal{F}^0$, and since the map $P$ commutes with left translations,

$$
\int_{G/H} P\phi(p)\, d\mu_{\Pi(x)f}(p) = \int_G \|f(x^{-1}y)\|_\sigma^2 \phi(y)\, dy
$$

$$
= \int_G \|f(y)\|_\sigma^2 L_{x^{-1}}\phi(y)\, dy = \int_G P(L_{x^{-1}}\phi)\, d\mu_f = \int_G P\phi(xp)\, d\mu_f(p).
$$

Thus $\mu_{\Pi(x)f}$ is the translate of $\mu_f$ by $x$, so

$$
\|\Pi(x)f\|^2 = \mu_{\Pi(x)f}(G/H) = \mu_f(x^{-1}(G/H)) = \mu_f(G/H) = \|f\|^2.
$$

In short, $\Pi(x)$ is an isometry, so it extends to a unitary operator on $\mathcal{F}$. The resulting unitary representation of $G$ on $\mathcal{F}$ is the **intrinsic version of the induced representation** $\mathrm{ind}_H^G(\sigma)$. (Note that if $G/H$ has an invariant measure $\mu$, then by Theorem 2.51, $\mathcal{F}^0 = \mathcal{F}_0$ and $d\mu_f(xH) = \|f(x)\|_\sigma^2\, d\mu(xH)$, so the present definitions of $\mathcal{F}$ and the induced representation coincide with the ones given earlier.)

To complete the picture, we show that $\Pi$ is unitarily equivalent to the representation $\Pi_\mu$ defined in terms of a strongly quasi-invariant measure $\mu$. Let $\rho$ be the rho-function associated to $\mu$. Then $f \mapsto \sqrt{\rho}\, f$ maps $\mathcal{F}_0$ onto $\mathcal{F}^0$; we claim that this map extends to a unitary map from $\mathcal{F}_\mu$ to $\mathcal{F}$ that intertwines $\Pi_\mu$ and $\Pi$.

Suppose $f \in \mathcal{F}_0$. By Lemma 2.49, we can choose $\Psi \in C_c(G)$ such that $P\Psi = 1$ on a neighborhood of $q(\mathrm{supp}\, f)$, and we can then choose $\phi \in C_c(G/H)$ such that $\phi = 1$ on $q(\mathrm{supp}\, f)$ and $P\Psi = 1$ on $\mathrm{supp}\,\phi$. Let

$$
\Phi(x) = \frac{\phi(q(x))\Psi(x)}{P\Psi(q(x))}.
$$

Then $\Phi \in C_c(G)$, $P\Phi = \phi$, and $P[\Phi\|f(\cdot)\|_\sigma^2] = \|f(\cdot)\|_\sigma^2$ (the quantity on the right being regarded as a function on $G/H$). Hence,

$$
\|f\|_{\mathcal{F}_\mu}^2 = \int_{G/H} \|f(x)\|_\sigma^2\, d\mu(xH) = \int_G \Phi(x)\|f(x)\|_\sigma^2 \rho(x)\, dx
$$

$$
= \int_{G/H} \phi\, d\mu_{\sqrt{\rho}\, f} = \|\sqrt{\rho}\, f\|_{\mathcal{F}}^2,
$$

since $\phi = 1$ on $\mathrm{supp}\,\mu_{\sqrt{\rho}\, f}$. Hence $f \mapsto \sqrt{\rho}\, f$ extends to a unitary map from $\mathcal{F}_\mu$ to $\mathcal{F}$. Finally,

$$
[\Pi(x)(\sqrt{\rho}\, f)](y) = \sqrt{\rho(x^{-1}y)}\, f(x^{-1}y)
$$

$$
= \sqrt{\rho(y)}\sqrt{\frac{\rho(x^{-1}y)}{\rho(y)}}\, f(x^{-1}y) = \sqrt{\rho(y)}[\Pi_\mu(x)f](y),
$$

so $f \mapsto \sqrt{\rho}\, f$ intertwines $\Pi_\mu$ and $\Pi$.

In view of the remark following the construction of $\Pi_\mu$, the space $\mathcal{F}$ (defined as the completion of $\mathcal{F}^0$) can be identified with a space of $\mathcal{H}_\sigma$-valued functions on $G$, in which two functions are identified if they are equal locally a.e. — namely, $\mathcal{F} = \{\sqrt{\rho}\, f : f \in \mathcal{F}_\mu\}$ — and the integral $\int \phi(x) f(x)\, dx$ is well defined for $\phi \in C_c(G)$ and $f \in \mathcal{F}$.

For later use, we now derive some facts about a useful subset of $\mathcal{F}^0$, namely the set of functions $f_\alpha$ obtained by taking $\alpha$ to be of the form $\alpha(x) = \phi(x)v$ ($\phi \in C_c(G)$, $v \in \mathcal{H}_\sigma$) in (6.5). That is, if $\phi \in C_c(G)$ and $v \in \mathcal{H}_\sigma$, we define

$$(6.7) \qquad f_{\phi,v}(x) = \int_H \sqrt{\frac{\Delta_G(\eta)}{\Delta_H(\eta)}}\, \phi(x\eta)\sigma(\eta)v\, d\eta$$

**6.8 Proposition.** *Let $\mathcal{D}$ be a dense subspace of $\mathcal{H}_\sigma$.*
  a. $\{f_{\phi,v}(x_0) : \phi \in C_c(G),\ v \in \mathcal{D}\}$ *is dense in $\mathcal{H}_\sigma$ for all $x_0 \in G$.*
  b. *The linear span of $\{f_{\phi,v} : \phi \in C_c(G),\ v \in \mathcal{D}\}$ is dense in $\mathcal{F}$.*

*Proof.* By the Tietze extension theorem, there is a family $\{\psi_U\}$ in $C_c(G)$ such that $\{\psi_U | H\}$ is an approximate identity on $H$. If we then set $\phi_U(x) = \psi_U(x_0^{-1}x)$, it is easily verified that $f_{\phi_U,v}(x_0) \to v$ as $U \to \{1\}$. This proves (a).

To deal with (b), we first observe that for any compact $K \subset G$ there is a constant $C_K > 0$ such that $\sup_{x\in G} \|f_\alpha(x)\|_\sigma \leq C_K \sup_{x\in G} \|\alpha(x)\|_\sigma$ for all $\alpha \in C_c(G, \mathcal{H}_\sigma)$ supported in $K$, where $f_\alpha$ is given by (6.5). Indeed, if $x \in K$, the integral over $H$ in (6.5) can be replaced by the integral over the compact set $H \cap (K^{-1}K)$, which yields $\|f_\alpha(x)\|_\sigma \leq C_K \sup_{y\in G} \|\alpha(y)\|_\sigma$ for $x \in K$. But then the same estimate holds for $x \in KH$ since $f_\alpha(x\xi) = \sigma(\xi^{-1})f_\alpha(x)$, and $f_\alpha(x) = 0$ for $x \notin KH$.

Now, to prove (b) it suffices to show that any $f \in \mathcal{F}^0$ can be approximated by linear combinations of $f_{\phi,v}$'s, and by Proposition 6.4 and the above remarks it is enough to show that any $\alpha \in C_c(G, \mathcal{H}_\sigma)$ can be uniformly approximated by functions of the form $\beta(x) = \sum_1^n \phi_j(x)v_j$ where $v_j \in \mathcal{D}$ and $\phi_j \in C_c(G)$ is supported in a fixed compact neighborhood $N$ of $\operatorname{supp}\alpha$ for each $j$. But this is easy: given $\epsilon > 0$, a standard partition of unity argument shows that there exist continuous $\phi_1, \ldots, \phi_n \in C_c(G)$ supported in $N$ such that $0 \leq \sum_1^n \phi_j \leq 1$, $\sum_1^n \phi_j = 1$ on $\operatorname{supp}\alpha$, and $\|\alpha(x) - \alpha(y)\|_\sigma < \frac{1}{2}\epsilon$ for $x, y \in \operatorname{supp}\phi_j$. For each $j$, pick $x_j \in \operatorname{supp}\phi_j$ and pick $v_j \in \mathcal{D}$ such that $\|\alpha(x_j) - v_j\|_\sigma < \frac{1}{2}\epsilon$. If $\beta(x) = \sum_1^n \phi_j(x)v_j$, then

$$\sup_x \|\alpha(x) - \beta(x)\|_\sigma = \sup_x \left\| \sum \phi_j(x)[\alpha(x) - v_j] \right\|_\sigma < \sup_x \sum \phi_j(x)\epsilon = \epsilon,$$

so we are done. $\qquad\qquad\square$

For future reference we record the following simple but important observations, whose proof we leave as an exercise for the reader.

**6.9 Proposition.** *If $\sigma$ and $\sigma'$ are equivalent representations of $H$, then $\mathrm{ind}_H^G(\sigma)$ and $\mathrm{ind}_H^G(\sigma')$ are equivalent representations of $G$. If $\{\sigma_i\}$ is any family of representations of $H$, then $\mathrm{ind}_H^G(\bigoplus \sigma_i)$ is equivalent to $\bigoplus \mathrm{ind}_H^G(\sigma_i)$.*

We conclude this section by giving a more geometric interpretation of the space $\mathcal{F}_0$. This interpretation will not play much of a role here, but it is of great importance in applications because it is the way in which many induced representations arise naturally in practice.

Define an equivalence relation on $G \times \mathcal{H}_\sigma$ by

$$(x, u) \sim (x\xi, \sigma(\xi^{-1})u) \text{ for all } \xi \in H,$$

and let $V$ be the set of equivalence classes. The projection $(x, u) \mapsto x$ of $G \times \mathcal{H}_\sigma$ onto $G$ induces a projection $p : V \to G/H$. For each $\overline{x} \in G/H$, let $V_{\overline{x}} = p^{-1}(\{\overline{x}\})$ be the fiber over $\overline{x}$. Each $V_{\overline{x}}$ has a natural structure as a Hilbert space isomorphic to $\mathcal{H}_\sigma$. Namely, if we fix $x \in q^{-1}(\overline{x})$, the map taking $u \in \mathcal{H}_\sigma$ to the equivalence class of $(x, u)$ is a bijection from $\mathcal{H}_\sigma$ to $V_{\overline{x}}$, by means of which we can transfer the Hilbert space structure of $\mathcal{H}_\sigma$ to $V_{\overline{x}}$. If we choose a different $x$, say $x' = x\xi$, we get the same Hilbert space structure on $V_{\overline{x}}$ since $\sigma(\xi)$ is unitary. Moreover, if $f \in \mathcal{F}_0$ we have $(x, f(x)) \sim (x\xi, f(x\xi))$ for all $\xi \subset H$, so $f$ can be regarded as a map from $G/H$ to $V$ whose value at any $\overline{x} \in G/H$, namely the equivalence class of $(x, f(x))$ for $x \in q^{-1}(\overline{x})$, lies in the fiber over $\overline{x}$.

Therefore, $V$ is a vector bundle over $G/H$ and $\mathcal{F}_0$ can be regarded as the space of compactly supported continuous sections of $V$. (We are glossing over one technicality, the question of local triviality, because it plays no role here; but see §6.9.) Conversely, suppose $V$ is a homogeneous Hermitian vector bundle over $G/H$: this means that the fibers are Hilbert spaces and that the action of $G$ on $G/H$ lifts to an action of $G$ as bundle automorphisms of $V$ that are unitary on the fibers. The latter action, restricted to $H$, preserves the fiber $V_0$ over the identity coset and defines a unitary representation $\sigma$ of $H$ on $V_0$. The preceding calculations, done in reverse, show that the continuous sections of $V$ can be identified with continuous mappings $f : G \to V_0$ such that $f(x\xi) = \sigma(\xi^{-1})f(x)$, and hence that the set of compactly supported ones can be identified with the space $\mathcal{F}_0$. Finally, if $\mu$ is a strongly quasi-invariant measure on $G/H$, $\mathcal{F}_\mu$ can be identified with the space of sections of $V$ that are square-integrable with respect to $\mu$.

Replacing $\mathcal{F}_0$ by $\mathcal{F}^0$ amounts to taking the tensor product of $V$ with the bundle of "half-densities" on $G/H$, that is, the line bundle associated to the one-dimensional (non-unitary) representation $\rho(\xi) =$

$\sqrt{\Delta_G(\xi)/\Delta_H(\xi)}$ of $H$. When $G/H$ admits an invariant measure, this is the trivial line bundle; otherwise, it precisely compensates for the absence of an invariant measure.

In short: *representations of $G$ induced from $H$ are the representations of $G$ on sections of homogeneous Hermitian vector bundles over $G/H$.* For example, if $G$ is a Lie group and $H$ is a compact subgroup, then $H$ acts linearly on the tangent space $V_0$ to the identity coset in $G/H$, and since $H$ is compact there is an inner product on $V_0$ that is $H$-invariant. This inner product can be transported to all the other tangent spaces by the action of $G$, thereby making the tangent bundle of $G$, or rather its complexification, into a homogeneous Hermitian vector bundle. The invariant inner product defines a unitary representation $\sigma$ of $H$ on the complexification of $V_0$, and $\mathrm{ind}_H^G(\sigma)$ is the natural representation of $G$ on the complex vector fields on $G/H$.

## 6.2    The Frobenius Reciprocity Theorem

Suppose $G$ is a locally compact group and $H$ is a closed subgroup. Any unitary representation of $G$ can be restricted to $H$, and any unitary representation of $H$ can be induced up to $G$. There is a remarkable relationship between these procedures, which for compact groups is embodied in the following theorem.

**6.10 Theorem** (The Frobenius Reciprocity Theorem). *Let $G$ be a compact group, $H$ a closed subgroup, $\pi$ an irreducible unitary representation of $G$, and $\sigma$ an irreducible unitary representation of $H$. Then*

$$\mathcal{C}(\pi, \mathrm{ind}_H^G(\sigma)) \cong \mathcal{C}(\pi|H, \sigma) \quad and \quad \mathrm{mult}(\pi, \mathrm{ind}_H^G(\sigma)) = \mathrm{mult}(\sigma, \pi|H).$$

*Proof.* It suffices to prove the first assertion, as the second one then follows from Proposition 5.4 and the fact that the map $S \mapsto S^*$ is an antilinear isomorphism from $\mathcal{C}(\pi|H, \sigma)$ to $\mathcal{C}(\sigma, \pi|H)$.

Let $\Pi = \mathrm{ind}_H^G(\sigma)$. Since $G/H$ admits an invariant measure (because $\Delta_G = \Delta_H = 1$) and $d_\sigma = \dim \mathcal{H}_\sigma$ is finite (Theorem 5.2), the simplest construction of $\Pi$ will work. Namely, let $L^2(G, \mathcal{H}_\sigma)$ be the space of square-integrable $\mathcal{H}_\sigma$-valued functions on $G$. The space $\mathcal{F}$ on which $\Pi$ acts is the subspace of $L^2(G, \mathcal{H}_\sigma)$ consisting of those $f$ such that $f(x\xi) = \sigma(\xi^{-1})f(x)$ for $x \in G$ and $\xi \in H$, and $\Pi$ is a subrepresentation of the left regular representation of $G$ on $L^2(G, \mathcal{H}_\sigma)$. Moreover, the latter is just the direct sum of $d_\sigma$ copies of the left regular representation of $G$ on $L^2(G)$. If $T \in \mathcal{C}(\pi, \Pi)$, then, the range of $T$ lies in the direct sum of

$d_\sigma$ copies of $\mathcal{E}_{\overline{\pi}}$ by the Peter-Weyl theorem, and the elements of $\mathcal{E}_{\overline{\pi}}$ are all continuous. Hence, if $f$ is in the range of $T$ it makes sense to evaluate $f$ pointwise.

Let $E : C(G, \mathcal{H}_\sigma) \to \mathcal{H}_\sigma$ be evaluation at 1: $Ef = f(1)$. We claim that the map $T \mapsto ET$ is an isomorphism from $\mathcal{C}(\pi, \Pi)$ to $\mathcal{C}(\pi|H, \sigma)$. First, if $T \in \mathcal{C}(\pi, \Pi)$, $v \in \mathcal{H}_\pi$, and $\xi \in H$,

$$\sigma(\xi)ETv = \sigma(\xi)[Tv(1)] = Tv(\xi^{-1}) = [\Pi(\xi)Tv](1)$$
$$= T[\pi(\xi)v](1) = ET\pi(\xi)v,$$

so $ET \in \mathcal{C}(\pi|H, \sigma)$. Next, if $ET = 0$, then for any $x \in G$ and $v \in \mathcal{H}_\pi$,

$$0 = [T\pi(x^{-1})v](1) = [\Pi(x^{-1})Tv](1) = Tv(x),$$

so $T = 0$. Thus $T \mapsto ET$ is injective, and it remains to show that it is surjective. If $S \in \mathcal{C}(\pi|H, \sigma)$, define $T : \mathcal{H}_\pi \to L^2(G, \mathcal{H}_\sigma)$ by $Tv(x) = S[\pi(x^{-1})v]$. We claim that $T \in \mathcal{C}(\pi, \Pi)$ and $S = ET$. Indeed, if $x \in G$ and $\xi \in H$,

$$Tv(x\xi) = S[\pi(\xi^{-1})\pi(x^{-1})v] = \sigma(\xi^{-1})S[\pi(x^{-1})v] = \sigma(\xi^{-1})Tv(x),$$

so $Tv \in \mathcal{F}$. Moreover,

$$[\Pi(y)Tv](x) = Tv(y^{-1}x) = S[\pi(x^{-1})\pi(y)v] = T[\pi(y)v](x),$$

so $T \in \mathcal{C}(\pi, \Pi)$. Finally, $ETv = Tv(1) = Sv$, so $S = ET$. $\qquad \square$

The Frobenius Reciprocity Theorem, in connection with Proposition 6.9, furnishes a powerful tool for determining the irreducible components of an induced representation of a compact group. Here are a few simple examples.

*Example 1.* If $\sigma$ is the trivial representation of the trivial subgroup $H = \{1\}$, then $\mathrm{ind}_H^G(\sigma)$ is the ordinary left regular representation of $G$. If $[\pi] \in \widehat{G}$, $\mathrm{mult}(\sigma, \pi|H)$ clearly equals $d_\pi$, so Frobenius reciprocity recaptures part of the Peter-Weyl theorem: each $[\pi] \in \widehat{G}$ occurs in the regular representation with multiplicity equal to its dimension.

*Example 2.* Take $G$ to be $SO(3)$, $H$ the subgroup that leaves the point $(1, 0, 0)$ fixed, and $\sigma$ the trivial representation of $H$ on $\mathbb{C}$. Then $G/H$ can be identified with the unit sphere $S^2 \subset \mathbb{R}^3$, and $\mathrm{ind}_H^G(\sigma)$ is the natural representation of $SO(3)$ on $L^2(S^2)$ by rotations. To analyze this situation, let us replace $SO(3)$ by its double cover $SU(2)$. The calculations leading to Theorem 5.44, together with (5.46), show that the subgroup $\widetilde{H}$ of $SU(2)$ corresponding to $H$ is the group

$\{F(\theta) : \theta \in \mathbb{R}\}$ given by (5.29). Moreover, with the notation of §5.4, $\pi_m(F(\theta))(z^j w^{m-j}) = e^{i(m-2j)\theta} z^j w^{m-j}$, so the characters of $\widetilde{H} \cong \mathbb{T}$ that occur in $\pi_m|H$ are $e^{i(m-2j)\theta}$, $0 \le j \le m$. Thus, the trivial representation occurs in $\pi_m|H$ with multiplicity 1 if $m$ is even and does not occur if $m$ is odd. Frobenius reciprocity therefore tells us that the natural representation of $SU(2)$ on $L^2(S^2)$ is the direct sum of subrepresentations equivalent to $\pi_{2k}$, $k = 0, 1, 2, \ldots$, each occurring with multiplicity one. By Corollary 5.48, this means that each irreducible representation $\rho_k$ of $SO(3)$ occurs in its representation on $L^2(S^2)$ with multiplicity one. The subspaces $\mathcal{H}_k$ of $L^2(S^2)$ on which these representations occur are the spaces of spherical harmonics of degree $k$ on $S^2$. (See, for example, Stein and Weiss [131].)

*Example 3.* Let $G$ and $H$ be as in Example 2. $H$ acts on the tangent plane to $S^2$ at the point $(1, 0, 0)$ (essentially the $yz$-plane in $xyz$-space) by rotations,

$$(6.11) \qquad (y, z) \mapsto (y \cos \theta - z \sin \theta, \ y \sin \theta + z \cos \theta).$$

Let $\sigma$ be the unitary representation of $H$ on $\mathbb{C}^2$ defined by (6.11); then, as explained at the end of §6.1, the induced representation $\mathrm{ind}_H^G(\sigma)$ is the natural representation of $SO(3)$ on the complex vector fields on $S^2$. $\sigma$ is reducible: the invariant subspaces of (6.11) are the spans of $(1, \pm i)$, with eigenvalues $e^{\mp i\theta}$. Hence, by Proposition 6.9, $\mathrm{ind}_H^G(\sigma)$ is the direct sum of the representations induced by the characters $e^{\pm i\theta}$ of $H$. Since passing from $H$ to its double cover $\widetilde{H}$ in $SU(2)$ amounts to replacing $\theta$ by $2\theta$, the same considerations as in Example 2 show that each irreducible representation $\rho_k$ of $SO(3)$ as in Corollary 5.48 occurs with multiplicity 2 in $\mathrm{ind}_H^G(\sigma)$ — once for the character $e^{i\theta}$ of $H$, and once for $e^{-i\theta}$ — except for the trivial representation $\rho_0$, which does not occur at all.

We can identify the irreducible subspaces as follows. First, the Euclidean metric allows us to identify vector fields on $S^2$ with differential 1-forms. The exterior derivative $d$ maps functions to 1-forms and commutes with the action of $SO(3)$, so by Schur's lemma it is either zero or an isomorphism on each irreducible subspace. The only functions $f$ such that $df = 0$ are the constants, corresponding to the trivial representation $\rho_0$. Hence, for $k > 0$, the image $d(\mathcal{H}_k)$ of the spherical harmonics of degree $k$ under $d$ is a space of 1-forms on which $SO(3)$ acts by $\rho_k$. Moreover, if $\omega$ is the 2-form on $S^2$ giving the Euclidean element of area, the map $f \mapsto f\omega$ is a bijection from functions to 2-forms that commutes with the action of $SO(3)$, so the space of 2-forms decomposes as $\bigoplus_0^\infty \mathcal{H}_k \omega$. The co-differential $d^*$ maps 2-forms to 1-forms and commutes with the action of $SO(3)$, and its kernel is $\mathcal{H}_0\omega$, so for each $k > 0$, $d^*(\mathcal{H}_k\omega)$ is another space of 1-forms on which $SO(3)$ acts by $\rho_k$. The spaces $d(\mathcal{H}_k)$ and $d^*(\mathcal{H}_k\omega)$ are mutually orthogonal since $d^2 = 0$, so the irreducible decomposition of the space of 1-forms is $\bigoplus_1^\infty \left[ d(\mathcal{H}_k) \oplus d^*(\mathcal{H}_k\omega) \right]$.

Similar considerations, together with a knowledge of the representations of $SO(n)$, enable one to write down the decomposition of the spaces of differential forms on the unit sphere in $\mathbb{R}^n$ under the natural action of $SO(n)$, for any $n$; see Folland [43].

## 6.3 Pseudomeasures and Induction in Stages

In §3.3 we explored the relationship between functions of positive type on a group $G$ and unitary representations of $G$. We now present a generalization of these ideas that is useful for the theory of induced representations.

Let $X$ be a locally compact Hausdorff space. If $K$ is a compact subset of $X$, let $C_K(X)$ be the space of continuous functions on $X$ supported in $K$; this is a Banach space under the uniform norm. The space $C_c(X)$, being the union of these Banach spaces, inherits a natural inductive limit topology. The definition and properties of this topology can be found in Bourbaki [19, §II.4.4]. (See also Rudin [123], where the similar space $C_c^\infty(\mathbb{R}^n)$ is discussed.) All we need to know about it are the following facts concerning continuity of maps, in which $X$ and $Y$ denote locally compact Hausdorff spaces and the topology on the spaces $C_K(X)$ is that of the uniform norm.

- A linear functional on $C_c(X)$ is continuous if and only if its restriction to $C_K(X)$ is continuous for each compact $K \subset X$.

- A linear map $T : C_c(X) \to C_c(Y)$ is continuous if for every compact $F \subset X$ there is a compact $K \subset Y$ such that $T$ maps $C_F(X)$ boundedly into $C_K(Y)$.

- A bilinear map $B : C_c(X) \times C_c(X) \to C_c(Y)$ is continuous if for every compact $F, F' \subset X$ there is a compact $K \subset Y$ such that $B$ maps $C_F(X) \times C_{F'}(X)$ boundedly into $C_K(Y)$.

- A map $\phi : X \to C_c(Y)$ is continuous if for every compact $F \subset X$ there is a compact $K \subset Y$ such that $\phi$ maps $F$ continuously into $C_K(Y)$.

We shall call a continuous linear functional on $C_c(X)$ a **pseudomeasure**. (There seems to be no standard name for these functionals in the literature except among the disciples of Bourbaki, who call them simply "measures.") For example, every positive Radon measure $\mu$ on $X$ defines a pseudomeasure by $\mu(f) = \int f \, d\mu$, and the Riesz representation theorem says that every positive pseudomeasure arises in this fashion. More generally, if $\lambda$ is a positive Radon measure and $\phi$ is a locally $\lambda$-integrable complex function, $\mu(f) = \int f\phi \, d\lambda$ is a pseudomeasure. A similar but slightly more complicated description is also available for general pseudomeasures, as follows.

Suppose that for each compact $K \subset X$ we are given a complex Radon measure $\mu_K$ on $K$ such that if $K' \subset K$, $\mu_{K'}$ is the restriction of $\mu_K$ to the Borel subsets of $K'$. Such a collection $\{\mu_K\}$ defines a pseudomeasure $\mu$ by $\mu(f) = \int f \, d\mu_K$ for all $f \in C_K(X)$. Conversely, if $\mu$ is a pseudomeasure and $K$ is a compact set in $X$, choose a compact neighborhood $V$ of $K$. The restriction of $\mu$ to $C_V(X)$ can be extended via the Hahn-Banach theorem to a continuous linear functional on $C(V)$, which is given by a complex Radon measure on $V$. Let $\mu_K$ be the restriction of this measure to the Borel subsets of $K$. It is easily verified that $\mu_K$ is independent of the choice of $V$ and the Hahn-Banach extension, and it is then immediate that $\mu_{K'}$ is the restriction of $\mu_K$ to subsets of $K'$ when $K' \subset K$ and that $\mu(f) = \int f \, d\mu_K$ for all $f \in C_K(X)$.

In short, pseudomeasures can be handled much like measures, and such results as Fubini's theorem can be applied, as long as one only integrates functions of compact support. We shall use the notations $\mu(f)$ and $\int f \, d\mu$ indifferently for the action of a pseudomeasure $\mu$ on a function $f \in C_c(X)$.

Now let $G$ be a locally compact group. It is easily checked that the involution $f \mapsto f^*$ (where $f^*(x) = \Delta(x^{-1})\overline{f(x^{-1})}$) and convolution $(f,g) \mapsto f * g$ are continuous operations on $C_c(G)$, so they make $C_c(G)$ into a topological $*$-algebra. A pseudomeasure $\mu$ is said to be of **positive type** if it is positive with respect to this algebra structure, that is, if

$$\mu(f^* * f) \geq 0 \text{ for all } f \in C_c(G).$$

For example, if $\phi \in L^\infty(G)$ is a function of positive type and $\lambda$ is left Haar measure, then $\mu(f) = \int f\phi \, d\lambda$ is a pseudomeasure of positive type. (These are the pseudomeasures of positive type that extend continuously to $L^1(G)$.) Another example is provided by the point mass at the identity of $G$.

The construction that led from functions of positive type to unitary representations in §3.3 works equally well for pseudomeasures. Namely, suppose $\mu$ is a pseudomeasure of positive type. Then

$$\langle f, g \rangle_\mu = \mu(g^* * f)$$

is a positive semi-definite Hermitian form on $C_c(G)$. Upon factoring out the nullspace $\{f \in C_c(G) : \mu(f^* * f) = 0\}$ one obtains an inner product space that can be completed to form a Hilbert space $\mathcal{H}_\mu$. Since

$$(6.12) \quad (L_x f)^* * (L_x f)(y) = \int \Delta(z^{-1})\overline{f(x^{-1}z^{-1})}f(x^{-1}z^{-1}y)\,dz$$

$$= \int \overline{f(x^{-1}z)}f(x^{-1}zy)\,dz = \int \overline{f(z)}f(zy)\,dz$$

$$= \int f^*(z)f(z^{-1}y)\,dz = f^* * f(y),$$

the map $f \mapsto L_x f$ induces a unitary operator $\pi_\mu(x)$ on $\mathcal{H}_\mu$. One clearly has $\pi_\mu(x)\pi_\mu(y) = \pi_\mu(xy)$, and $x \mapsto \pi_\mu(x)$ is strongly continuous because $x \mapsto L_x f$ is continuous from $G$ to $C_c(G)$ and $\mu$ is continuous on $C_c(G)$. Hence $\pi_\mu$ is a unitary representation of $G$ on $\mathcal{H}_\mu$.

When $\mu$ comes from a function of positive type as described above, this construction coincides with the one in §3.3. (The reader may verify that $L^1(G)$ could be replaced by $C_c(G)$ there without changing anything essential.) However, functions of positive type yield only cyclic representations, as Proposition 3.20 shows, whereas pseudomeasures can yield more general representations. For example, let $\mu$ be the point mass at the identity; then

$$\mu(f^* * f) = \int \Delta(y^{-1})\overline{f(y^{-1})}f(y^{-1})\,dy = \int \overline{f(y)}f(y)\,dy,$$

so $\mathcal{H}_\mu = L^2(G)$ and $\pi_\mu$ is just the left regular representation of $G$. By Proposition 4.72, this representation is not cyclic when $G$ is Abelian and $\widehat{G}$ is not $\sigma$-compact.

The reason for the suitability of pseudomeasures in the theory of induced representations is that one can easily relate pseudomeasures on a closed subgroup to pseudomeasures on the whole group. Namely, if $H$ is a closed subgroup of $G$, any pseudomeasure $\mu$ on $H$ defines a pseudomeasure $\widetilde{\mu}$ on $G$ by $\widetilde{\mu}(f) = \mu(f|H)$. We shall call $\widetilde{\mu}$ the **injection** of $\mu$ into $G$. With this in mind, we come to the main result of this section.

**6.13 Theorem.** *Suppose $G$ is a locally compact group and $H$ is a closed subgroup, with modular functions $\Delta_G$ and $\Delta_H$. Let $\mu$ be a pseudomeasure of positive type on $H$, let $\sigma_\mu$ be the associated unitary representation of $H$, and let $\nu$ be the injection of $\sqrt{\Delta_G/\Delta_H}\,\mu$ into $G$, i.e., the pseudomeasure on $G$ defined by*

$$\nu(f) = \int_H \sqrt{\frac{\Delta_G(\xi)}{\Delta_H(\xi)}}\,f(\xi)\,d\mu(\xi).$$

*Then $\nu$ is of positive type, and the associated unitary representation $\pi_\nu$ of $G$ is unitarily equivalent to the induced representation $\Pi = \mathrm{ind}_H^G(\sigma_\mu)$.*

*Proof.* If $f \in C_c(G)$, define a continuous map $Uf : G \to C_c(H)$ by

$$[Uf(x)](\xi) = \sqrt{\frac{\Delta_G(\xi)}{\Delta_H(\xi)}} \, f(x\xi),$$

and (by Lemma 2.49) choose $h \in C_c(G)$ such that $\int_H h(y\eta) \, d\eta = 1$ for $y \in (\operatorname{supp} f)H$. We then have

$$\nu(f^* * f) = \iint \overline{f(y)} f(y\xi) \sqrt{\frac{\Delta_G(\xi)}{\Delta_H(\xi)}} \, dy \, d\mu(\xi)$$

$$= \iiint h(y\eta) \overline{f(y)} f(y\xi) \sqrt{\frac{\Delta_G(\xi)}{\Delta_H(\xi)}} \, d\eta \, dy \, d\mu(\xi).$$

Substituting $y\eta^{-1}$ for $y$ and then $\eta^{-1}$ for $\eta$, we obtain

$$\nu(f^* * f) = \iiint h(y) \overline{f(y\eta)} f(y\eta\xi) \sqrt{\frac{\Delta_G(\xi)}{\Delta_H(\xi)} \frac{\Delta_G(\eta)}{\Delta_H(\eta)}} \, d\eta \, dy \, d\mu(\xi)$$

$$= \iint h(y) \left[ (Uf(y))^* * Uf(y) \right](\xi) \, dy \, d\mu(\xi)$$

$$= \int h(y) \mu \left[ (Uf(y))^* * Uf(y) \right] dy.$$

This shows immediately that the positivity of $\mu$ implies the positivity of $\nu$. Moreover, if we regard $Uf(y) \in C_c(H)$ as an element of the Hilbert space $\mathcal{H}_\mu$ for the representation $\sigma_\mu$, then: (i) its norm is given by $\|Uf(y)\|_\mu^2 = \mu \left[ (Uf(y))^* * Uf(y) \right]$; (ii) $Uf$ is an element of the Hilbert space $\mathcal{F}$ for the intrinsic version of the induced representation $\Pi$; (iii) in view of (6.6) and the fact that $(\operatorname{supp} Uf)H = (\operatorname{supp} f)H$, we have

$$\nu(f^* * f) = \int h(y) \mu \left[ (Uf(y))^* * Uf(y) \right] dy = \|Uf\|_{\mathcal{F}}^2.$$

In other words, $U$ defines an isometry from the Hilbert space $\mathcal{H}_\nu$ for $\pi_\nu$ into the Hilbert space $\mathcal{F}$. Also, we clearly have

$$U[\pi_\nu(x)f](y) = \sqrt{\frac{\Delta_G}{\Delta_H}} f(x^{-1} y(\cdot)) = [\Pi(x) Uf](y),$$

so $U$ intertwines $\pi_\nu$ and $\Pi$.

It therefore remains to show that $U$ is surjective, and it is enough to show that the range of $U$ is dense. Given $\phi \in C_c(G)$ and $v \in C_c(H)$, define $f \in C_c(G)$ by

$$f(x) = \int_H \sqrt{\frac{\Delta_G(\eta)}{\Delta_H(\eta)}} \phi(x\eta) v(\eta^{-1}) \, d\eta.$$

Then

$$[Uf(x)](\xi) = \sqrt{\frac{\Delta_G(\xi)}{\Delta_H(\xi)}} \int_H \sqrt{\frac{\Delta_G(\eta)}{\Delta_H(\eta)}} \phi(x\xi\eta) v(\eta^{-1}) \, d\eta$$

$$= \int_H \sqrt{\frac{\Delta_G(\eta)}{\Delta_H(\eta)}} \phi(x\eta) v(\eta^{-1}\xi) \, d\eta.$$

But this is $[f_{\phi,v}(x)](\xi)$ in the notation of (6.7), with $\sigma = \sigma_\mu$, so the range of $U$ is dense by Proposition 6.8. $\qquad \square$

As an immediate corollary of Theorem 6.13, we obtain the theorem on "induction in stages," one of the fundamental results of the subject.

**6.14 Theorem.** *Suppose $H$ is a closed subgroup of $G$, $K$ is a closed subgroup of $H$, and $\sigma$ is a unitary representation of $K$. Then the representations $\operatorname{ind}_K^G(\sigma)$ and $\operatorname{ind}_H^G(\operatorname{ind}_K^H(\sigma))$ are unitarily equivalent.*

*Proof.* First suppose that $\sigma$ arises from a pseudomeasure $\mu$ of positive type on $K$. Let $\nu$ be the injection of $\sqrt{\Delta_H/\Delta_K}\,\mu$ into $H$. The desired result follows immediately from Theorem 6.13 and the obvious fact that the injections of $\sqrt{\Delta_G/\Delta_K}\,\mu$ and $\sqrt{\Delta_G/\Delta_H}\,\nu$ into $G$ are equal.

Now, any representation of $K$ is a direct sum of cyclic representations by Proposition 3.3, and every cyclic representation is equivalent to one that arises from a pseudomeasure (in fact, a function) of positive type by Corollary 3.24. An application of Proposition 6.9 therefore completes the proof. $\qquad \square$

## 6.4 Systems of Imprimitivity

Let $G$ be a locally compact group. A **system of imprimitivity** on $G$ is an ordered triple $\Sigma = (\pi, S, P)$ consisting of:

I1. a unitary representation $\pi$ of $G$ on a Hilbert space $\mathcal{H}_\pi$,

I2. a $G$-space $S$, and

I3. a regular $\mathcal{H}_\pi$-projection-valued measure $P$ on $S$ such that

$$(6.15) \qquad \pi(x)P(E)\pi(x)^{-1} = P(xE) \text{ for all } x \in G \text{ and } E \subset S.$$

(Recall that we have defined a $G$-space to be a locally compact Hausdorff space equipped with a continuous left $G$-action. One can increase the generality here a bit by taking $S$ to be a measurable space on which $G$ acts measurably. However, we shall have no reason to consider extensions of this sort.)

The definition we have just given is standard, but there is another equivalent one that is often technically easier to work with. The projection-valued measure $P$ determines a nondegenerate $*$-representation $M$ of the C$*$ algebra $C_0(S)$ on $\mathcal{H}_\pi$, namely $M(\phi) = \int \phi \, dP$. Since (6.15) can be re-expressed as

$$(6.16) \qquad \pi(x) \, dP(s) \, \pi(x)^{-1} = dP(xs),$$

we clearly have

$$(6.17) \qquad \pi(x)M(\phi)\pi(x^{-1}) = M(L_x\phi), \text{ where } L_x\phi(s) = \phi(x^{-1}s).$$

Conversely, suppose $M$ is a nondegenerate $*$-representation of $C_0(S)$ on $\mathcal{H}_\pi$ satisfying (6.17). By Corollary 1.55, $M$ determines a regular $\mathcal{H}_\pi$-projection-valued measure $P$ on $S$ such that $M(\phi) = \int \phi \, dP$ for all $\phi \in C_0(S)$. This $P$ satisfies 6.15. Indeed, if $x \in G$, $E \mapsto \pi(x)P(E)\pi(x)^{-1}$ is the projection-valued measure associated to the $*$-representation $\phi \mapsto \pi(x)M(\phi)\pi(x)^{-1}$, whereas $E \mapsto P(xE)$ is the projection-valued measure associated to the $*$-representation $\phi \mapsto M(L_x\phi)$. Since these representations are equal, (6.15) follows from the uniqueness of the projection-valued measure in Corollary 1.55.

In short, a system of imprimitivity can be thought of as an ordered triple $(\pi, S, M)$ where $\pi$ and $S$ are as in (I1) and (I2), and

I3$'$. $M$ is a nondegenerate $*$-representation of $C_0(S)$ on $\mathcal{H}_\pi$ satisfying (6.17).

We shall employ whichever of these definitions seems more convenient, and we shall simply use the notational distinction between the letters $P$ and $M$ to indicate which one we are using. (This practice, while potentially hazardous, will cause no confusion for us.)

A representation $\pi$ is called **imprimitive** if it belongs to a nontrivial system of imprimitivity, that is, one in which $S$ is not a single point. Every reducible representation is imprimitive. Indeed, let $\mathcal{A}$ be a commutative C$*$ subalgebra of $\mathcal{C}(\pi)$ (e.g., the algebra generated by a self-adjoint element of $\mathcal{C}(\pi)$), and let $S$ be the spectrum of $\mathcal{A}$. If $M : C_0(S) \to \mathcal{A}$ is the inverse of the Gelfand transform on $\mathcal{A}$, we have $M(\phi) \in \mathcal{C}(\pi)$ for all $\phi \in C_0(S)$, i.e., $\pi(x)M(\phi)\pi(x)^{-1} = M(\phi)$ for all $x \in G$. $(\pi, S, M)$ is therefore a system of imprimitivity if we endow $S$ with the trivial $G$-action ($xs = s$ for all $x$ and $s$).

More interestingly, every induced representation is imprimitive. Let $H$ be a closed subgroup of $G$ and let $q : G \to G/H$ be the quotient map. If $\sigma$ is a representation of $H$, the Hilbert space $\mathcal{F}$ for $\Phi = \mathrm{ind}_H^G(\sigma)$ is the completion of the space $\mathcal{F}^0$ of continuous $\mathcal{H}_\sigma$-valued functions on $G$ satisfying (6.3) and such that $q(\mathrm{supp}\, f)$ is compact. If $\phi \in C_0(G/H)$ and $f \in \mathcal{F}^0$, it is obvious that $(\phi \circ q)f \in \mathcal{F}^0$ and that $\|(\phi \circ q)f\|_{\mathcal{F}} \leq \|\phi\|_{\sup}\|f\|_{\mathcal{F}}$. Hence, if we set

$$M(\phi)f = (\phi \circ q)f,$$

$M$ is a $*$-representation of $C_0(G/H)$ on $\mathcal{F}$ that is clearly nondegenerate, and

$$\Phi(x)M(\phi)\Phi(x)^{-1}f(y) = M(\phi)\Phi(x)^{-1}f(x^{-1}y)$$
$$= \phi(q(x^{-1}y))f(y) = M(L_x\phi)f(y).$$

Thus $(\Phi, G/H, M)$ is a system of imprimitivity, called the **canonical system of imprimitivity** associated to $\Phi$.

It is clear what the associated projection-valued measure for this system of imprimitivity is: it is given by $P(E)f = (\chi_E \circ q)f$. From our definition of $\mathcal{F}$ as the completion of $\mathcal{F}^0$, however, it is not immediately obvious that multiplication by $\chi_E \circ q$ preserves $\mathcal{F}$. (It certainly doesn't preserve $\mathcal{F}^0$ in general!) Our use of the representation $M$ rather than the projection-valued measure $P$ neatly finesses this technical problem.

Various notions associated to unitary representations have analogues for systems of imprimitivity. For example, two systems of imprimitivity $(\pi, S, M)$ and $(\pi', S, M')$ (with the same $S$) are **(unitarily) equivalent** if there is a unitary $U : \mathcal{H}_\pi \to \mathcal{H}_{\pi'}$ such that $U\pi(x) = \pi'(x)U$ for all $x \in G$ and $UM(\phi) = M'(\phi)U$ for all $\phi \in C_0(S)$. (This condition on $M, M'$ is equivalent to the following condition on the associated projection-valued measures: $UP(E) = P'(E)U$ for all $E \subset S$.) If $\Sigma = (\pi, S, M)$ (or $(\pi, S, P)$) is a system of imprimitivity, a closed subspace $\mathcal{M}$ of $\mathcal{H}_\pi$ is called **invariant** under $\Sigma$ if it is invariant under all the operators $\pi(x)$ and $M(\phi)$ (or, equivalently, under all the operators $\pi(x)$ and $P(E)$). Also, if $\{(\pi_i, S, M_i)\}_{i \in I}$ is a family of systems of imprimitivity, all with the same $S$, their **direct sum** is the system $(\pi, S, M)$ where $\pi(x) = \bigoplus \pi_i(x)$ and $M(\phi) = \bigoplus M_i(\phi)$, acting on the Hilbert space $\bigoplus \mathcal{H}_{\pi_i}$. Reciprocally, if $(\pi, S, M)$ is a system of imprimitivity and $\mathcal{H}_\pi$ is the orthogonal sum of invariant closed subspaces $\mathcal{M}_i$, we can regard $(\pi, S, M)$ as the direct sum of the systems $(\pi_i, S, M_i)$, where $\pi_i(x) = \pi(x)|\mathcal{M}_i$ and $M_i(\phi) = M(\phi)|\mathcal{M}_i$.

Just as unitary representations of $G$ give rise to representations of the group algebra $L^1(G)$, systems of imprimitivity $(\pi, S, P)$ on $G$ give rise to representations of an algebra $L(S \times G)$ that we now describe. Let $S$ be any $G$-space. As a vector space, $L(S \times G)$ is simply $C_c(S \times G)$. The product is a type of convolution,

$$f * g(s,x) = \int_G f(s,y)g(y^{-1}s, y^{-1}x)\, dy,$$

and the involution is given by

$$f^*(s,x) = \overline{f(x^{-1}s, x^{-1})}\Delta(x^{-1}),$$

where $\Delta$ is the modular function of $G$. When $L(S \times G)$ is equipped with the usual inductive limit topology of $C_c(S \times G)$, the operations $(f,g) \mapsto f * g$ and $f \mapsto f^*$ are easily seen to be continuous and to satisfy all the usual algebraic properties, so that $L(S \times G)$ is a topological $*$-algebra. (The interested reader may also verify that the completion of $L(S \times G)$ with respect to the norm

$$\|f\| = \int_G \sup_{s \in S} |f(s,x)|\, dx$$

is a Banach $*$-algebra. We shall have no need to consider this larger algebra, although it provides a more exact analogue of the group algebra $L^1(G)$.)

Suppose now that $\Sigma = (\pi, S, M)$ is a system of imprimitivity on $G$. If $f \in L(S \times G)$, $M[f(\cdot, x)]$ is a bounded operator on $\mathcal{H}_\pi$ that is compactly supported and continuous in the norm topology as a function of $x$. If $v \in \mathcal{H}_\pi$, then, $M[f(\cdot, x)]\pi(x)v$ is a continuous, compactly supported $\mathcal{H}_\pi$-valued function of $x$, and we can integrate it to obtain a vector

$$T_\Sigma(f)v = \int_G M[f(\cdot, x)]\pi(x)v\, dx.$$

If the projection of supp $f \subset S \times G$ onto $G$ is contained in a compact set $K$, we have

$$\|T_\Sigma(f)v\| \le |K| \sup_{x \in G} \|M[f(\cdot, x)]\|\|v\| \le |K|\|f\|_{\sup}\|v\|.$$

Hence $T_\Sigma$ is a continuous linear map from $L(S \times G)$ into $\mathcal{L}(\mathcal{H}_\pi)$.

In terms of the projection-valued measure $P$ associated to $M$, we have $M[f(\cdot, x)] = \int f(s,x)\, dP(s)$, and hence

(6.18)    $$T_\Sigma(f) = \int_G M[f(\cdot, x)]\pi(x)\, dx = \int_G \int_S f(s,x)\, dP(s)\, \pi(x)\, dx.$$

In view of (6.16), however, we also have

$$M[f(\cdot, x)] = \int f(xs, x)\, dP(xs) = \pi(x) \int f(xs, x)\, dP(s)\, \pi(x)^{-1}$$

and hence

$$(6.19) \qquad T_\Sigma(f)v = \int_G \pi(x) \int_S f(xs, x) \, dP(s) \, dx.$$

(The use of the projection-valued measure $P$ rather than $M$ in (6.18) and (6.19) is mainly a convenience for keeping track of the variables.)

**6.20 Theorem.** *$T_\Sigma$ is a nondegenerate $*$-representation of $L(S \times G)$ on $\mathcal{H}_\pi$.*

*Proof.* First, by (6.19), we have

$$\begin{aligned} T_\Sigma(f^*) &= \int_G \pi(x) \int_S \overline{f(s, x^{-1})} \Delta(x^{-1}) \, dP(s) \, dx \\ &= \int_G \pi(x^{-1}) \int_S \overline{f(s, x)} \, dP(s) \, dx \\ &= \int_G \pi(x)^* M[f(\cdot, x)]^* \, dx = T_\Sigma(f)^*. \end{aligned}$$

Next, to show that $T_\Sigma(f*g) = T_\Sigma(f) T_\Sigma(g)$ we will need to apply Fubini's theorem to the operator-valued integral defining $T_\Sigma(f * g)$. To justify this, one can reduce to the scalar case by considering the scalar integral defining $\langle T_\Sigma(f * g)u, v \rangle$ for $u, v \in \mathcal{H}_\pi$; we leave the details to the reader. We have

$$T_\Sigma(f * g) = \int_G \int_S \int_G f(s, y) g(y^{-1}s, y^{-1}x) \, dy \, dP(s) \, \pi(x) \, dx.$$

Moving the $y$-integral to the outside and then substituting $yx$ for $x$ and $ys$ for $s$, we obtain

$$T_\Sigma(f * g) = \int_G \int_G \int_S f(ys, y) g(s, x) \, dP(ys) \, \pi(yx) \, dx \, dy.$$

Now an application of (6.16) yields

$$T_\Sigma(f * g) = \int_G \pi(y) \int_G \int_S f(ys, y) g(s, x) \, dP(s) \, \pi(x) \, dx \, dy,$$

so by the multiplicativity of the spectral functional calculus and (6.19),

$$\begin{aligned} &T_\Sigma(f * g) \\ &= \int_G \pi(y) \int_G \left[ \int_S f(ys, y) \, dP(s) \right] \left[ \int_S g(s, x) \, dP(s) \right] \pi(x) \, dx \, dy \\ &= \left[ \int_G \pi(y) \int_S f(ys, y) \, dP(s) \, dy \right] T_\Sigma(g) = T_\Sigma(f) T_\Sigma(g). \end{aligned}$$

Finally, if $v \neq 0 \in \mathcal{H}_\pi$, we can choose $g \in C_c(G)$ such that $\pi(g)v \neq 0$, and we can then choose $h \in C_c(S)$ such that $M(h)\pi(g)v \neq 0$. Let $f(s,x) = g(x)h(s)$; then $T_\Sigma(f)v = M(h)\pi(g)v \neq 0$, so $T_\Sigma$ is nondegenerate. $\qquad\square$

If $S$ is a $G$-space, a pseudomeasure $\mu$ on $S \times G$ is said to be of **positive type** if it is positive as a linear functional on the algebra $L(S \times G)$, that is, if $\mu(f^* * f) \geq 0$ for all $f \in L(S \times G)$.

For example, suppose $\Sigma$ is a system of imprimitivity and $T_\Sigma$ is the associated representation of $L(S \times G)$ defined by (6.18). Then for any $v \in \mathcal{H}_\pi$ the pseudomeasure $\mu$ defined by

$$(6.21) \qquad \mu(f) = \langle T_\Sigma(f)v, v \rangle$$

is of positive type, for

$$\langle T_\Sigma(f^* * f)v, v \rangle = \langle T_\Sigma(f)^* T_\Sigma(f)v, v \rangle = \|T_\Sigma(f)v\|^2 \geq 0.$$

Conversely, any pseudomeasure $\mu$ of positive type on $S \times G$ gives rise to a system of imprimitivity in much the same way that a pseudomeasure of positive type on $G$ gives rise to a unitary representation. First, the sesquilinear form

$$\langle f, g \rangle_\mu = \mu(g^* * f)$$

on $L(S \times G)$ is positive semidefinite, so by factoring out its nullspace and completing the resulting inner product space, we obtain a Hilbert space $\mathcal{H}_\mu$. Next, for $x \in G$ we define the operator $\pi(x)$ on $L(S \times G)$ by

$$(6.22) \qquad \pi(x)f(s,y) = f(x^{-1}s, x^{-1}y).$$

A calculation entirely similar to (6.12) shows that

$$[\pi(x)f]^* * [\pi(x)f] = f^* * f,$$

so $\pi(x)$ induces an isometry $\pi_\mu(x)$ on $\mathcal{H}_\mu$. We clearly have $\pi_\mu(x^{-1}) = \pi_\mu(x)^{-1}$, and the obvious continuity of the map $x \mapsto \pi(x)f$ from $G$ to $L(S \times G)$ easily implies the strong continuity of $\pi_\mu(x)$ in $x$. In short, $\pi_\mu$ is a unitary representation of $G$ on $\mathcal{H}_\mu$.

Next, if $\phi \in C_0(S)$, we define an operator $M(\phi)$ on $L(S \times G)$ by

$$(6.23) \qquad M(\phi)f(s,y) = \phi(s)f(s,y).$$

**6.24 Proposition.** *For $\phi \in C_0(S)$, the operator $M(\phi)$ on $L(S \times G)$ induces a bounded operator $M_\mu(\phi)$ on $\mathcal{H}_\mu$ satisfying (6.17). $M_\mu$ is a nondegenerate $*$-representation of $C_0(S)$ on $\mathcal{H}_\mu$.*

*Proof.* We obviously have $M(\phi\psi) = M(\phi)M(\psi)$. Also,
(6.25)

$$g^* * [M(\phi)f](s,x) = \int \overline{g(y^{-1}s, y^{-1})} \Delta(y^{-1}) \phi(y^{-1}s) f(y^{-1}s, y^{-1}x) \, dy$$
$$= [M(\overline{\phi})g]^* * f(s,x)$$

and

$$\pi(x)M(\phi)\pi(x^{-1})f(s,y) = M(\phi)\pi(x^{-1})f(x^{-1}s, x^{-1}y) = \phi(x^{-1}s)f(s,y).$$

From these relations it will follow that $M_\mu(\phi\psi) = M_\mu(\phi)M_\mu(\psi)$ and $M_\mu(\phi)^* = M_\mu(\overline{\phi})$ and that (6.17) holds, and nondegeneracy is obvious, so it remains to prove the boundedness. If $f \in L(S \times G)$, by (6.25) we have

$$\|M(\phi)f\|_\mu = \langle M(|\phi|^2)f, f\rangle_\mu^{1/2} \le \|M(|\phi|^2)f\|_\mu^{1/2}\|f\|_\mu^{1/2},$$

so by induction,

$$\|M(\phi)f\|_\mu \le \|M(|\phi|^{2^n})f\|_\mu^{2^{-n}}\|f\|_\mu^{1-2^{-n}}.$$

But

$$\|M(|\phi|^{2^n})f\|_\mu^2 - \int_{S \times G} \int_G \overline{f(ys,y)}|\phi(ys)|^{2^{n+1}} f(ys, yx) \, dy \, d\mu(s,x)$$
$$\le \|\phi\|_{\sup}^{2^{n+1}} \int_{S \times G} \int_G |f(ys,y)f(ys,yx)| \, dy \, d|\mu|(s,x)$$
$$= C_f \|\phi\|_{\sup}^{2^{n+1}},$$

so

$$\|M(\phi)f\|_\mu \le C_f^{2^{-n}} \|\phi\|_{\sup}\|f\|_\mu^{1-2^{-n}},$$

and letting $n \to \infty$ we obtain $\|M(\phi)f\|_\mu \le \|\phi\|_{\sup}\|f\|_\mu$.   □

In short, starting with a pseudomeasure $\mu$ of positive type on $S \times G$, we have constructed a system of imprimitivity $(\pi_\mu, S, M_\mu)$, which we shall call the system **derived from** $\mu$.

To complete the circle, we have the following theorem, an analogue of Corollary 3.24. A system of imprimitivity $\Sigma = (\pi, S, M)$ will be called **cyclic** if there is a vector $v \in \mathcal{H}_\pi$ such that $\{T_\Sigma(f)v : f \in L(S \times G)\}$ is dense in $\mathcal{H}_\pi$, where $T_\Sigma$ is defined by (6.18), in which case $v$ is called a **cyclic vector** for $\Sigma$. (This is in a slightly different spirit than our definition of cyclicity for unitary representations of $G$; it corresponds to cyclicity of the associated representation of $L^1(G)$. However, it is an easy exercise to show that the latter two notions are equivalent.)

**6.26 Theorem.** *Suppose $\Sigma = (\pi, S, M)$ is a cyclic system of imprimitivity, with cyclic vector $v$. Let $T_\Sigma$ be the associated representation of $L(S \times G)$ defined by (6.18), let $\mu$ be the pseudomeasure of positive type on $S \times G$ defined by $\mu(f) = \langle T_\Sigma(f)v, v \rangle$, and let $\Sigma_\mu = (\pi_\mu, S, M_\mu)$ be the system of imprimitivity derived from $\mu$. Then $\Sigma$ and $\Sigma_\mu$ are equivalent.*

*Proof.* If $f \in L(S \times G)$ we have

$$\|f\|_\mu^2 = \mu(f^* * f) = \langle T_\Sigma(f^* * f)v, v \rangle = \|T_\Sigma(f)v\|^2,$$

so the map $f \mapsto T_\Sigma(f)v$ induces an isometry from $\mathcal{H}_\mu$ into $\mathcal{H}_\pi$. Since $v$ is a cyclic vector, the range of $U$ is dense in $\mathcal{H}_\pi$, so $U$ is actually unitary. By (6.19), we have

$$T_\Sigma(\pi_\mu(x)f) = \int_G \pi(y) \int_S f(x^{-1}ys, x^{-1}y) \, dP(s) \, dy$$

$$= \int_G \pi(xy) \int_S f(ys, y) \, dP(s) \, dy = \pi(x)T_\Sigma(f),$$

from which it follows that $U$ intertwines $\pi_\mu$ and $\pi$. Moreover, if $\phi \in C_0(S)$,

$$T_\Sigma[M_\mu(\phi)f] = \int_G M[\phi f(\cdot, x)]\pi(x) \, dx$$

$$= \int_G M(\phi)M[f(\cdot, x)]\pi(x) \, dx = M(\phi)T_\Sigma(f),$$

and hence $UM_\mu(\phi) = M(\phi)U$ for $\phi \in C_0(S)$.          $\square$

We therefore have a natural correspondence between cyclic systems of imprimitivity $(\pi, S, M)$ on $G$ and pseudomeasures of positive type on $S \times G$. These results apply to arbitrary systems of imprimitivity via the following result, an analogue of Proposition 3.3.

**6.27 Proposition.** *Every system of imprimitivity $\Sigma = (\pi, S, M)$ is a direct sum of cyclic systems of imprimitivity.*

*Proof.* Since $T_\Sigma(f)^* = T_\Sigma(f^*)$, the orthogonal complement of a subspace of $\mathcal{H}_\pi$ that is invariant under all the operators $T_\Sigma(f)$, $f \in L(S \times G)$ is likewise invariant. A routine application of Zorn's lemma then shows that there is a maximal collection $\{v_i\}$ of unit vectors in $\mathcal{H}_\pi$ such that the closures $\mathcal{M}_i$ of the spaces $\{T_\Sigma(f)v_i : f \in L(S \times G)\}$ are mutually orthogonal. We then have $\mathcal{H}_\pi = \bigoplus \mathcal{M}_i$ by maximality, and it remains to show that each of the spaces $\mathcal{M}_i$ is invariant under all the operators

$\pi(x)$ and $M(\phi)$. But by (6.19),

$$\pi(x)T_\Sigma(f) = \int_G \pi(xy) \int_S f(ys, y) \, dP(s) \, dy$$

$$= \int_G \pi(y) \int_S f(x^{-1}ys, x^{-1}y) \, dP(s) \, dy = T_\Sigma(f_x),$$

where $f_x(s, y) = f(x^{-1}s, x^{-1}y)$, so $\mathcal{M}_i$ is invariant under $\pi(x)$. Moreover, if $\phi \in C_0(S)$,

$$M(\phi)T_\Sigma(f) = \int_G M(\phi)M[f(\cdot, x)]\pi(x) \, dx = T_\Sigma(\phi f)$$

(where $(\phi f)(s, x) = \phi(s)f(s, x)$), so $\mathcal{M}_i$ is invariant under every $M(\phi)$.
$\square$

## 6.5   The Imprimitivity Theorem

A system of imprimitivity $(\pi, S, M)$ on $G$ is called **transitive** if $S$ is a homogeneous space of $G$, that is, if $S = G/H$ for some closed subgroup $H$ of $G$. For example, the canonical system of imprimitivity associated with any induced representation is transitive. The principal object of this section is to prove that every transitive system of imprimitivity is of this form, and moreover to give a useful characterization of the operators that commute with the operators in a transitive system of imprimitivity. Since we shall need the latter result to prove uniqueness in the former one, we present it first.

Let $\Sigma = (\pi, S, M)$ (or $(\pi, S, P)$) be a system of imprimitivity. The **commutant** of $\Sigma$ is the set $\mathcal{C}(\Sigma)$ of all $T \in \mathcal{L}(\mathcal{H}_\pi)$ that commute with all the operators $\pi(x)$ and $M(\phi)$ (or equivalently, with all $\pi(x)$ and $P(E)$). Just as with unitary representations, a closed subspace $\mathcal{M}$ is invariant under $\Sigma$ if and only if the orthogonal projection onto $\mathcal{M}$ belongs to $\mathcal{C}(\Sigma)$ (see Proposition 3.4).

Suppose $H$ is a closed subgroup of $G$ and $\sigma$ is a unitary representation of $H$. Let $\Pi = \mathrm{ind}_H^G(\sigma)$ be the intrinsic version of the representation induced from $\sigma$, and let $\Sigma = (\Pi, G/H, M)$ be the canonical system of imprimitivity associated to $\Pi$. As usual, we denote the Hilbert space on which $\Pi$ acts by $\mathcal{F}$ and the space of continuous $\mathcal{H}_\sigma$-valued functions of which it is the completion by $\mathcal{F}^0$, and we denote the quotient map from $G$ to $G/H$ by $q$. If $T \in \mathcal{C}(\sigma)$, define the operator $\widetilde{T}$ on $\mathcal{F}^0$ by

$$[\widetilde{T}f](x) = T[f(x)].$$

$\widetilde{T}$ maps $\mathcal{F}^0$ into itself since $T$ commutes with every $\sigma(\xi)$, and $\widetilde{T}$ extends to a bounded operator on $\mathcal{F}$ such that $\|\widetilde{T}\| \leq \|T\|$ since $\|\widetilde{T}f(x)\|_\sigma \leq \|T\|\|f(x)\|_\sigma$ for all $x \in G$. Moreover, it is obvious that $\widetilde{T}$ commutes with left translations and with multiplication by functions in $C_0(S)$; in other words, $\widetilde{T} \in \mathcal{C}(\Sigma)$. Our first main result is the following.

**6.28 Theorem.** *With notation as above, the map $T \mapsto \widetilde{T}$ is an isometric $*$-isomorphism from $\mathcal{C}(\sigma)$ to $\mathcal{C}(\Sigma)$.*

*Proof.* It is trivial to verify that $(ST)\tilde{\ } = \widetilde{S}\widetilde{T}$ and that $(T^*)\tilde{\ } = (\widetilde{T})^*$, and we have already observed that $\|\widetilde{T}\| \leq \|T\|$. We next show that $\|\widetilde{T}\| \geq \|T\|$, so that $T \mapsto \widetilde{T}$ is an isometry.

Given $\epsilon > 0$, we can choose a unit vector $v \in \mathcal{H}_\sigma$ such that $\|Tv\|_\sigma \geq (1-\epsilon)\|T\|$. By Proposition 6.8(a), there is an $f \in \mathcal{F}^0$ such that $\|f(1)\|_\sigma < 1$ and $\|f(1) - v\|_\sigma < \epsilon$. Let $U$ be an open neighborhood of 1 in $G$ such that $\|f(x)\|_\sigma < 1$ and $\|f(x) - f(1)\|_\sigma < \epsilon$ for $x \in U$. Pick $\psi \neq 0$ in $C_c(G/H)$ such that $\operatorname{supp}\psi \subset q(U)$, and let $g(x) = \psi(q(x))f(x)$.

If $x$ is such that $g(x) \neq 0$, there exist $y \in U$ and $\xi \in H$ with $x = y\xi$. We have $\|f(y) - v\|_\sigma < 2\epsilon$, so

$$\|Tf(y)\|_\sigma \geq \|Tv\|_\sigma - \|T\|\|f(y) - v\|_\sigma > (1-3\epsilon)\|T\| > (1-3\epsilon)\|T\|\|f(y)\|_\sigma,$$

and hence $\|Tg(y)\|_\sigma > (1 - 3\epsilon)\|T\|\|g(y)\|_\sigma$. But then, by (6.3),

$$\|Tg(x)\|_\sigma = \sqrt{\frac{\Delta_H(\xi)}{\Delta_G(\xi)}} \|Tg(y)\|_\sigma$$

$$> \sqrt{\frac{\Delta_H(\xi)}{\Delta_G(\xi)}}(1 - 3\epsilon)\|T\|\|g(y)\|_\sigma = (1 - 3\epsilon)\|T\|\|g(x)\|_\sigma.$$

It follows that $\|Tg(x)\|_\sigma > (1 - 3\epsilon)\|T\|\|g(x)\|_\sigma$ for all $x \in G$ and hence that $\|\widetilde{T}g\|_\mathcal{F} > (1-3\epsilon)\|T\|\|g\|_\mathcal{F}$. Since $\epsilon$ is arbitrary, we have $\|\widetilde{T}\| \geq \|T\|$.

Now comes the hard part: proving that every operator in $\mathcal{C}(\Sigma)$ is of the form $\widetilde{T}$. For this we shall need two lemmas, in which — for the first and only time — we need the fact that elements $f$ of $\mathcal{F}$ can be realized as measurable functions such that the integral $\int \phi(x)f(x)\,dx$ is well-defined for $\phi \in C_c(G)$. (See the paragraphs preceding (6.3) and (6.7) in §6.1.)

**6.29 Lemma.** *If $\mathcal{N}$ is a closed subspace of $\mathcal{F}$ that is invariant under $\Sigma$, $\mathcal{F}^0 \cap \mathcal{N}$ is dense in $\mathcal{N}$.*

*Proof.* Let $\{\psi_U\}$ be an approximate identity in $C_c(G)$, and for $f \in \mathcal{N}$, consider $\psi_U * f$. If we write $\psi_U * f(x) = \int \psi_U(xy)f(y^{-1})\,dy$, we easily see that $\psi_U * f$ is continuous. On the other hand, since $\Pi(y)f(x) =$

$f(y^{-1}x)$, we can write $\psi_U * f = \int \psi_U(y)\Pi(y)f\,dy$, interpreted as a $\mathcal{F}$-valued integral (see Appendix 4), whence $\psi_U * f \in \mathbb{N}$ since $\Pi(y)f \in \mathbb{N}$ for all $y$. Moreover, the usual arguments show that $\psi_U * f \to f$ as $U \to \{1\}$. Thus if $f \in \mathbb{N}$ and $\epsilon > 0$ there is a continuous $g \in \mathbb{N}$ such that $\|g-f\| < \epsilon$. But also there exists $\phi \in C_c(G/H)$ such that $\|M(\phi)g - g\| < \epsilon$ (take $\phi = 1$ on a sufficiently large compact set), and $M(\phi)g \in \mathcal{F}^0$. $\qquad\square$

**6.30 Lemma.** *If $\mathcal{M}$ is a closed subspace of $\mathcal{H}_\sigma$, let*

$$\widetilde{\mathcal{M}} = \text{ closure in } \mathcal{F} \text{ of } \{f \in \mathcal{F}^0 : f(x) \in \mathcal{M} \text{ for all } x \in G\}.$$

*Then the correspondence $\mathcal{M} \mapsto \widetilde{\mathcal{M}}$ is a bijection between the set of $\sigma$-invariant closed subspaces of $\mathcal{H}_\sigma$ and the set of $\Sigma$-invariant closed subspaces of $\mathcal{F}$.*

*Proof.* It is clear that $\widetilde{\mathcal{M}}$ is invariant under translations and multiplication by functions in $C_0(G/H)$, i.e., invariant under $\Sigma$. In fact, when $\mathcal{M}$ is $\sigma$-invariant, $\widetilde{\mathcal{M}}$ is just the Hilbert space for the induced representation $\text{ind}_H^G(\sigma^\mathcal{M})$. If $\mathcal{M}_1$ and $\mathcal{M}_2$ are distinct closed $\sigma$-invariant subspaces of $\mathcal{H}_\sigma$, there is a vector $v$ in one but not the other, say $v \in \mathcal{M}_1 \setminus \mathcal{M}_2$. By Proposition 6.8(a) (with $\sigma$ replaced by $\sigma^{\mathcal{M}_1}$), there is a continuous $f \in \widetilde{\mathcal{M}}_1$ with $\|f(1) - v\|_\sigma$ arbitrarily small, and hence $f(x) \notin \mathcal{M}_2$ for $x$ in some neighborhood of 1. Then $f \in \widetilde{\mathcal{M}}_1 \setminus \widetilde{\mathcal{M}}_2$, so the map $\mathcal{M} \mapsto \widetilde{\mathcal{M}}$ is an injection.

On the other hand, suppose $\mathbb{N}$ is a $\Sigma$-invariant closed subspace of $\mathcal{F}$. If $f \in \mathbb{N}$ and $\phi \in C_c(G)$, let

$$v_{f,\phi} = \int \phi(x^{-1})f(x^{-1})\,dx,$$

and let $\mathcal{M}$ be the closed linear span of $\{v_{f,\phi} : f \in \mathbb{N},\ \phi \in C_c(G)\}$ in $\mathcal{H}_\sigma$. For any $\xi \in H$ we have

$$\sigma(\xi)v_{f,\phi} = \int \phi(x^{-1})\sigma(\xi)f(x^{-1})\,dx = \int \phi(x^{-1})\sqrt{\frac{\Delta_H(\xi)}{\Delta_G(\xi)}}f(x^{-1}\xi^{-1})\,dx$$

$$= \int \phi(x^{-1}\xi)\sqrt{\frac{\Delta_H(\xi)}{\Delta_G(\xi)}}f(x^{-1})\,dx = v_{f,\phi'}$$

where $\phi'(x) = \sqrt{\Delta_H(\xi)/\Delta_G(\xi)}\,\phi(x\xi)$, so $\mathcal{M}$ is $\sigma$-invariant.

We claim that $\mathbb{N} = \widetilde{\mathcal{M}}$. On the one hand, if $f \in \mathcal{F}^0 \cap \mathbb{N}$ and $w \in \mathcal{M}^\perp$, we have

$$0 = \langle v_{f,\phi}, w \rangle = \int \phi(x^{-1})\langle f(x^{-1}), w \rangle\,dx$$

for all $\phi \in C_c(G)$. This implies that $\langle f(x), w \rangle = 0$ for all $x \in G$ and $w \in \mathcal{M}^\perp$, so $f \in \widetilde{\mathcal{M}}$. By Lemma 6.29, then, $\mathcal{N} \subset \widetilde{\mathcal{M}}$. On the other hand, if $f \in \mathcal{N}$ and $\phi, \psi \in C_c(G)$, we can form an element $g_{f,\phi,\psi}$ of $\widetilde{\mathcal{M}}$ by the prescription (6.7):

$$g_{f,\phi,\psi}(x) = \int_H \sqrt{\frac{\Delta_G(\xi)}{\Delta_H(\xi)}} \psi(x\xi)\sigma(\xi)v_{f,\phi}\,d\xi.$$

By Proposition 6.8(b) (with $\sigma$ replaced by $\sigma^{\mathcal{M}}$), such functions are dense in $\widetilde{\mathcal{M}}$, so to prove that $\widetilde{\mathcal{M}} \subset \mathcal{N}$ it suffices to show that $g_{f,\phi,\psi} \in \mathcal{N}$. But

$$g_{f,\phi,\psi}(x) = \int_H \int_G \sqrt{\frac{\Delta_G(\xi)}{\Delta_H(\xi)}} \psi(x\xi)\phi(y^{-1})\sigma(\xi)f(y^{-1})\,dy\,d\xi$$

$$= \int_H \int_G \psi(x\xi)\phi(y^{-1})f(y^{-1}\xi^{-1})\,dy\,d\xi$$

$$= \int_H \int_G \psi(x\xi)\phi(y^{-1}x\xi)f(y^{-1}x)\,dy\,d\xi.$$

In other words, if we set $h(q(x), y) = \int_H \psi(x\xi)\phi(y^{-1}x\xi)\,d\xi$, we have

$$g_{f,\phi,\psi}(x) = \int_G h(q(x), y)f(y^{-1}x)\,dy = [T_\Sigma(h)f](x),$$

where $T_\Sigma$ is defined by (6.18). Since $f \in \mathcal{N}$ and $\mathcal{N}$ is $\Sigma$-invariant, $T_\Sigma f \in \mathcal{N}$, and we are done. $\qquad\square$

Now we can complete the proof of Theorem 6.28. If $\mathcal{M}$ is a closed $\sigma$-invariant subspace of $\mathcal{H}_\sigma$ and $P$ is the orthogonal projection onto $\mathcal{M}$, clearly $\widetilde{P}$ is the orthogonal projection onto $\widetilde{\mathcal{M}}$ in the notation of Lemma 6.30. That lemma therefore shows that every orthogonal projection in $\mathcal{C}(\Sigma)$ is of the form $\widetilde{P}$ for some projection $P \in \mathcal{C}(\sigma)$. Moreover, every self-adjoint element of $\mathcal{C}(\Sigma)$ is the norm limit of linear combinations of such projections, by the spectral theorem, and $\mathcal{C}(\sigma)\widetilde{\ }$ is norm-closed since $T \mapsto \widetilde{T}$ is an isometry, so every self-adjoint element of $\mathcal{C}(\Sigma)$ belongs to $\mathcal{C}(\sigma)\widetilde{\ }$. Finally, every $T \in \mathcal{C}(\Sigma)$ is a linear combination of self-adjoint elements ($T = A + iB$ where $A = \frac{1}{2}(T + T^*)$ and $B = \frac{1}{2i}(T - T^*)$), so $\mathcal{C}(\Sigma) \subset \mathcal{C}(\sigma)\widetilde{\ }$. $\qquad\square$

Now we come to the second major result of this section, and indeed the centerpiece of this whole chapter:

**6.31 Theorem** (The Imprimitivity Theorem). *Suppose $G$ is a locally compact group, $H$ is a closed subgroup, and $S = G/H$. Let $\Sigma = (\pi, S, M)$*

be a transitive system of imprimitivity on $G$. Then there is a unitary representation $\sigma$ of $H$ such that $\Sigma$ is equivalent to the canonical system of imprimitivity associated to $\text{ind}_H^G(\sigma)$. (In particular, $\pi$ is equivalent to $\text{ind}_H^G(\sigma)$.) Moreover, $\sigma$ is uniquely determined up to equivalence by $\Sigma$.

*Proof.* Let us first establish the uniqueness of $\sigma$. Let $\sigma_1$ and $\sigma_2$ be unitary representations of $H$, let $\Pi_j = \text{ind}_H^G(\sigma_j)$, let $\Sigma_j = (\Pi_j, S, M_j)$ be the associated canonical system of imprimitivity on the Hilbert space $\mathcal{F}_j$, and suppose $U : \mathcal{F}_1 \to \mathcal{F}_2$ is a unitary equivalence of $\Sigma_1$ and $\Sigma_2$. Consider the direct sum $\Sigma = (\Pi_1 \oplus \Pi_2, S, M_1 \oplus M_2)$, the system induced from $\sigma_1 \oplus \sigma_2$. Define an operator $V$ on $\mathcal{F}_1 \oplus \mathcal{F}_2$ by $V(f_1, f_2) = (0, Uf_1)$. Then $V^*(f_1, f_2) = (U^*f_2, 0)$, so since $U$ is unitary, $V^*V$ is the orthogonal projection onto $\mathcal{F}_1$ while $VV^*$ is the orthogonal projection onto $\mathcal{F}_2$. On the other hand, it is trivial to check that $V \in \mathcal{C}(\Sigma)$, so by Theorem 6.28 there is an operator $T \in \mathcal{C}(\sigma_1 \oplus \sigma_2)$ such that $V = \widetilde{T}$. Since $T \mapsto \widetilde{T}$ is a *-isomorphism, $T^*T$ and $TT^*$ are the orthogonal projections onto $\mathcal{H}_{\sigma_1}$ and $\mathcal{H}_{\sigma_2}$ respectively. But this implies that $T_0 = T|\mathcal{H}_{\sigma_1}$ is a unitary isomorphism from $\mathcal{H}_{\sigma_1}$ to $\mathcal{H}_{\sigma_2}$, and the fact that $T \in \mathcal{C}(\sigma_1 \oplus \sigma_2)$ implies that $T_0$ intertwines $\sigma_1$ and $\sigma_2$. Thus $\sigma_1$ and $\sigma_2$ are equivalent.

Before proceeding to the general proof of existence, we remark that a simple argument is available when $S = G/H$ is discrete. In this case, the compact sets in $G/H$ are finite, so the (inner) regularity of the projection-valued measure $P$ of $\Sigma$ implies that $P(\{s\}) \neq 0$ for some, and hence every, singleton set $\{s\} \subset S$. Let $\iota$ denote the coset of the identity on $G/H$. Then $\pi(\xi)P(\{\iota\})\pi(\xi)^{-1} = P(\{\xi\iota\}) = P(\{\iota\})$ for $\xi \in H$, so the range $\mathcal{M}$ of $P(\{\iota\})$ is invariant under $\pi|H$, and $\sigma = (\pi|H)^{\mathcal{M}}$ is a unitary representation of $H$. Given $v \in \mathcal{H}$, define a function $f_v : G \to \mathcal{M}$ by

$$f_v(x) = P(\{\iota\})\pi(x)^{-1}v = \pi(x)^{-1}P(\{x\iota\})v.$$

Then, for $\xi \in H$,

$$f_v(x\xi) = \pi(x\xi)^{-1}P(\{x\xi\iota\})v = \pi(\xi)^{-1}\pi(x)^{-1}P(\{x\iota\})v = \sigma(\xi)^{-1}f_v(x),$$

and

$$\sum_{x\iota \in G/H} \|f_v(x)\|^2 = \sum_{x\iota \in G/H} \|P(\{x\iota\})v\|^2 = \|v\|^2$$

since $\mathcal{H}_\pi$ is the direct sum of the ranges of the projections $P(\{x\iota\})$. It is now easy to see that $v \mapsto f_v$ is a unitary map from $\mathcal{H}_\pi$ to the Hilbert space $\mathcal{F}$ for $\text{ind}_H^G(\sigma)$ that defines an equivalence between $\Sigma$ and the canonical system of imprimitivity for $\text{ind}_H^G(\sigma)$. (Obviously, counting measure is an invariant measure on $G/H$, so there is no need to worry about modular functions.) Equivalently, if one wishes to think of the elements of $\mathcal{F}$ as sections of a vector bundle over $G/H$, the bundle in question is the one whose fiber over each $s \in G/H$ is the range of $P(\{s\})$.

When $G/H$ is not discrete, $P(\{\iota\})$ will normally be zero, so we must work harder to construct the representation $\sigma$. Here is the strategy: In the first place, since every system of imprimitivity is a direct sum of cyclic ones (Proposition 6.27), and inducing commutes with direct sums (Proposition 6.9), it suffices to assume that $\Sigma$ is cyclic. But then, by Theorem 6.26, we may assume that $\Sigma$ is the system of imprimitivity derived from a pseudomeasure $\mu$ of positive type on $S \times G$ (where $S = G/H$). We then proceed as follows.

i. We shall define a positive semidefinite Hermitian form $\langle \cdot, \cdot \rangle_\lambda$ on $C_c(G)$ and thence obtain a Hilbert space $\mathcal{H}_\lambda$ by factoring out the nullspace and completing the resulting inner product space. We shall also define an action of $H$ on $C_c(G)$ that will turn into a unitary representation $\sigma$ of $H$ on $\mathcal{H}_\lambda$.

ii. The Hilbert space for $\mathrm{ind}_H^G(\sigma)$ is fashioned out of continuous functions on $G$ with values in $\mathcal{H}_\lambda$, and thus ultimately out of continuous functions on $G$ with values in $C_c(G)$. On the other hand, the Hilbert space for the system of imprimitivity derived from $\mu$ is fashioned out of the space $L(S \times G)$. We shall therefore construct a unitary equivalence between the two out of a linear map $U$ from $L(S \times G)$ to $C(G, C_c(G))$.

It turns out that we need the map $U$ in (ii) to prove the positivity of the Hermitian form in (i), so the argument is a bit convoluted. But let us begin.

Suppose then that $\mu$ is a pseudomeasure of positive type on $S \times G$, where $S = G/H$. If $F \in C_c(G \times G)$, define $\widetilde{F} \in L(S \times G)$ by

$$\widetilde{F}(q(y), x) = \int_H f(x^{-1}y\xi, y\xi) \Delta_G(y\xi)^{-1} \, d\xi.$$

It is easily verified that the map $F \mapsto \widetilde{F}$ is continuous from $C_c(G \times G)$ to $L(S \times G) = C_c(S \times G)$ (see the remarks at the beginning of §6.3), so we can use it to pull the pseudomeasure $\mu$ back to $G \times G$. We denote the resulting pseudomeasure on $G \times G$ by $\lambda$:

$$\lambda(F) = \mu(\widetilde{F}).$$

We now define a sesquilinear form on $C_c(G)$ by

$$\langle f, g \rangle_\lambda = \lambda(f \otimes \bar{g}) = \int_{G \times G} f(x)\overline{g(y)} \, d\lambda(x, y)$$

$$= \int_{S \times G} \int_H f(x^{-1}y\xi)\overline{g(y\xi)}\Delta_G(y\xi)^{-1} \, d\xi \, d\mu(q(y), x).$$

This is the form described in (i); now we need the map $U : L(S \times G) \to$

$C(G, C_c(G))$ of (ii). It is given by

$$[Uf(x)](y) = f(q(x), xy^{-1}).$$

Again, it is easy to check that $Uf$ is indeed continuous from $G$ to $C_c(G)$.

**6.32 Lemma.** *Suppose $f \in L(S \times G)$ and $\phi$ is a nonnegative function in $C_c(G)$. Define $\phi' \in C_c(S)$ by $\phi'(q(x)) = \int_H \phi(x\xi)\,d\xi$, and define $g \in L(S \times G)$ by $g(s, x) = \phi'(s)^{1/2}f(s, x)$. Then*

$$\int_G \phi(x)\langle Uf(x), Uf(x)\rangle_\lambda\,dx = \mu(g^* * g).$$

*Proof.* By definition of $\langle \cdot, \cdot \rangle_\lambda$,

$$\langle Uf(x), Uf(x)\rangle_\lambda$$
$$= \iint [Uf(x)](z^{-1}y\xi)\overline{[Uf(x)](y\xi)}\Delta_G(y\xi)^{-1}\,d\xi\,d\mu(q(y), z)$$
$$= \iint f(q(x), x(y\xi)^{-1}z)\overline{f(q(x), x(y\xi)^{-1})}\Delta_G(y\xi)^{-1}\,d\xi\,d\mu(q(y), z).$$

Therefore,

$$\int \phi(x)\langle Uf(x), Uf(x)\rangle_\lambda\,dx$$
$$= \iiint \phi(x)f(q(x), x(y\xi)^{-1}z)\overline{f(q(x), x(y\xi)^{-1})}\Delta_G(y\xi)^{-1}dx\,d\xi\,d\mu(q(y), z)$$
$$= \iiint \phi(x\xi)f(q(x), xy^{-1}z)\overline{f(q(x), xy^{-1})}\Delta_G(y)^{-1}\,dx\,d\xi\,d\mu(q(y), z)$$
$$= \iint \phi'(q(x))f(q(x), xy^{-1}z)\overline{f(q(x), xy^{-1})}\Delta_G(y)^{-1}\,dx\,d\mu(q(y), z)$$

Since $g^*(q(y), yx^{-1}) = \overline{g(q(x), xy^{-1})}\Delta_G(xy^{-1})$, this equals

$$\iint g(q(x), xy^{-1}z)g^*(q(y), yx^{-1})\Delta_G(x^{-1})\,dx\,d\mu(q(y), z)$$
$$= \iint g^*(s, w)g(w^{-1}s, w^{-1}z)\,dw\,d\mu(s, z)$$
$$= \mu(g^* * g),$$

by the substitution $w = yx^{-1}$. $\qquad\qquad\square$

**6.33 Corollary.** *The form $\langle \cdot, \cdot \rangle_\lambda$ is positive semidefinite on $C_c(G)$.*

*Proof.* Since $\mu$ is of positive type, the lemma shows that

$$\int_G \phi(x)\langle Uf(x), Uf(x)\rangle_\lambda \, dx \geq 0$$

for all nonnegative $\phi \in C_c(G)$. Since $\langle Uf(x), Uf(x)\rangle_\lambda$ is continuous in $x$, this implies that $\langle Uf(x), Uf(x)\rangle_\lambda \geq 0$ for all $f \in L(S \times G)$ and all $x \in G$. But it is an easy exercise to see that, for any $x$, $f \mapsto Uf(x)$ maps $L(S \times G)$ onto $C_c(G)$, and the result follows. $\qquad\square$

In view of Corollary 6.33, we obtain a Hilbert space $\mathcal{H}_\lambda$ from $C_c(G)$ by passing to a quotient and completing it with respect to $\langle \cdot, \cdot \rangle_\lambda$, as promised. Next, for $\xi \in H$ we define a map $\sigma(\xi) : C_c(G) \to C_c(G)$ by

$$[\sigma(\xi)f](x) = \sqrt{\frac{\Delta_H(\xi)}{\Delta_G(\xi)}} f(x\xi).$$

We have

$$\langle \sigma(\xi)f, \sigma(\xi)g\rangle_\lambda = \iint \frac{\Delta_H(\xi)}{\Delta_G(\xi)} f(x^{-1}y\eta\xi)\overline{g(y\eta\xi)}\Delta_G(y\eta)^{-1}d\eta\, d\mu(q(y), x)$$

$$= \iint f(x^{-1}y\eta)\overline{g(y\eta)}\Delta_G(y\eta)^{-1}\, d\eta\, d\mu(q(y), x) = \langle f, g\rangle_\lambda,$$

so $\sigma(\xi)$ is an isometry with respect to $\langle \cdot, \cdot \rangle_\lambda$. Moreover, it is clear that $\sigma(\xi\eta) = \sigma(\xi)\sigma(\eta)$, $\sigma(x^{-1}) = \sigma(x)^{-1}$, and $x \mapsto \sigma(x)f$ is continuous from $G$ to $C_c(G)$ for each $f \in C_c(G)$. It follows easily that the operators $\sigma(\xi)$ yield a unitary representation (still denoted by $\sigma$) of $H$ on $\mathcal{H}_\lambda$.

Now we can complete the proof of the imprimitivity theorem. We observe that if $f \in L(S \times G)$,

$$[Uf(x\xi)](y) = f(q(x\xi), x\xi y^{-1}) = f(q(x), x(y\xi^{-1})^{-1})$$

$$= [Uf(x)](y\xi^{-1}) = \sqrt{\frac{\Delta_H(\xi)}{\Delta_G(\xi)}}[\sigma(\xi^{-1})Uf(x)](y),$$

so if we think of $Uf(x)$ as being an element of $\mathcal{H}_\lambda$, we see that $Uf$ is an element of the Hilbert space $\mathcal{F}$ for the induced representation $\Pi = \mathrm{ind}_H^G(\sigma)$. Moreover, if $f \in L(S \times G)$, let $V$ be a compact set in $S$ such that $\operatorname{supp} f \subset V \times G$, and choose $\phi \geq 0$ in $C_c(G)$ such that $\int_H \phi(x\xi) \, d\xi = 1$ for $x \in q^{-1}(V)$ (Lemma 2.49). Then $\operatorname{supp} Uf \subset q^{-1}(V)$, and $g = f$ in the notation of Lemma 6.32. Hence, by that lemma and (6.6),

$$\|Uf\|_{\mathcal{F}}^2 = \int \phi(x)\langle Uf(x), Uf(x)\rangle_\lambda \, dx = \mu(f^* * f),$$

so $U$ defines an isometry from the Hilbert space $\mathcal{H}_\mu$ (on which the system of imprimitivity $(\pi_\mu, S, M_\mu)$ derived from $\mu$ lives) to $\mathcal{F}$. It satisfies

$$[U[\pi_\mu(x)f](y)](z) = [\pi_\mu(x)f](q(y), yz^{-1}) = f(q(x^{-1}y), x^{-1}yz^{-1})$$
$$= [Uf(x^{-1}y)](z) = [[\Pi(x)Uf](y)](z)$$

and for $\phi \in C_0(S)$,

$$[U[M_\mu(\phi)f](y)](z) = \phi(q(y))f(q(y), yz^{-1}) = [[M(\phi)Uf](y)](z),$$

where $M$ is the representation of $C_0(S)$ in the canonical system of imprimitivity associated to $\Pi$. In other words, $U$ intertwines the two systems of imprimitivity.

It remains only to show that $U$ maps $\mathcal{H}_\mu$ onto $\mathcal{F}$, and it is enough to show that its range is dense. Given $\phi, \psi \in C_c(G)$, define $g \in L(S \times G)$ by

$$g(q(x), y) = \int_H \phi(x\xi)\psi(y^{-1}x\xi)\,d\xi.$$

Then

$$[Ug(x)](y) = g(q(x), xy^{-1}) = \int_H \phi(x\xi)\psi(y\xi)\,d\xi$$
$$= \int_H \sqrt{\frac{\Delta_G(\xi)}{\Delta_H(\xi)}}\,\phi(x\xi)[\sigma(\xi)\psi](y)\,d\xi.$$

But this says that $Ug = f_{\phi,\psi}$ in the notation of (6.7), so the desired result follows from Proposition 6.8(b). $\qquad\square$

## 6.6  Introduction to the Mackey Machine

The imprimitivity theorem is the foundation of a method, due to Mackey, for analyzing the representations of a group $G$ in terms of the representations of a closed normal subgroup $N$ and various subgroups of $G/N$. An exposition of this general theory is beyond the scope of this book, and we shall restrict ourselves to some special cases where things work out rather simply. Namely, we shall assume that $N$ is Abelian and satisfies a regularity condition to be explained below, and we shall obtain definitive results only in the case where $G$ is the semi-direct product of $N$ and another closed subgroup $H$. We shall sketch the more general picture in §6.9.

Suppose then that $G$ is a locally compact group and $N$ is a nontrivial closed Abelian normal subgroup of $G$. $G$ acts on $N$ by conjugation, and this induces an action of $G$ on the dual group $\widehat{N}$, $(x, \nu) \mapsto x\nu$, defined by

$$(6.34) \qquad \langle n, x\nu \rangle = \langle x^{-1}nx, \nu \rangle \qquad (x \in G, \ \nu \in \widehat{N}, \ n \in N).$$

For each $\nu \in \widehat{N}$, we denote by $G_\nu$ the stabilizer of $\nu$,

$$G_\nu = \{x \in G : x\nu = \nu\},$$

which is a closed subgroup of $G$, and we denote by $\mathcal{O}_\nu$ the orbit of $\nu$:

$$\mathcal{O}_\nu = \{x\nu : x \in G\}.$$

The action of $G$ on $\widehat{N}$ is never transitive (for one thing, $\mathcal{O}_1 = \{1\}$), and the structure of the orbits can be very complicated. We shall say that $G$ acts **regularly** on $\widehat{N}$ if the following two conditions are satisfied.

R1. The orbit space is **countably separated**, that is, there is a countable family $\{E_j\}$ of $G$-invariant Borel sets in $\widehat{N}$ such that each orbit in $\widehat{N}$ is the intersection of all the $E_j$'s that contain it.

R2. For each $\nu \in \widehat{N}$, the natural map $xG_\nu \mapsto x\nu$ from $G/G_\nu$ to $\mathcal{O}_\nu$ is a homeomorphism.

When $G$ is $\sigma$-compact, (R2) is equivalent to

R2′. Each orbit in $\widehat{N}$ is relatively open in its closure.

Indeed, it is an easy exercise to see that (R2′) holds if and only if each orbit is locally compact in the relative topology, and this is equivalent to (R2) by Proposition 2.46.

*Remark.* When $G$ is second countable, (R1) and (R2) are actually equivalent, and they are both implied by

R3. There is a Borel set in $\widehat{N}$ that intersects each orbit in exactly one point.

The former assertion was proved by Glimm [52] and the latter one by Mackey [92, Theorem 5.2]. We shall not prove them here, for in most cases it is easy to see directly whether (R1) and (R2) (or (R2′)) hold. In §§6.7–8 we shall work out several examples, including some where (R1) and (R2) fail.

Now suppose $\pi$ is a unitary representation of $G$. By Theorem 4.45, there is a unique regular projection-valued measure $P$ on $\widehat{N}$ such that

$$(6.35) \qquad \pi(n) = \int \langle n, \nu \rangle \, dP(\nu) \text{ for } n \in N.$$

**6.36 Proposition.** $(\pi, \widehat{N}, P)$ *is a system of imprimitivity. If $\pi$ is irreducible, then $P$ is ergodic in the sense that if $E \subset \widehat{N}$ is a $G$-invariant Borel set then $P(E)$ is either $0$ or $I$.*

*Proof.* If $x \in G$, $E \mapsto \pi(x)P(E)\pi(x)^{-1}$ is the projection-valued measure associated with the representation $n \mapsto \pi(x)\pi(n)\pi(x^{-1})$ of $N$, whereas by (6.34), $E \mapsto P(xE)$ is the projection-valued measure associated to the representation $n \mapsto \pi(xnx^{-1})$. These representations are equal; hence so are the measures, by the uniqueness in Theorem 4.45.

If $\pi$ is irreducible and $E \subset N$ is $G$-invariant, then $\pi(x)P(E)\pi(x)^{-1} = P(E)$ for all $x \in G$, so $P(E) \in \mathcal{C}(\pi)$. By Schur's lemma, $P(E) = 0$ or $I$. $\qquad\square$

**6.37 Proposition.** *If $\pi$ is irreducible and $G$ acts regularly on $\widehat{N}$, there is an orbit $\mathcal{O} \subset \widehat{N}$ such that $P(\mathcal{O}) = I$.*

*Proof.* Let $\{E_j\}_1^\infty$ be a countable separating family for the orbit space, as in condition (R1). If $\mathcal{O}$ is an orbit, we have $\mathcal{O} = \bigcap_{j \in J} E_j$ for some $J \subset \mathbb{Z}^+$, so $P(\mathcal{O})$ is the projection onto the intersection of the ranges of the $P(E_j)$, $j \in J$. Each $P(E_j)$ is either $0$ or $I$ by Proposition 6.36; hence either $P(E_j) = I$ for all $j \in J$, in which case $P(\mathcal{O}) = I$, or $P(E_j) = 0$ for some $j \in J$, in which case $P(\mathcal{O}) = 0$. Therefore, if $P(\mathcal{O}) = 0$ for every orbit $\mathcal{O}$, for each $\mathcal{O}$ there is a $j(\mathcal{O})$ such that $\mathcal{O} \subset E_{j(\mathcal{O})}$ and $P(E_{j(\mathcal{O})}) = 0$. But then $\widehat{N} = \bigcup_{\mathcal{O}} E_{j(\mathcal{O})}$, so $P(\widehat{N}) = 0$, an absurdity. Hence $P(\mathcal{O}) = I$ for some $\mathcal{O}$. $\qquad\square$

Suppose the conditions of Proposition 6.37 hold. The orbit $\mathcal{O}$ such that $P(\mathcal{O}) = I$ is of course unique, since $P(N \setminus \mathcal{O}) = 0$. Fix an element $\nu$ of $\mathcal{O}$. By the regularity condition (R2), $\mathcal{O}$ can be identified with the homogeneous space $G/G_\nu$. The projection-valued measure $P$ can therefore be regarded as living on $G/G_\nu$ rather than on $\widehat{N}$, and $(\pi, G/G_\nu, P)$ is then a *transitive* system of imprimitivity. Consequently, by the imprimitivity theorem, there is a representation $\sigma$ of $G_\nu$, necessarily irreducible since $\pi$ is, such that $(\pi, G/G_\nu, P)$ is unitarily equivalent to the canonical system of imprimitivity associated to $\mathrm{ind}_{G_\nu}^G(\sigma)$.

**6.38 Proposition.** $\sigma(n) = \langle n, \nu \rangle I$ *for all $n \in N$.*

*Proof.* We can assume that $(\pi, G/G_\nu, P)$ actually equals the canonical system of imprimitivity for $\mathrm{ind}_{G_\nu}^G(\sigma)$, acting on the usual space $\mathcal{F}$. When we transfer $P$ from $\widehat{N}$ to $G/G_\nu$, the formula $\pi(n) = \int \langle n, \cdot \rangle \, dP$ for $n \in N$ becomes $\pi(n) = \int \langle n, x\nu \rangle \, dP(q(x))$, where $q : G \to G/G_\nu$ is the quotient map. Now, from the definition of the canonical system of imprimitivity, for $\phi \in C_0(G/G_\nu)$ and $f \in \mathcal{F}^0$ we have $(\int \phi \, dP)f = M(\phi)f = (\phi \circ q)f$. The formula $(\int \phi \, dP)f = (\phi \circ q)f$ continues to hold for any bounded continuous function $\phi$ on $G/G_\nu$, as one sees by unraveling the definitions in the spectral functional calculus, and in particular it holds for $\phi(q(x)) = \langle n, x\nu \rangle$. Hence, for $f \in \mathcal{F}^0$,

$$[\pi(n)f](x) = \langle n, x\nu \rangle f(x) = \langle x^{-1}nx, \nu \rangle f(x).$$

On the other hand, we observe that since $N$ is normal in $G$ and in $G_\nu$, the spaces $G/N$ and $G_\nu/N$ both possess $G$-invariant measures, namely, their respective Haar measures. Hence, by Theorem 2.51, $\Delta_G|N = \Delta_{G_\nu}|N = \Delta_N$ $(= 1$, since $N$ is Abelian). Taking this into account, if $n \in N$ and $f \in \mathcal{F}^0$ we have

$$[\pi(n)f](x) = f(n^{-1}x) = f(x(x^{-1}n^{-1}x)) = \sigma(x^{-1}nx)f(x).$$

Comparing the last two displayed equations, we see that $\sigma(x^{-1}nx)f(x) = \langle x^{-1}nx, \nu \rangle f(x)$. In particular, $\sigma(n)f(1) = \langle n, \nu \rangle f(1)$. But $\{f(1) : f \in \mathcal{F}^0\}$ is dense in $\mathcal{H}_\sigma$ by Proposition 6.8(a), so the result follows.          $\square$

We summarize our results so far in a theorem.

**6.39 Theorem.** *Suppose $G$ acts regularly on $\widehat{N}$. If $\pi$ is an irreducible unitary representation of $G$, there exist $\nu \in \widehat{N}$ and an irreducible representation $\sigma$ of $G_\nu$ with $\sigma(n) = \langle n, \nu \rangle I$ for $n \in N$ such that $\pi$ is unitarily equivalent to $\mathrm{ind}_{G_\nu}^G(\sigma)$.*

In Theorem 6.39, the orbit $\mathcal{O}_\nu$ is uniquely determined by $\pi$, but the choice of $\nu$ in this orbit is arbitrary. If $\nu'$ is another element of this orbit, say $\nu' = x\nu$, the groups $G_\nu$ and $G_{\nu'}$ are isomorphic — in fact, $G_{\nu'} = xG_\nu x^{-1}$. Moreover, the correspondence $\sigma \leftrightarrow \sigma'$, where $\sigma'(y) = \sigma(x^{-1}yx)$, is a bijection between the representations of $G_\nu$ and those of $G_{\nu'}$, and the map $Uf(y) = f(x^{-1}yx)$ defines a unitary equivalence between $\mathrm{ind}_{G_\nu}^G(\sigma)$ and $\mathrm{ind}_{G_\nu}^G(\sigma')$. Hence the non-uniqueness of $\nu$ is of no essential significance.

Now we prove the converse of Theorem 6.39:

**6.40 Theorem.** *Suppose $G$ acts regularly on $\widehat{N}$. If $\nu \in \widehat{N}$ and $\sigma$ is an irreducible representation of $G_\nu$ such that $\sigma(n) = \langle n, \nu \rangle I$ for $n \in N$, then $\pi = \mathrm{ind}_{G_\nu}^G(\sigma)$ is irreducible. If $\sigma'$ is another such representation of $G_\nu$ such that $\mathrm{ind}_{G_\nu}^G(\sigma)$ and $\mathrm{ind}_{G_\nu}^G(\sigma')$ are unitarily equivalent, then $\sigma$ and $\sigma'$ are unitarily equivalent.*

*Proof.* Let $\Sigma = (\pi, G/G_\nu, P)$ be the canonical system of imprimitivity associated with $\pi$. We identify $G/G_\nu$ with $\mathcal{O}_\nu$ and regard $P$ as a projection-valued measure on $\widehat{N}$ by setting $P(E) = P(E \cap \mathcal{O}_\nu)$ for

$E \subset N$. (It is an easy exercise to check that this transferred $P$ is still regular.) For $n \in N$ and $f \in \mathcal{F}^0$ we have

$$\pi(n)f(x) = f(n^{-1}x) = f(xx^{-1}n^{-1}x)$$
$$= \sigma(x^{-1}nx)f(x) = \langle x^{-1}nx, \nu \rangle f(x) = \langle n, x\nu \rangle f(x),$$

from which it follows as in the proof of Proposition 6.38 that $\pi(n) = \int_{\widehat{N}} \langle n, \cdot \rangle \, dP$. Thus $P$ is the projection-valued measure associated to $\pi|N$ by Theorem 4.45.

This implies both assertions of the theorem. First, if $T \in \mathcal{C}(\pi)$, $T$ commutes with every $P(E)$ by Theorem 1.44, and hence $T \in \mathcal{C}(\Sigma)$. But by Theorem 6.28 and Schur's lemma, $\mathcal{C}(\Sigma) \cong \mathcal{C}(\sigma) = \mathbb{C}I$. Thus $\mathcal{C}(\pi) = \mathbb{C}I$ and $\pi$ is irreducible. Second, any unitary equivalence between $\pi$ and $\pi' = \operatorname{ind}_{G_\nu}^G(\sigma')$ is actually a unitary equivalence between their canonical systems of imprimitivity, so if such an equivalence exists then $\sigma$ and $\sigma'$ are equivalent by the uniqueness in the imprimitivity theorem.  $\square$

These results are not yet satisfactory, because $G_\nu$ may be rather large and its representations not much easier to analyze than those of $G$ itself. In fact, it can happen that $G_\nu = G$, in which case the preceding results are vacuous except for the fact that $\pi|N = \nu I$. In particular, this happens when $\nu = 1$; but in that case we actually can reduce the analysis to a smaller group, namely $G/N$. Indeed, if the orbit associated to $\pi$ in Proposition 6.37 is $\{1\}$, it simply means that $\pi$ is trivial on $N$ and so factors through $G/N$. In many cases something similar happens even when $\nu \neq 1$.

**6.41 Proposition.** *Suppose that $\nu \in \widehat{N}$ can be extended to a representation of $G_\nu$, i.e., that there exists a continuous homomorphism $\widetilde{\nu} : G_\nu \to \mathbb{T}$ such that $\widetilde{\nu}|N = \nu$. If $\rho$ is any irreducible representation of $G_\nu/N$, the formula $\sigma(y) = \widetilde{\nu}(y)\rho(yN)$ defines an irreducible representation of $G_\nu$ on $\mathcal{H}_\rho$ such that $\sigma(n) = \langle n, \nu \rangle I$ for $n \in N$. Moreover, every such representation of $G_\nu$ arises in this way.*

*Proof.* The first assertion is obvious. If $\sigma$ is any irreducible representation of $G_\nu$ such that $\sigma(n) = \langle n, \nu \rangle I$ for $n \in N$, let $\sigma'(y) = \widetilde{\nu}(y)^{-1}\sigma(y)$. Then $\sigma'$ is an irreducible representation of $G_\nu$ that is trivial on $N$, so $\sigma'(y) = \rho(yN)$ where $\rho$ is an irreducible representation of $G_\nu/N$, and $\sigma(y) = \widetilde{\nu}(y)\rho(yN)$.  $\square$

The most important case where the phenomenon in Proposition 6.41 is guaranteed to occur is that in which $G$ is the semi-direct product of $N$ and another closed subgroup $H$. We recall that, in general, a topological group $G$ is the **semi-direct product** of two closed subgroups $N$ and $H$ if $N$ is normal in $G$ and the map $(n, h) \mapsto nh$ from $N \times H$ to $G$

is a homeomorphism; in this case we write $G = N \ltimes H$.[1] Thus, when $G = N \ltimes H$ every element of $G$ can be written uniquely as $nh$ with $n \in N$ and $h \in H$, and when this is done the group law takes the form

$$(6.42) \qquad (n_1 h_1)(n_2 h_2) = (n_1[h_1 n_2 h_1^{-1}])(h_1 h_2).$$

Returning to our situation in which $N$ is Abelian, for $\nu \in \widehat{N}$ we define the **little group** $H_\nu$ associated to $\nu$ to be

$$H_\nu = G_\nu \cap H.$$

Since $G_\nu \supset N$, we then have $G_\nu = N \ltimes H_\nu$ and $H_\nu \cong G_\nu/N$. The character $\nu$ always extends to a homomorphism $\widetilde{\nu} : G_\nu \to \mathbb{T}$ by the formula $\widetilde{\nu}(nh) = \nu(n) = \langle n, \nu \rangle$. Indeed, by (6.42),

$$\widetilde{\nu}((n_1 h_1)(n_2 h_2)) = \langle n_1(h_1 n_2 h_1^{-1}), \nu \rangle = \langle n_1, \nu \rangle \langle h_1 n_2 h_1^{-1}, \nu \rangle,$$

and since $h_1 \in H_\nu$,

$$\widetilde{\nu}((n_1 h_1)(n_2 h_2)) = \langle n_1, \nu \rangle \langle n_2, \nu \rangle = \widetilde{\nu}(n_1 h_1)\widetilde{\nu}(n_2 h_2).$$

The prescription of Proposition 6.41 then takes the following form. If $\nu \in \widehat{N}$ and $\rho$ is an irreducible representation of $H_\nu$, we obtain an irreducible representation of $G_\nu$, which we denote by $\nu\rho$, by setting

$$(\nu\rho)(nh) = \langle n, \nu \rangle \rho(h),$$

and every irreducible representation $\sigma$ of $G_\nu$ such that $\sigma(n) = \langle n, \nu \rangle I$ for $n \in N$ is of this form. Moreover, since $(\nu\rho)|H_\nu = \rho$, $\nu\rho$ is equivalent to $\nu\rho'$ if and only if $\rho$ is equivalent to $\rho'$.

One further simple observation: since $N$ acts trivially on $\widehat{N}$, the $G$-orbit of $\nu \in \widehat{N}$ is the same as its $H$-orbit, and if $\nu' = x\nu$ ($x \in H$) belongs to this orbit, the little groups of $\nu$ and $\nu'$ are related by $H_{\nu'} = x H_\nu x^{-1}$. In particular, they are isomorphic.

To sum up, if we combine Theorems 6.39 and 6.40, Proposition 6.41, and the remarks following Theorem 6.39, we see that we have completely classified the irreducible representations of $G = N \ltimes H$ in terms of the irreducible representations of $N$ (i.e., the characters $\nu \in \widehat{N}$) and the irreducible representations of their little groups $H_\nu$:

---

[1] Some people write this as $N \rtimes H$. Either way, the notation is meant to recall the notation $N \lhd G$, indicating that $N$ is a normal subgroup of $G$. Our notation identifies the normal factor by putting the triangle in the product sign next to it but gives the triangle the wrong orientation with respect to it; the alternative makes the triangle point the right way but puts it on the wrong side. In any case, $N$ is a normal subgroup of $G$, not $H$, so the parallel with "$N \lhd G$" is inexact.

**6.43 Theorem.** *Suppose* $G = N \ltimes H$, *where* $N$ *is Abelian and* $G$ *acts regularly on* $\widehat{N}$. *If* $\nu \in \widehat{N}$ *and* $\rho$ *is an irreducible representation of* $H_\nu$, *then* $\mathrm{ind}_{G_\nu}^{G}(\nu\rho)$ *is an irreducible representation of* $G$, *and every irreducible representation of* $G$ *is equivalent to one of this form. Moreover,* $\mathrm{ind}_{G_\nu}^{G}(\nu\rho)$ *and* $\mathrm{ind}_{G_{\nu'}}^{G}(\nu'\rho')$ *are equivalent if and only if* $\nu$ *and* $\nu'$ *belong to the same orbit, say* $\nu' = x\nu$, *and* $h \mapsto \rho(h)$ *and* $h \mapsto \rho'(xhx^{-1})$ *are equivalent representations of* $H_\nu$.

## 6.7 Examples: The Classics

In this section we present several examples, or classes of examples, to show how the general theory of the preceding sections applies in some important specific cases.

**1. The ax + b group.** Let $G$ be the $ax+b$ group, whose underlying manifold is $(0, \infty) \times \mathbb{R}$ and whose group law is

$$(a,b)(a',b') = (aa', b + ab').$$

We have $G = N \ltimes H$ where $N = \{(1,b) : b \in \mathbb{R}\}$ and $H = \{(a,0) : a > 0\}$. We identify $N$ with $\mathbb{R}$ via the correspondence $(1,b) \leftrightarrow b$, and then identify $\widehat{N}$ with $\mathbb{R}$ via the pairing $\langle b, \beta \rangle = e^{2\pi i \beta b}$. Simple calculations show that

$$(a,b)^{-1}(1,b')(a,b) = (1, b'/a),$$

and hence that the action of $G$ on $\widehat{N}$ is given by

$$(a,b)\beta = \beta/a.$$

There are only three orbits: $(0, \infty)$, $(-\infty, 0)$, and $\{0\}$, so it is trivial to verify that $G$ acts regularly on $\widehat{N}$.

The irreducible representations of $G$ are therefore described by Theorem 6.43. The ones associated to the orbit $\{0\}$ are the characters of $G/N \cong H$ lifted to $G$:

$$\pi_\lambda^0(a,b) = a^{i\lambda} \qquad (\lambda \in \mathbb{R}).$$

To analyze the orbit $(0, \infty)$, we fix the base point $1 \in (0, \infty)$. The little group $H_1$ is trivial, so up to equivalence there is only one irreducible representation $\pi^+$ associated to $(0, \infty)$, the one obtained by inducing the character $\langle b, 1 \rangle = e^{2\pi i b}$ from $N$ up to $G$. Likewise, there is one representation $\pi^-$ associated to the orbit $(-\infty, 0)$, the one obtained by inducing the character $\langle b, -1 \rangle = e^{-2\pi i b}$ from $N$ up to $G$.

Let us take a closer look at $\pi^+$. The homogeneous space $G/G_1 = G/N \cong H$ admits a $G$-invariant measure, namely the Haar measure $da/a$ on $H$. The Hilbert space $\mathcal{F}$ for $\pi^+$ therefore consists of functions $f : G \to \mathbb{C}$ such that

(6.44)        $f(a, b + ab') = f((a, b)(1, b')) = e^{-2\pi i b'} f(a, b),$

with norm

$$\|f\|^2 = \int_0^\infty |f(a, b)|^2 \, \frac{da}{a}.$$

If we take $b = 0$ in (6.44) and then write $b/a$ in place of $b'$, we see that $f(a, b) = e^{-2\pi i b/a} f(a, 0)$. Functions in $\mathcal{F}$ are thus completely determined by their values on $H$, and if we set

$$f_0(t) = f(t, 0),$$

the correspondence $f \mapsto f_0$ is a unitary map from $\mathcal{F}$ to $L^2((0, \infty), dt/t)$ whose inverse is given by $f(a, b) = e^{-2\pi i b/a} f_0(a)$. Now, the representation $\pi^+$ is given on $\mathcal{F}$ by

$$[\pi^+(a, b)f](c, d) = f((a, b)^{-1}(c, d)) = f(a^{-1}c, \, a^{-1}(d - b)),$$

and we have

$$[\pi^+(a, b)f](c, d) = e^{-2\pi i d/c}[\pi^+(a, b)f]_0(c),$$
$$f(a^{-1}c, \, a^{-1}(d - b)) = e^{2\pi i(b-d)/c} f_0(a^{-1}c).$$

Hence, if we conjugate $\pi^+$ by the map $f \mapsto f_0$, we obtain an equivalent representation $\widetilde{\pi}^+$ on $L^2((0, \infty), dt/t)$, namely

$$\widetilde{\pi}^+(a, b)f_0(t) = e^{2\pi i b/t} f_0(a^{-1}t).$$

We now make one final change of variables:

$$s = t^{-1}, \qquad g(s) = s^{-1/2} f_0(s^{-1}).$$

The map $f_0 \mapsto g$ transforms $L^2((0, \infty), dt/t)$ into $L^2((0, \infty), ds)$ and $\widetilde{\pi}^+(a, b)$ into

$$[\pi_+(a, b)g](s) = a^{1/2} e^{2\pi i b s} g(as) \qquad (g \in L^2((0, \infty), ds)).$$

A similar analysis shows that $\pi^-$ is equivalent to the representation $\pi_-$ of $G$ given by the same formula as $\pi_+$, but acting on $L^2((-\infty, 0), ds)$:

$$[\pi_-(a, b)g](s) = a^{1/2} e^{2\pi i b s} g(as) \qquad (g \in L^2((-\infty, 0), ds)).$$

The interest of these representations is that their direct sum $\widehat{\pi} = \pi_+ \oplus \pi_-$ is given by $[\widehat{\pi}(a,b)g](s) = a^{1/2} e^{2\pi i b s} g(as)$ on $L^2(\mathbb{R}, ds)$, and conjugation by the Fourier transform $\mathcal{F}$ on $\mathbb{R}$ yields the representation $\pi(a,b) = \mathcal{F}\widehat{\pi}(a,b)\mathcal{F}^{-1}$ that is given by

$$\pi(a,b)f(x) = a^{1/2} f(a^{-1}(x-b)).$$

This is just the representation of $G$ on $L^2(\mathbb{R})$ associated to the natural action of $G$ on $\mathbb{R}$. ($x \mapsto a^{-1}(x-b)$ is of course the inverse of the transformation $x \mapsto ax + b$.)

As an exercise, the reader may perform a similar analysis for the extended $ax + b$ group $\widetilde{G}$ in which $a$ is allowed to be either positive or negative. The result is that there are twice as many one-dimensional representations of $\widetilde{G}$, $(a,b) \mapsto |a|^{i\lambda}$ and $(a,b) \mapsto (\mathrm{sgn}\, a)|a|^{i\lambda}$, but that the two infinite-dimensional representations $\pi^{\pm}$ of $G$ coalesce into a single irreducible representation of $\widetilde{G}$, which may be realized as the natural action of $\widetilde{G}$ on $L^2(\mathbb{R})$.

**2. The Euclidean groups.** Let $E(n)$ be the group of rigid motions of $\mathbb{R}^n$, that is, the group generated by translations and rotations. $E(n)$ is the semi-direct product of the group of translations and the group of rotations, $E(n) = \mathbb{R}^n \ltimes SO(n)$, with group law given by

(6.45) $$(b,T)(b',T') = (b + Tb',\, TT').$$

Here $N \cong \mathbb{R}^n$ and, as usual, we identify $\widehat{N}$ with $\mathbb{R}^n$ by the pairing $\langle b, \beta \rangle = e^{2\pi i \beta \cdot b}$. It is then an easy exercise to see that the action of $G$ on $\widehat{N}$ is given by $(b,T)\beta = T\beta$, so the orbits are the spheres centered at the origin,

$$\mathcal{O}_r = \{\beta : |\beta| = r\} \qquad (r \geq 0).$$

It follows easily that $G$ acts regularly on $\widehat{N}$, so Theorem 6.43 applies. (For the countable separation property, one can take the $G$-invariant sets to be the annuli $\{\beta : s < |\beta| < t\}$ with $s$ and $t$ rational.) As a base point on the orbit $\mathcal{O}_r$ we can take the point $r\mathbf{1} = (r, 0, \ldots, 0)$; then the little group $H_{r\mathbf{1}}$ is $SO(n-1)$ (considered as the subgroup of $SO(n)$ that leaves the first coordinate fixed), except when $r = 0$, in which case it is $SO(n)$. Hence the irreducible representations of $E(n)$ are (i) the irreducible representations of $SO(n)$ lifted to $E(n)$, and (ii) the representations induced from products of nontrivial characters of $\mathbb{R}^n$ and irreducible representations of $SO(n-1)$. In short, one can describe all the irreducible representations of the Euclidean groups in terms of the irreducible representations of the rotation groups $SO(k)$. We have determined the latter for $k = 1$ ($SO(1) = \{1\}$), $k = 2$ ($SO(2) \cong \mathbb{T}$), and $k = 3$ or $4$ (see §5.5), and of course they are also well-known for higher values of $k$ (see, e.g., Boerner [16] or Weyl [147]).

**3. The Poincaré group.** Let us denote coordinates on $\mathbb{R}^4$ by $(x_0, x_1, x_2, x_3)$. The set of linear transformations of $\mathbb{R}^4$ that leave invariant the Lorentz (pseudo)-inner product

$$L(x, y) = x_0 y_0 - x_1 y_1 - x_2 y_2 - x_3 y_3,$$

or equivalently the set of $4 \times 4$ real matrices $T$ such that $T^* L T = L$ where $L$ is the diagonal matrix with diagonal entries $(1, -1, -1, -1)$, is denoted by $O(3, 1)$ and called the (4-dimensional) **Lorentz group**. $O(3, 1)$ admits two nontrivial homomorphisms into the 2-element group $\{\pm 1\}$, namely $T \mapsto \det T$ and $T \mapsto \operatorname{sgn} T_{00}$ (where $T_{00}$ is the $(0,0)$ entry of the matrix $T$). The intersection of the kernels of these homomorphisms, which is also the connected component of the identity in $O(3, 1)$, is called the **restricted Lorentz group** and is denoted by $SO_o(3, 1)$.

The (4-dimensional, restricted) **Poincaré group** $\mathcal{P}$ is the group formed from translations and restricted Lorentz transformations in the same way as the Euclidean group $E_4$ is formed from translations and rotations:

$$\mathcal{P} = \mathbb{R}^4 \ltimes SO_o(3, 1); \qquad (b, T)(b', T') = (b + Tb', TT').$$

As in the case of the Euclidean group, the normal subgroup $N$ is $\mathbb{R}^4$, but this time it is convenient to identify $\widehat{N}$ with $\mathbb{R}^4$ by using the Lorentz inner product rather than the Euclidean one: $\langle b, \beta \rangle = e^{2\pi i L(b, \beta)}$. With this convention, the action of $G$ on $\widehat{N}$ is given by $(b, T)\beta = T\beta$ as before.

We sketch the analysis of the orbits in $\widehat{N}$, leaving the verifications to the reader. The level sets

$$M_\lambda = \{\beta : L(\beta, \beta) = \lambda\} \qquad (\lambda \in \mathbb{R})$$

of the Lorentz form are $G$-invariant, but they are not all orbits. Instead, the situation is as follows. For $\lambda < 0$, $M_\lambda$ is a hyperboloid of one sheet, and it is an orbit. For $\lambda > 0$, $M_\lambda$ is a hyperboloid of two sheets, each of which is an orbit. For $\lambda = 0$, $M_0$ is a cone; the origin $\{0\}$ and the two connected half-cones in $M_0 \setminus \{0\}$ are each orbits. In short, the orbits are

$$\mathcal{O}_\lambda = M_\lambda \ (\lambda < 0), \qquad \mathcal{O}_\lambda^{\pm} = \{\beta \in M_\lambda : \pm\beta_0 > 0\} \ (\lambda \geq 0), \qquad \{0\}.$$

From this it is an easy exercise to see that $G$ acts regularly on $\widehat{N}$. (For countable separation, start with the family of invariant sets $U_{ab} = \bigcup_{a < \lambda < b} M_\lambda$ with $a$ and $b$ rational, and elaborate it a bit.)

As a representative point in the orbit $\mathcal{O}_\lambda$ with $\lambda < 0$ we may take $(0, 0, 0, |\lambda|^{1/2})$, and the corresponding little group is $SO_o(2, 1)$, the 3-dimensional restricted Lorentz group acting on $(x_0, x_1, x_2)$-space. As a representative point in $\mathcal{O}_\lambda^{\pm}$ with $\lambda > 0$ we may take $(\pm\lambda^{1/2}, 0, 0, 0)$, and the corresponding little group is $SO(3)$, acting on $(x_1, x_2, x_3)$-space. The little groups of the orbits $\mathcal{O}_0^{\pm}$ are not so transparent, but they turn out to be isomorphic to the Euclidean group $E(2)$; we shall prove this shortly in Proposition 6.46. And, of course, the little group for the orbit $\{0\}$ is $SO_o(3,1)$.

In view of Theorem 6.43, the irreducible representations of the Poincaré group can be described in terms of the irreducible representations of $SO(3)$, $E(2)$, $SO_o(2,1)$, and $SO_o(3,1)$. We have catalogued the irreducible representations of $SO(3)$ in §5.4, and the irreducible representations of $E(2)$ are obtained by another application of Theorem 6.43 as described above: they are the representations induced from nontrivial characters of the subgroup $\mathbb{R}^2$, together with the characters of the quotient group $E(2)/\mathbb{R}^2 \cong SO(2)$ lifted to $E(2)$.

The groups $SO_o(2,1)$ and $SO_o(3,1)$ have no nontrivial closed normal subgroups, so the Mackey machine cannot be applied to them, but the classification of their irreducible representations is well understood because of their intimate relation with $SL(2,\mathbb{R})$ and $SL(2,\mathbb{C})$. To be precise, consider the linear isomorphism $M$ from $\mathbb{R}^4$ to the space $\mathcal{H}$ of $2 \times 2$ Hermitian matrices given by

$$M(x) = \begin{pmatrix} x_0 + x_1 & x_2 - ix_3 \\ x_2 + ix_3 & x_0 - x_1 \end{pmatrix}.$$

If $A \in SL(2,\mathbb{C})$ and $H \in \mathcal{H}$ then $AHA^* \in \mathcal{H}$, so the map $\rho(A)x = M^{-1}[AM(x)A^*]$ defines a linear action of $SL(2,\mathbb{C})$ on $\mathbb{R}^4$. It is not hard to verify that the kernel of $\rho$ is $\pm I$ and that the image of $SL(2,\mathbb{C})$ under $\rho$ is precisely $SO_o(3,1)$. (The key point is that $L(x,x) = \det M(x)$.) Moreover, if we identify $SO_o(2,1)$ with the subgroup of $SO_o(3,1)$ leaving the point $(0,0,0,1)$ fixed, the inverse image of $SO_o(2,1)$ under $\rho$ is precisely $SL(2,\mathbb{R})$. (Use the fact that if $J = \begin{pmatrix} 0 & 1 \\ -1 & 0 \end{pmatrix}$ then $J = iM(0,0,0,1)$ and $AJA^T = J$ for all $A \in SL(2,\mathbb{C})$.) Hence

$$SO_o(3,1) \cong SL(2,\mathbb{C})/(\pm I), \qquad SO_o(2,1) \cong SL(2,\mathbb{R})/(\pm I),$$

and the representations of $SO_o(3,1)$ and $SO_o(2,1)$ are essentially the representations of $SL(2,\mathbb{C})$ and $SL(2,\mathbb{R})$ that are trivial on $-I$. We shall describe the irreducible representations of $SL(2,\mathbb{R})$ in §7.6, and the corresponding description for $SL(2,\mathbb{C})$ can be found in Knapp [79].

At this point we can attend to a bit of unfinished business:

**6.46 Proposition.** *The little groups of the orbits $\mathcal{O}_0^{\pm}$ are isomorphic to $E(2)$.*

*Proof.* We consider $\mathcal{O}_0^+$; the proof for $\mathcal{O}_0^-$ is similar. As a representative point in $\mathcal{O}_0^+$ we take the point $p = (1,1,0,0)$, so that $M(p) = \begin{pmatrix} 2 & 0 \\ 0 & 0 \end{pmatrix}$. It is easy to check that if $A = \begin{pmatrix} a & b \\ c & d \end{pmatrix} \in SL(2, \mathbb{C})$, we have $AM(p)A^* = M(p)$ if and only if $|a| = 1$, $c = 0$, and $d = a^{-1} = \bar{a}$. Let $G$ be the group of such matrices. Thus, with $z = ab$, $G$ is the set of matrices of the form

$$m(z, a) = \begin{pmatrix} a & \bar{a}z \\ 0 & \bar{a} \end{pmatrix} \qquad (z \in \mathbb{C}, \ |a| = 1),$$

and it is easily verified that

$$m(z, a)m(w, b) = m(z + a^2 w, \, ab).$$

On comparing this with (6.45), we see that the map

$$m(x + iy, \, e^{i\theta}) \mapsto \big((x, y), \text{ rotation through angle } 2\theta\big)$$

is a 2-to-1 homomorphism from $G$ to $\mathbb{R}^2 \ltimes SO(2) = E(2)$, with kernel $\{\pm I\} = \{m(0, \pm 1)\}$. On the other hand, the restriction of $\rho : SL(2, \mathbb{C}) \to SO_o(3, 1)$ to $G$ is a 2-to-1 homomorphism from $G$ to the little group of $p$, with the same kernel. The latter groups are therefore isomorphic.  $\square$

The Poincaré group is the group of transformations of space-time that arises in special relativity, and its representations are of importance in relativistic quantum mechanics. In particular, the representation derived from the orbit $\mathcal{O}_\lambda^+$ and the representation $\rho_k$ of the little group $SO(3)$ (cf. Corollary 5.48) is connected with the description of particles of mass $m = \sqrt{\lambda}$ and spin $k$. (For particles of half-integer spin one must pass to the double cover of the Poincaré group, isomorphic to $\mathbb{R}^4 \ltimes SL(2, \mathbb{C})$ as indicated above, where the corresponding little group is $SU(2)$.) Likewise, the representation derived from the orbit $\mathcal{O}_0^+$ and the character $(b, T_\theta) \mapsto e^{ik\theta}$ of the little group $E(2)$ (where $T_\theta$ is rotation through the angle $\theta$) has to do with particles of mass zero and spin $k$. For more details, see Simms [128] or Varadarajan [138].

**4. The Heisenberg groups.** For $n \geq 1$, the **Heisenberg group** $H_n$ is the group whose underlying space is $\mathbb{R}^n \times \mathbb{R}^n \times \mathbb{R}$ and whose group law is

$$(6.47) \quad (x, \xi, t)(x', \xi', t') = \big(x + x', \ \xi + \xi', \ t + t' + \tfrac{1}{2}(x \cdot \xi' - \xi \cdot x')\big).$$

With the notation $X = (x, \xi, t)$, it is easily verified that the sets

$$N_1 = \{X : x = 0\}, \quad N_2 = \{X : \xi = 0\},$$
$$K_1 = \{X : \xi = 0, \ t = 0\}, \quad K_2 = \{X : x = 0, \ t = 0\}$$

are Abelian subgroups of $H_n$ such that

$$H_n = N_1 \ltimes K_1 = N_2 \ltimes K_2.$$

Theorem 6.43 can therefore be applied to determine the irreducible representations of $H_n$. We leave this as an exercise for the reader; instead, we shall give a different argument based directly on the imprimitivity theorem that yields a stronger result.

For each nonzero real number $h$ there is a representation $\rho_h$ of $H_n$ on $L^2(\mathbb{R}^n)$ defined by

$$[\rho_h(x,\xi,t)f](y) = e^{2\pi i h t + \pi i h \xi \cdot x} e^{-2\pi i h \xi \cdot y} f(y-x).$$

(These differ from the representations called $\rho_h$ in Folland [42] but are equivalent to them via the map $Uf(y) = |h|^{1/2} f(-hy)$. They are equivalent to representations of $H_n$ induced from one-dimensional representations of the subgroup $N_1$, as the reader may verify; cf. the second remark preceding (6.3).) Let

$$\rho_h'(x) = \rho_h(x,0,0), \qquad \rho_h''(\xi) = \rho_h(0,\xi,0).$$

Then $\rho_h'$ and $\rho_h''$ are representations of $\mathbb{R}^n$ on $L^2(\mathbb{R}^n)$; in fact, $\rho_h'$ is the regular representation of $\mathbb{R}^n$, and $\rho_h''$ is the regular representation conjugated by the Fourier transform and rescaling by the factor $h$. Since

$$(6.48) \qquad (x,0,0)(0,\xi,0) = (x,\xi,\tfrac{1}{2}x\cdot\xi) = (0,0,x\cdot\xi)(0,\xi,0)(x,0,0),$$

we have

$$(6.49) \qquad \rho_h'(x)\rho_h''(\xi) = e^{2\pi i h x \cdot \xi}\rho_h''(\xi)\rho_h'(x).$$

These are the integrated form of the canonical commutation relations of quantum mechanics, in which $h$ is Planck's constant; hence the name "Heisenberg group." (See Folland [42] or Mackey [99] for more details.) The crucial point is that the relations (6.49) essentially determine the representation $\rho_h$ uniquely.

**6.50 Theorem** (The Stone-von Neumann Theorem).

a. *Suppose $\pi'$ and $\pi''$ are unitary representations of $\mathbb{R}^n$ on a Hilbert space $\mathcal{H}$ that satisfy*

$$(6.51) \qquad \pi'(x)\pi''(\xi) = e^{2\pi i h x \cdot \xi}\pi''(\xi)\pi'(x) \qquad (x,\xi \in \mathbb{R}^n)$$

*for some $h \neq 0$. Then there is a unitary isomorphism from $\mathcal{H}$ to a direct sum of copies of $L^2(\mathbb{R}^n)$ such that $U\pi'(x)U^{-1} = \rho_h'(x)$ and $U\pi''(\xi)U^{-1} = \rho_h''(\xi)$ on each copy.*

b. *If $\pi$ is a unitary representation of $H_n$ such that $\pi(0,0,t) = e^{2\pi i h t} I$ where $h \neq 0$, then $\pi$ is unitarily equivalent to a direct sum of copies of $\rho_h$.*

*Proof.* We first observe that assertions (a) and (b) are equivalent. Indeed, if $\pi$ is a representation of $H_n$ such that $\pi(0,0,t) = e^{2\pi i h t} I$, then the representations $\pi'(x) = \pi(x,0,0)$ and $\pi''(\xi) = \pi(0,\xi,0)$ of $\mathbb{R}^n$ satisfy (6.51), because of (6.48). Conversely, if $\pi'$ and $\pi''$ are representations of $\mathbb{R}^n$ satisfying (6.51), it is easy to check that

$$\pi(x,\xi,t) = e^{2\pi i h t - \pi i h x \cdot \xi}\pi'(x)\pi''(\xi) = e^{2\pi i h t + \pi i h x \cdot \xi}\pi''(\xi)\pi'(x)$$

is a representation of $H_n$, and it satisfies $\pi(0,0,t) = e^{2\pi i h t} I$.

Moreover, it suffices to consider $h = 1$. In (a) one simply replaces the representation $\pi''$ by the representation $\xi \mapsto \pi''(\xi/h)$, and in (b) one replaces $\pi$ by $\pi \circ \alpha$, where the automorphism $\alpha$ of $H_n$ is given by $\alpha(x,\xi,t) = (x,\xi/h,t/h)$.

Suppose then that $\pi'$ and $\pi''$ satisfy (6.51) with $h = 1$. By Theorem 4.45, there is a unique regular projection-valued measure on $\mathbb{R}^n$ such that $\pi''(\xi) = \int e^{-2\pi i \xi \cdot y}\, dP(y)$ for $\xi \in \mathbb{R}^n$. Then, for each $x \in \mathbb{R}^n$, the projection-valued measures associated to the representations $\xi \mapsto \pi'(x)\pi''(\xi)\pi'(x)^{-1}$ and $\xi \mapsto e^{2\pi i x \cdot \xi}\pi''(\xi)$ are $E \mapsto \pi'(x)P(E)\pi'(x)^{-1}$ and $E \mapsto P(x+E)$ (since $\int e^{2\pi i \xi \cdot (x-y)}\, dP(y) = \int e^{-2\pi i \xi \cdot y}\, dP(x+y)$). But these two representations are equal, so $\pi'(x)P(E)\pi'(x)^{-1} = P(x + E)$. In other words, $(\pi', \mathbb{R}^n, P)$ is a transitive system of imprimitivity on $\mathbb{R}^n$. By the imprimitivity theorem, it is equivalent to the system induced by a representation of the trivial subgroup $\{0\}$ on some Hilbert space, necessarily a direct sum of copies of the (unique) representation $\tau$ of $\{0\}$ on $\mathbb{C}$. But $\mathrm{ind}_{\{0\}}^{\mathbb{R}^n}(\tau)$ is just the left regular representation of $\mathbb{R}^n$, namely $\rho_1'$, and the associated projection-valued measure is $P_\tau(E) = $ multiplication by $\chi_E$, so that

$$\int e^{-2\pi i \xi \cdot y}\, dP_\tau(y) = \text{multiplication by } e^{-2\pi i \xi \cdot y} = \rho_1''(\xi).$$

This proves (a) for the case $h = 1$ and hence establishes the theorem. $\square$

*Remark.* This argument also shows that the representations $\rho_h$ are irreducible. Indeed, it is easy to check that $\mathcal{C}(\rho_1) = \mathcal{C}(\rho_1')\cap\mathcal{C}(\rho_1'')$. In the notation established above, we have $\mathcal{C}(\rho_1') \cap \mathcal{C}(\rho_1'') = \mathcal{C}(\rho_1', \mathbb{R}^n, P_\tau)$, and by Theorem 6.28 this is isomorphic to $\mathcal{C}(\tau) = \mathbb{C}I$. So $\rho_1$ is irreducible, and hence so is every $\rho_h$. This irreducibility can also be established by more elementary means, as can the Stone-von Neumann theorem itself; cf. Folland [42, Proposition 1.43 and Theorem 1.50].

**6.52 Corollary.** *Every irreducible representation of $H_n$ is equivalent to one and only one of the following:*
   i. $\rho_h$ $(h \neq 0)$,

ii. the representation $\pi_{b,\beta}$ $(b,\beta \in \mathbb{R}^n)$ on $\mathbb{C}$ defined by $\pi_{b,\beta}(x,\xi,t) = e^{2\pi i(b\cdot x+\beta\cdot\xi)}$.

*Proof.* We observe that the center of $H_n$ is $Z = \{(0,0,t) : t \in \mathbb{R}\}$. If $\pi$ is irreducible, by Schur's lemma $\pi|Z$ must act as scalar multiples of $I$, so $\pi(0,0,t) = e^{2\pi i h t}I$ for some $h \in \mathbb{R}$. If $h \neq 0$ then $\pi$ is equivalent to $\rho_h$ by Theorem 6.50. If $h = 0$ then $\pi$ factors through $H_n/Z \cong \mathbb{R}^n \times \mathbb{R}^n$ and so is of the form $\pi_{b,\beta}$. Of course the representations $\rho_h$ and $\pi_{b,\beta}$ are all inequivalent (the former are already inequivalent on $Z$). $\qquad\square$

## 6.8 More Examples, Good and Bad

In this section we present some further examples to illustrate various phenomena, some pleasant and some pathological, that can arise in applying the theory of induced representations.

**1. The Mautner group.** The **Mautner group** $M$ is the simplest example of a connected Lie group with a pathological representation theory. Its underlying set is $\mathbb{C}^2 \times \mathbb{R}$, and its group law is given by

$$(z,w,t)(z',w',t') = (z + e^{it}z', \, w + e^{2\pi it}w', \, t+t') \qquad (z,w \in \mathbb{C}, \, t \in \mathbb{R}).$$

From this it is easily checked that $M$ is a semidirect product of $\mathbb{C}^2$ (the normal factor) and $\mathbb{R}$.

We identify $(\mathbb{C}^2)\widehat{\phantom{x}}$ with $\mathbb{C}^2$ via the pairing

$$\langle (z,w), \, (\zeta,\omega) \rangle = e^{2\pi i \, \mathrm{Re}(z\overline{\zeta}+w\overline{\omega})}.$$

(This is the usual identification of $(\mathbb{R}^4)\widehat{\phantom{x}}$ with $\mathbb{R}^4$, written in complex coordinates.) A simple calculation then shows that the action of $M$ on $(\mathbb{C}^2)\widehat{\phantom{x}}$ is given by

$$(z,w,t)(\zeta,\omega) = (e^{it}\zeta, \, e^{2\pi it}\omega).$$

The origin $\{(0,0)\}$ and the circles $\{(0,\omega) : |\omega| = r\}$ and $\{(\zeta,0) : |\zeta| = r\}$ $(r > 0)$ are orbits. The tori

$$T_{rs} = \big\{(\zeta,\omega) : |\zeta| = r, \, |\omega| = s\big\} \qquad (r,s > 0)$$

are invariant under $M$, but they are not orbits: the orbits in $T_{rs}$ are curves that wind around $T_{rs}$ without ever closing up, each of which is dense in $T_{rs}$. (If we identify $T_{rs}$ with $\mathbb{R}^2/\mathbb{Z}^2$ in the obvious way, the inverse image of the orbit through $(re^{2\pi ia}, se^{2\pi ib})$ in $\mathbb{R}^2$ is the union of the lines with slope $2\pi$ and $y$-intercept of the form $b - a + j + 2\pi k$ where $j, k \in \mathbb{Z}$; this set is dense in the plane.)

From this it is clear that the regularity condition (R2) or (R2$'$) fails. We can also see that (R1) fails, as follows. For simplicity we consider the torus $T_{11}$, which is just the standard 2-torus $\mathbb{T}^2$, and we denote normalized Haar measure on $T_{11}$ by $\mu$.

**6.53 Lemma.** *If $f \in L^2(\mu)$ is invariant under the action of $M$, then $f$ is constant $\mu$-a.e.*

*Proof.* We expand $f$ in a Fourier series:

$$f(\zeta, \omega) = \sum c_{mn} \zeta^m \omega^n, \text{ where } c_{mn} = \int f(\zeta, \omega) \overline{\zeta^m \omega^n} \, d\mu(\zeta, \omega).$$

The invariance means that $e^{it(m+2\pi n)} c_{mn} = c_{mn}$ for all $t$, and hence $c_{mn} = 0$ unless $m = n = 0$. $\quad\square$

If $E$ is an $M$-invariant Borel set in $T_{11}$, by applying Lemma 6.53 to $\chi_E$ we see that $\mu(E)$ is either 0 or 1. If there were a countable separating family for the orbits (in $\mathbb{C}^2$ and *a fortiori* in $T_{11}$), the argument that proves Proposition 6.37 would show that there is an orbit $\mathcal{O} \subset T_{11}$ such that $\mu(\mathcal{O}) = 1$. But $\mathcal{O}$ intersects each circle $\zeta = \text{const.}$ in a countable set, so Fubini's theorem implies that $\mu(\mathcal{O}) = 0$. Hence condition (R1) fails.

Finally, it is easy to produce an irreducible representation $\pi$ of $M$ such that the projection-valued measure associated to $\pi|\mathbb{C}^2$ is not concentrated on an orbit. Indeed, let $\mu$ be Haar measure on $T_{11}$ as above, and define the representation $\pi$ on $L^2(\mu)$ by

$$\pi(z, w, t) f(\zeta, \omega) = e^{2\pi i \, \mathrm{Re}(z\bar{\zeta} + w\bar{\omega})} f(e^{-it}\zeta, e^{-2\pi it}\omega).$$

We leave it to the reader to check that $\pi$ is indeed a representation and that the projection-valued measure associated to $\pi|\mathbb{C}^2$ is just $P(E) =$ multiplication by $\chi_{E \cap T_{11}}$, so that $P$ has the same nullsets as $\mu$ and hence is not concentrated on an orbit. To see that $\pi$ is irreducible, suppose $T \in \mathcal{C}(\pi)$. Since $T$ commutes with $\pi(z, w, 0)$ for all $z$ and $w$, it commutes with multiplication by any $g \in C(T_{11})$. But then $Tg = T(g \cdot 1) = g \cdot T1$ for $g \in C(T_{11})$, and it follows that $T$ is multiplication by $f = T1$. But then, since $T$ commutes with $\pi(0, 0, t)$ for $t \in \mathbb{R}$, Lemma 6.53 implies that $f$ is constant $\mu$-a.e., and so $T = cI$.

**2. The discrete Heisenberg group and its central quotients.**
We begin by presenting a slightly different parametrization of the Heisenberg group $H_1$ from the preceding section. Namely, let $\widetilde{H}_1$ be $\mathbb{R}^3$

equipped with the group law

$$(x, \xi, t)(x', \xi', t') = (x + x', \, \xi + \xi', \, t + t' + x\xi'),$$
$$(x, \xi, t)^{-1} = (-x, -\xi, -t + x\xi).$$

It is easily verified that the map $(x, \xi, t) \mapsto (x, \xi, t + \frac{1}{2}x\xi)$ is an isomorphism from $H_1$ to $\widetilde{H}_1$. The advantage of $\widetilde{H}_1$ is that the lattice of points whose coordinates are all integers is a subgroup. We call it the **discrete Heisenberg group** and denote it by H:

(6.54)
$$H = \mathbb{Z}^3 \text{ with group law}$$
$$(j, k, l)(j', k', l') = (j + j', \, k + k', \, l + l' + jk').$$

(Of course there are many other discrete subgroups of $\widetilde{H}_1$, and one can also consider discrete subgroups of higher-dimensional Heisenberg groups; see Folland [46] for more about them. But H will suffice to illustrate the phenomena that we wish to exhibit.)

The center and the commutator subgroup of H are both equal to

$$Z = \big\{(0, 0, l) : l \in \mathbb{Z}\big\}.$$

We shall also be concerned with the quotients of H by the subgroups of Z. Namely, for each positive integer $q$ we have the group

$$H_q = H/q Z = \big\{(j, k, l) : j, k \in \mathbb{Z}, \, l \in \mathbb{Z}/q\mathbb{Z}\big\}.$$

The group law is again given by (6.54), except that the addition in the last coordinate is computed modulo $q$, and the center and commutator subgroup of $H_q$ are

$$Z_q = Z/q Z = \big\{(0, 0, l) : l \in \mathbb{Z}/q\mathbb{Z}\big\}.$$

We wish to investigate the irreducible representations of H and $H_q$, the latter being essentially identical to the irreducible representations of H that are trivial on $q Z$. The first step in classifying these representations is the observation that, by Schur's lemma, if $\pi$ is an irreducible representation of H, its restriction to Z acts by scalar multiples of the identity; thus, $\pi(0, 0, l) = \chi(l)I$ for some character $\chi$ of Z, which we call the **central character** of $\pi$. More concretely, we have $\chi(l) = e^{2\pi i \omega l}$ for some real number $\omega$ determined modulo $\mathbb{Z}$. We regard $\omega$ as an element of $\mathbb{R}$ or of $\mathbb{R}/\mathbb{Z}$ as convenience dictates, and we adopt the notation

$$\chi_\omega(l) = e^{2\pi i \omega l}.$$

The central characters of irreducible representations of $H_q$ are of the same form except that $\omega$ must be of the form $p/q$ for some $p \in \mathbb{Z}$ determined modulo $q$.

The Mackey machine is ideally suited to describing the irreducible representations of the groups $H_q$. Indeed, $H_q$ is the semi-direct product of the Abelian subgroups

$$N = \{(0, k, l) : k \in \mathbb{Z},\ l \in \mathbb{Z}/q\mathbb{Z}\}, \quad H = \{(j, 0, 0) : j \in \mathbb{Z}\}.$$

$\widehat{N}$ consists of the characters

$$\nu_{\beta,\gamma}(0, k, l) = e^{2\pi i[\beta k + (\gamma/q)l))} \quad (\beta \in \mathbb{R}/\mathbb{Z},\ \gamma \in \mathbb{Z}/q\mathbb{Z}).$$

Since $(j, 0, 0)(0, k, l)(j, 0, 0)^{-1} = (0, k, l + jk)$, the action of $H$ on $\widehat{N}$ is given by

$$[(j, 0, 0)\nu_{\beta,\gamma}](0, k, l) = e^{2\pi i[\beta k + (\gamma/q)(l+jk)]},$$

that is,

(6.55) $$(j, 0, 0)\nu_{\beta,\gamma} = \nu_{\beta+(\gamma/q)j,\gamma}.$$

Thus the orbits are finite sets (remember that $(\gamma/q)j$ is taken mod 1), and it follows easily that the action of $H$ is regular. Thus, by Theorem 6.43, the ingredients for constructing an irreducible representation of $H_q$ are a character $\nu_{\beta,\gamma}$ of $N$ and a character of the corresponding little group in $H$. Given such a pair of characters, it is easy to see on *a priori* grounds — and will follow from the explicit calculations that we are about to perform — that the corresponding representation of $H_q$ will have central character $\chi_{p/q}$ if and only if $\nu_{\beta,\gamma}(0, 0, l) = \chi_{p/q}(l)$, that is, $\gamma = p$.

To find all the irreducible representations of $H_q$ (for all $q$) it is enough to find the ones whose central character $\chi_{p/q}$ is faithful, that is, for which $p$ and $q$ are relatively prime, for the ones where $\gcd(p, q) = d > 1$ are just representations of $H_{q/d}$ lifted to $H_q$. If $p$ and $q$ are relatively prime, the little group of $\nu_{\beta,p}$ is $qH = \{(j, 0, 0) : q|j\}$, and its characters are naturally parametrized by real numbers modulo rational numbers with denominator $q$:

$$\rho_\alpha(j, 0, 0) = e^{2\pi i \alpha j} \quad (j \in q\mathbb{Z},\ \alpha \in \mathbb{R}/q^{-1}\mathbb{Z}).$$

The irreducible representations of $H_q$ with central character $\chi_{p/q}$ ($\gcd(p, q) = 1$) are therefore obtained by inducing a character $\rho_\alpha \nu_{\beta,p}$ of $N \ltimes qH$ (this group is Abelian, i.e., the product is actually direct) up to $H_q$ to obtain a representation that we shall call $\pi_{\alpha,\beta}^{(p,q)}$. Since an element $(j, k, l) \in H_q$ belongs to $N \ltimes qH$ if and only if $q$ divides $j$, we see that $N \ltimes qH$ is of index $q$ in $H_q$ and hence that $\pi_{\alpha,\beta}^{(p,q)}$ is $q$-dimensional.

According to the definition at the beginning of §6.1, it acts on the space $\mathcal{F}_0$ of all functions $f : \mathsf{H}_q \to \mathbb{C}$ that satisfy

$$f((j,k,l)(r,s,t)) = e^{-2\pi i[\alpha r + \beta s + (p/q)t]} f(j,k,l) \text{ for } r \in q\mathbb{Z}$$

by left translation. However, since $(j,k,l) = (j,0,0)(0,k,l-jk)$, functions in $\mathcal{F}_0$ are determined by their restrictions to $H$,

$$f(j,k,l) = e^{-2\pi i[\beta k + (p/q)(l-jk)]} f(j,0,0),$$

and the action on these restrictions is given by

$$\begin{aligned}
\pi_{\alpha,\beta}^{(p,q)}(j,k,l)f(m,0,0) &= f((-j,-k,-l+jk)(m,0,0)) \\
&= f(m-j,-k,-l+jk) \\
&= f((m-j,0,0)(0,-k,-l+mk)) \\
&= e^{2\pi i[\beta k + (p/q)(l-mk)]} f(m-j,0,0).
\end{aligned}$$

Moreover, these restrictions have the following quasi-periodicity property:

$$f(j+mq,0,0) = f((j,0,0)(mq,0,0)) = e^{-2\pi i\alpha mq} f(j,0,0) \quad (m \in \mathbb{Z}).$$

They are therefore determined by their values $f(j,0,0)$ with $j = 0,1,\ldots,q-1$, in accordance with the fact that $\mathcal{F}_0$ is $q$-dimensional. However, the description of $\pi_{\alpha,\beta}^{(p,q)}$ is simpler if we do not make this last reduction. Rather, we take $\pi_{\alpha,\beta}^{(p,q)}$ — in its final version — to act on

$$(6.56) \qquad \mathcal{H}_\alpha^q = \{f : \mathbb{Z} \to \mathbb{C} : f(j+mq) = e^{-2\pi i\alpha mq} f(j) \text{ for } m \in \mathbb{Z}\}$$

(with norm $\|f\|^2 = \sum_{J+1}^{J+q} |f(j)|^2$, independent of $J$) by

$$(6.57) \qquad \pi_{\alpha,\beta}^{(p,q)}(j,k,l)f(m) = e^{2\pi i[\beta k + (p/q)(l-mk)]} f(m-j).$$

(It is curious that this formula for $\pi_{\alpha,\beta}^{(p,q)}$ depends only on $\beta$, whereas the $\alpha$-dependence comes in the space $\mathcal{H}_\alpha^q$ on which it acts.)

The representations $\pi_{\alpha,\beta}^{(p,q)}$ $(\alpha,\beta \in \mathbb{R})$ give a complete list of irreducible representations of $\mathsf{H}$ or $\mathsf{H}_q$ with central character $\chi_r$ when $r = p/q$ in lowest terms, up to equivalence. It is obvious from the definitions that $\pi_{\alpha,\beta}^{(p,q)}$ depends only on the equivalence classes of $\alpha$ in $\mathbb{R}/q^{-1}\mathbb{Z}$ and of $\beta$ in $\mathbb{R}/\mathbb{Z}$. However, $\pi_{\alpha,\beta}^{(p,q)}$ and $\pi_{\alpha,\beta'}^{(p,q)}$ are also equivalent (although not identical) when $\nu_{\beta,p}$ and $\nu_{\beta',p}$ belong to the same orbit in $\widehat{N}$. By (6.55), this happens precisely when $\beta' = \beta \mod (p/q)\mathbb{Z}$, that is — since the map $j \mapsto pj$ is an isomorphism on $\mathbb{Z}/q\mathbb{Z}$ ($p$ and $q$ being relatively prime) — when $\beta' = \beta$ mod $q^{-1}\mathbb{Z}$. (It is an easy exercise to check that when $\beta' = \beta + (p/q)j$, the intertwining map $T : \mathcal{H}_\alpha^q \to \mathcal{H}_\alpha^q$ from $\pi_{\alpha,\beta}^{(p,q)}$ to $\pi_{\alpha,\beta'}^{(p,q)}$ is given by $Tf(m) = f(m+j)$.) To sum up, we have proved the following:

**6.58 Theorem.** *Given positive integers $p, q$ with $\gcd(p,q) = 1$, the map $(\alpha, \beta) \mapsto [\pi_{\alpha,\beta}^{(p,q)}]$ defined by (6.56) and (6.57) is a bijection from the 2-torus $(\mathbb{R}/q^{-1}\mathbb{Z})^2$ to the set of equivalence classes of irreducible representations of $\mathsf{H}$ or $\mathsf{H}_q$ with central character $\chi_{p/q}$.*

This immediately yields a complete list of irreducible representations of $\mathsf{H}_q$:

**6.59 Corollary.** *Every irreducible representation of $\mathsf{H}_q$ belongs to an equivalence class $[\pi_{\alpha,\beta}^{(p',q')}]$ for some divisor $q'$ of $q$ (including $q$ and 1), some $p' \in \mathbb{Z}/q'\mathbb{Z}$ with $\gcd(p',q') = 1$, and some $\alpha, \beta \in \mathbb{R}/(q')^{-1}\mathbb{Z}$. These equivalence classes are all distinct.*

It is of interest that even though $\mathsf{H}_q$ is neither Abelian nor compact, its irreducible representations are all finite-dimensional; indeed, their dimensions have a finite upper bound (namely, $q$).

What remains is to analyze the irreducible representations of $\mathsf{H}$ with central character $\chi_\omega$ where $\omega$ is irrational, and here all hell breaks loose. For the remainder of this discussion we consider a fixed $\omega \in \mathbb{R} \setminus \mathbb{Q}$.

As with $\mathsf{H}_q$, $\mathsf{H}$ is a semi-direct product $N \ltimes H$ with

$$N = \{(0, k, l) : k, l \in \mathbb{Z}\}, \quad H = \{(j, 0, 0) : j \in \mathbb{Z}\},$$

$\widehat{N}$ consists of the characters

$$\nu_{\beta,\gamma}(k, l) = e^{2\pi i(\beta k + \gamma l)} \quad (\beta, \gamma \in \mathbb{R}/\mathbb{Z}),$$

and the action of $H$ on $\widehat{N}$ is given by

$$(j, 0, 0)\nu_{\beta,\gamma} = \nu_{\beta + j\gamma, \gamma}.$$

But this action is *not regular*. When $\gamma = \omega$, the orbit $\{(\beta + j\omega, \omega) : j \in \mathbb{Z}\}$ of $\beta$ in the parameter space $(\mathbb{R}/\mathbb{Z})^2$ is a dense set in the circle $\gamma = \omega$. An argument similar to the proof of Lemma 6.53 shows that these orbits are not countably separated; they are plainly not homeomorphic to $H/N \cong \mathbb{Z}$ or open in their closures; and not only is there no Borel cross-section for them, there is not even a Lebesgue measurable one. Indeed, cross-sections for these orbits in the circle $\mathbb{R}/\mathbb{Z}$ are essentially the canonical examples of Lebesgue non-measurable sets. (The usual examples are the cross-sections for the cosets of $\mathbb{Q}$ in $\mathbb{R}/\mathbb{Z}$, but the cosets of $\omega\mathbb{Z}$ work just as well; the crucial property that $\mathbb{Q}$ and $\omega\mathbb{Z}$ have in common is that their images in $\mathbb{R}/\mathbb{Z}$ are countably infinite.)

Although the Mackey machine in the form of Theorem 6.43 breaks down, we can still use parts of it to construct irreducible representations of H. The little group of $\nu_{\beta,\omega}$ is trivial, so it is just a question of inducing $\nu_{\beta,\omega}$ up to H to obtain a representation that we call $\pi_\beta^\omega$. A calculation like the one leading to (6.57) shows that this representation can be conveniently realized on $l^2 = l^2(\mathbb{Z})$ by

$$(6.60) \qquad \pi_\beta^\omega(j,k,l)f(m) = e^{2\pi i[\beta k + \omega(l - mk)]} f(m - j).$$

**6.61 Proposition.** *The representation $\pi_\beta^\omega$ of H defined by (6.60) is irreducible.*

*Proof.* Suppose $\mathcal{V}$ is a nontrivial invariant subspace of $l^2$, $0 \neq f \in \mathcal{V}$, and $g \perp \mathcal{V}$. Thus, for all $j,k,l \in \mathbb{Z}$,

$$0 = \langle \pi_\beta^\omega(j,k,l)f, g \rangle = e^{2\pi i(\beta k + \omega l)} \sum_m e^{-2\pi i \omega k m} f(m - j)\overline{g(m)}.$$

Since $f(\cdot - j)\overline{g(\cdot)} \in l^1$, the function $\phi(\theta) = \sum_m e^{-2\pi i m \theta} f(m - j)\overline{g(m)}$ is continuous on $\mathbb{R}/\mathbb{Z}$, and it vanishes at $\theta = \omega k$ for all $k$. Hence it vanishes identically, so its Fourier coefficients $f(m - j)\overline{g(m)}$ all vanish. But $f(m_0) \neq 0$ for at least one $m_0$, and by taking $j = m - m_0$ we see that $g(m) = 0$ for all $m$. Hence $\mathcal{V} = l^2$. $\qquad \square$

Thus we still obtain a nice family of irreducible representations of H. However, there are two problems with it:

1. The representations $\pi_\beta^\omega$ and $\pi_{\beta'}^\omega$ are equivalent if and only if $\beta$ and $\beta'$ belong to the same orbit $\{\beta + j\omega : j \in \mathbb{Z}\}$. Consequently, there is no Lebesgue measurable way to parametrize the equivalence classes.

2. There are *many* irreducible representations of H with central character $\chi_\omega$ that are not equivalent to any $\pi_\beta^\omega$.

Let us elaborate on these points. Concerning (1), if $\beta' = \beta + j\omega$, the map $Tf(m) = f(m + j)$ on $l^2$ intertwines $\pi_\beta^\omega$ and $\pi_{\beta'}^\omega$. On the other hand, the projection-valued measure $P_\beta^\omega$ on $\widehat{N}$ associated to $\pi_\beta^\omega$ by (6.35) is supported on the orbit $\{\nu_{\beta + j\omega, \omega} : j \in \mathbb{Z}\}$, so by the uniqueness in Theorem 4.45, different orbits give inequivalent representations.

Concerning (2), we now present a way to construct a much larger family of irreducible representations of H. First, some terminology. Define

$S : \mathbb{R}/\mathbb{Z} \to \mathbb{R}/\mathbb{Z}$ by $S(t) = t-\omega$, and suppose $\mu$ is a $\sigma$-finite Borel measure on $\mathbb{R}/\mathbb{Z}$. (In what follows, all measures in question will be assumed $\sigma$-finite. In fact, there is no harm in assuming that they are finite, as every $\sigma$-finite measure is equivalent to a finite one.) We say that $\mu$ is **ergodic** (under $S$) if for any $S$-invariant set $E \subset \mathbb{R}/\mathbb{Z}$, either $\mu(E) = 0$ or $\mu((\mathbb{R}/\mathbb{Z}) \setminus E) = 0$; and we say that $\mu$ is **quasi-invariant** (under $S$) if the measures $\mu_j(E) \equiv \mu(S^j(E)) = \mu(E - j\omega)$ $(j \in \mathbb{Z})$ are all equivalent.

**6.62 Lemma.** *If $\mu$ and $\mu'$ are ergodic and quasi-invariant, they are either equivalent or mutually singular.*

*Proof.* If $\mu$ and $\mu'$ are not equivalent, there is a set $E$ such that $\mu(E) = 0$ and $\mu'(E) > 0$, or vice versa; we assume the former. Let $F = \bigcup_{j \in \mathbb{Z}} S^j(E)$. Since $\mu$ is quasi-invariant, we have $\mu(S^j(E)) = 0$ for all $j$ and hence $\mu(F) = 0$; on the other hand, $\mu'(F) \geq \mu'(E) > 0$, so since $\mu'$ is ergodic, we have $\mu'((\mathbb{R}/\mathbb{Z}) \setminus F) = 0$. Thus $\mu$ and $\mu'$ are mutually singular. $\qquad \square$

**6.63 Theorem.** *Given an ergodic, quasi-invariant measure $\mu$ on $\mathbb{R}/\mathbb{Z}$, define the operators $\rho_\mu(j, k, l)$ on $L^2(\mu)$ by*

$$(6.64) \qquad \rho_\mu(j, k, l)f(t) = e^{2\pi i(kt+\omega l)} f(t + j\omega)\sqrt{(d\mu_j/d\mu)(t)}.$$

*Then $\rho_\mu$ is an irreducible representation of $\mathsf{H}$. Moreover, if $\mu'$ is another such measure, $\rho_\mu$ and $\rho_{\mu'}$ are equivalent if and only if $\mu$ and $\mu'$ are equivalent.*

*Proof.* We leave it to the reader to check that $\rho_\mu$ really is a unitary representation, supplying only the hint that $d\mu_j(t) = d\mu(t + j\omega)$.

The proof of irreducibility is similar to that of Proposition 6.61. If $\mathcal{V}$ is an invariant subspace of $L^2(\mu)$, $0 \neq f \in \mathcal{V}$, and $g \perp \mathcal{V}$, then for all $j, k, l$ we have

$$0 = \langle \rho_\mu(j, k, l)f, g \rangle = e^{2\pi i\omega l} \int e^{2\pi ikt} \sqrt{(d\mu_j/d\mu)(t)} \, f(t + j\omega)\overline{g(t)} \, d\mu(t).$$

By the Fourier uniqueness theorem and the fact that $d\mu_j/d\mu > 0$ $\mu$-a.e., the complex measure $f(t + j\omega)\overline{g(t)} \, d\mu(t)$ vanishes for all $j$, i.e., $f(t + j\omega)\overline{g(t)} = 0$ for a.e. $t$ (with respect to $\mu$) and all $j$. Thus, if $E = \{t : f(t) \neq 0\}$, we see that $g = 0$ a.e. on $S^j(E)$ for all $j$ and hence $g = 0$ a.e. on $\bigcup_j S^j(E)$. But since $\mu$ is ergodic, the complement of the latter set has measure zero, so $g = 0$ a.e. on $\mathbb{R}/\mathbb{Z}$; hence $\mathcal{V} = L^2(\mu)$.

If $\mu$ and $\mu'$ are equivalent, the map $f \mapsto f\sqrt{d\mu/d\mu'}$ is unitary from $L^2(\mu)$ to $L^2(\mu')$ and intertwines $\rho_\mu$ and $\rho_{\mu'}$. If not, then by Lemma 6.62, there are disjoint sets $E$ and $F$ with $E \cup F = \mathbb{R}/\mathbb{Z}$ and $\mu(F) = \mu'(E) = 0$. But then it is easily verified that the projection-valued measures $P_\mu$ and

$P_{\mu'}$ associated to $\rho_\mu$ and $\rho_{\mu'}$ by (6.35) are supported in $E \times \{\omega\}$ and $F \times \{\omega\}$ respectively, so $\rho_\mu$ and $\rho_{\mu'}$ are inequivalent (even on $N$) by the uniqueness in Theorem 4.45. $\qquad\qquad\qquad\qquad\qquad\qquad\qquad\qquad\qquad$ □

The immediate question is now: what are the ergodic quasi-invariant measures on $\mathbb{R}/\mathbb{Z}$? One obvious class of examples is the counting measures $\kappa_\beta$ on the $S$-orbits $\{\beta - m\omega : m \in \mathbb{Z}\}$; these just give us the representations $\pi_\beta^\omega$ that we obtained earlier. (The map $m \mapsto \beta - m\omega$ takes $\mathbb{Z}$ to the orbit of $\beta$ and $l^2(\mathbb{Z})$ to $L^2(\kappa_\beta)$, and it intertwines $\pi_\beta^\omega$ with $\rho_{\kappa_\beta}$.) Another obvious example is Lebesgue measure; this gives us a representation of H analogous to the representation of the Mautner group discussed after Lemma 6.53. But there are *uncountably* many others that are pairwise mutually singular — see Keane [74], Katznelson and Weiss [73], and Moran [107] for three different constructions of uncountable families of mutually singular ergodic, quasi-invariant measures — and classifying all of them (up to equivalence) seems to be a completely hopeless task.

Moreover, this is not the end of the story, for one can obtain many more representations via the following modification of (6.64). A **cocycle** or **multiplier** (for the action $(j, t) \mapsto t + j\omega$ of $\mathbb{Z}$ on $\mathbb{R}/\mathbb{Z}$) is a Borel measurable map $\sigma : \mathbb{Z} \times \mathbb{R}/\mathbb{Z} \to \mathbb{T}$ $(=U(1))$ that satisfies

$$(6.65) \qquad \sigma(j + j', t) = \sigma(j, t)\sigma(j', t + j\omega).$$

If $\mu$ is an ergodic quasi-invariant measure, two cocycles $\sigma$ and $\sigma'$ are $\mu$-**cohomologous** if there is a Borel measurable $\lambda : \mathbb{R}/\mathbb{Z} \to \mathbb{T}$ such that

$$(6.66) \qquad \sigma'(j, t) = \frac{\lambda(t + j\omega)}{\lambda(t)}\sigma(j, t) \text{ for all } j \text{ and } \mu\text{-almost all } t.$$

This is easily seen to be an equivalence relation, and it depends only on the equivalence class of $\mu$.

**6.67 Theorem.** *For any cocycle $\sigma$ and any ergodic quasi-invariant measure $\mu$, let*

$$(6.68) \qquad \rho_{\sigma,\mu}(j, k, l)f(t) = e^{2\pi i(kt + \omega l)}\sigma(j, t)f(t + j\omega)\sqrt{(d\mu_j/d\mu)(t)}.$$

*Then $\rho_{\sigma,\mu}$ is an irreducible representation of H. Moreover, $\rho_{\sigma,\mu}$ and $\rho_{\sigma',\mu'}$ are equivalent if and only if $\mu$ and $\mu'$ are equivalent and $\sigma$ and $\sigma'$ are $\mu$-cohomologous.*

*Proof.* The first statement and the fact that $\rho_{\sigma,\mu}$ and $\rho_{\sigma',\mu'}$ are inequivalent when $\mu$ and $\mu'$ are inequivalent are proved just as in Theorem 6.63. If $\mu$ and $\mu'$ are equivalent, we may assume that $\mu' = \mu$ and that

$\mu$ is finite so that $C(\mathbb{R}/\mathbb{Z}) \subset L^2(\mu)$. Suppose then that there is a unitary intertwining operator $T$ for $\rho_{\sigma,\mu}$ and $\rho_{\sigma',\mu}$. Taking $j = l = 0$, we see that $T$ commutes with multiplication by $e^{2\pi ikt}$ for all $k$, and hence (by taking linear combinations and limits) with multiplication by any $f \in C(\mathbb{R}/\mathbb{Z})$. Let $\lambda = T(1)$; then we have $T(f) = T(f \cdot 1) = fT(1) = \lambda f$ for all $f \in C(\mathbb{R}/\mathbb{Z})$ and hence for all $f \in L^2(\mu)$, and $|\lambda| = 1$ since $T$ is unitary. The fact that $T$ intertwines $\rho_{\sigma,\mu}(j, 0, 0)$ and $\rho_{\sigma',\mu}(j, 0, 0)$ then implies that $\sigma$ and $\sigma'$ are related by (6.66). Conversely, if (6.66) holds, the operator $Tf = \lambda f$ intertwines $\rho_{\sigma,\mu}$ and $\rho_{\sigma',\mu}$. $\qquad\square$

The simplest cocycles are the ones that are constant in $t$, namely, $\sigma_\alpha(j, t) = e^{2\pi i\alpha j}$ ($\alpha \in \mathbb{R}$). It is easy to check that $\sigma_\alpha$ and $\sigma_{\alpha'}$ are $\mu$-cohomologous with $\mu =$ Lebesgue measure if and only if $\alpha' - \alpha \in \mathbb{Z} + \omega\mathbb{Z}$. (Note that as with counting measure on orbits, there is no measurable cross-section for these cohomology classes.) Another simple family is given by $\sigma_{(n)}(j, t) = e^{2\pi i njt + \pi i n\omega j^2}$ ($n \in \mathbb{Z}$), no two of which are $\mu$-cohomologous with $\mu =$ Lebesgue measure. (One might think otherwise because $\sigma_{(n')}$ and $\sigma_{(n)}$ satisfy (6.66) with $\lambda(t) = e^{\pi i(n'-n)t^2/\omega}$, but this $\lambda$ is well defined only on $\mathbb{R}$, not on $\mathbb{R}/\mathbb{Z}$.) There are many others, and as with the ergodic quasi-invariant measures, the problem of exhibiting and classifying them up to $\mu$-cohomology is quite intractable. It is even worse than it might at first appear because the conditions "for $\mu$-almost all $t$" and "for $\mu'$-almost all $t$" as in (6.66) are entirely uncorrelated when $\mu$ and $\mu'$ are mutually singular.

There is yet one further generalization that yields even more new representations: instead of considering scalar-valued functions and cocycles, one can take the function $f$ in (6.68) to take values in a Hilbert space $\mathcal{H}$ (that is, the space on which $\rho_{\sigma,\mu}$ acts is $L^2(\mu, \mathcal{H})$) and the cocycle $\sigma$ to take values in the unitary group $\mathcal{U}(\mathcal{H})$. The cocycle relation is still given by (6.65), but the cohomology relation (6.66) should be restated as $\lambda(t)\sigma'(j, t) = \sigma(j, t)\lambda(t + j\omega)$, $\lambda$ again being $\mathcal{U}(\mathcal{H})$-valued. There is an additional condition for irreducibility of $\rho_{\sigma,\mu}$: $\sigma$ should not be $\mu$-cohomologous to a direct sum (where, if $\sigma_j$ is a $\mathcal{U}(\mathcal{H}_j)$-valued cocycle for $j = 1, 2$, the direct sum $\sigma_1 \oplus \sigma_2$ is defined in the obvious way as a $\mathcal{U}(\mathcal{H}_1 \oplus \mathcal{H}_2)$-valued cocycle). It follows from results of Ramsay [115] that this class of representations is, finally, exhaustive: every irreducible representation of $\mathsf{H}$ with central character $\chi_\omega$ is equivalent to one of them. We shall not insist on the details, as the main point should by now be abundantly clear: when $\omega$ is irrational, the set of equivalence classes of irreducible representations of $\mathsf{H}$ with central character $\chi_\omega$ is, in a quite literal sense, incalculably large.

## 6.9   Notes and References

Induced representations of finite groups were first studied by Frobenius [47]. For general locally compact groups, the notion of induced representation was formalized by Mackey [88], [90]; the "intrinsic version" we use in this chapter is due to Blattner [13].

The Hilbert space $\mathcal{F}$ or $\mathcal{F}_\mu$ on which $\mathrm{ind}_H^G(\sigma)$ acts can be precisely identified as the space of all $\mathcal{H}_\sigma$-valued functions on $G$ that are (in a suitable sense) measurable and square-integrable and that satisfy the appropriate covariance equation with respect to $H$. We have omitted this point since we have no need for the result, but see Blattner [13] or Gaal [48].

With regard to the realization of induced representations on sections of vector bundles: the usual definition of a vector bundle requires that the bundle be locally trivial, that is, that it be locally isomorphic to the product of the base space and the vector space on which the fibers are modelled. It is easy to see that this will happen for the homogeneous vector bundles over $G/H$ on which induced representations live if the fibration of $G$ itself over $G/H$ is locally trivial, i.e., if the quotient map $q : G \to G/H$ admits a local continuous right inverse near some (and hence any) point of $G/H$. When $G$ is a Lie group, the cosets of $H$ form a foliation of $G$, and one easily obtains a smooth local inverse to $q$ near $q(x)$ by considering a submanifold of $G$ passing through $x$ that is transverse to the cosets. However, the following example shows that local triviality of $G$ over $G/H$ is false in general, even for separable compact groups. Let $G = \mathbb{T}^\omega$ be the product of a countable number of circles, and let $H = \{\pm 1\}^\omega$ be the subgroup consisting of points whose coordinates are all $\pm 1$. Since $\mathbb{T}/\{\pm 1\}$ is isomorphic to $\mathbb{T}$, $G/H$ is isomorphic to $G$, and a continuous local inverse to the quotient map on an open set $U \subset G$ is just a map from $U$ to $G$ that is a continuous branch of the square root function in each coordinate. But this is impossible, for any open set contains a set of the form $\prod_1^\infty U_j$ where each $U_j$ is open in $\mathbb{T}$ and all but finitely many are equal to $\mathbb{T}$, and there is no continuous square root function on all of $\mathbb{T}$. (Nonetheless, see Appendix C of Fell and Doran [40], where it is proved that Banach bundles over locally compact spaces always have lots of continuous sections.)

The Frobenius reciprocity theorem for compact groups is due to Weil [146, §23]. Mautner [103] and Mackey [91] have proved versions of the Frobenius Reciprocity Theorem for noncompact groups, but they are much more technical both to state and to prove since representations must generally be decomposed as direct integrals rather than direct sums of irreducible components. A condition on groups $G$ and closed subgroups $H$ known as "weak Frobenius reciprocity" was introduced by Fell [39] and subsequently studied by several authors; see Fell and Doran [41] and the references given there.

The theorem on induction in stages and the imprimitivity theorem were first proved by Mackey [88], [93], [98] for second countable groups. Loomis [85] extended the imprimitivity theorem to arbitrary locally compact groups by a proof involving functions of positive type. The arguments we have presented here, and in particular the ideas of using pseudomeasures of positive type and the algebra $L(S \times G)$, are due to Blattner [14]. Our exposition also owes much to some unpublished notes of J. M. G. Fell; in particular, these are the source for our proof of Theorem 6.28. Another proof of the imprimitivity theorem has been given by Ørsted [110]. There are also generalizations of the imprimitivity theorem to the settings of representations of C* algebras and representations of Banach *-algebraic bundles, due respectively to Rieffel [119] and Fell and Doran [41].

The historical genesis of the imprimitivity theorem is of some interest. Mackey was led to it by a threefold generalization of the Stone-von Neumann theorem (6.50), the two intermediate stages being the main theorems in Mackey [87]. Suppose $G$ is a separable locally compact Abelian group, and $\pi$ and $\rho$ are representations of $G$ and $\widehat{G}$, respectively, on a Hilbert space $\mathcal{H}$ that satisfy

(6.69)
$$\rho(\xi)\pi(x) = \langle x, \xi \rangle \pi(x)\rho(\xi).$$

The first theorem in [87] is that $\pi$ and $\rho$ are simultaneously equivalent to a direct sum of copies of the representations $\pi_0(x) = L_x$ (the left regular representation) and $\rho_0(\xi)f = \xi f$. Moreover, if $P$ is the projection-valued measure on $G$ ($= \widehat{\widehat{G}}$) associated to $\rho$ as in Theorem 4.45, (6.69) implies that $(\pi, G, P)$ is a system of imprimitivity, and the result just stated says that $(\pi, G, P)$ is equivalent to a direct sum of copies of the canonical system of imprimitivity associated to the left regular representation ($= \operatorname{ind}_{\{1\}}^G(1)$). The second theorem in [87] is that in this form, the result remains valid when $G$ is an arbitrary separable locally compact group. The full imprimitivity theorem is obtained by generalizing this from systems of imprimitivity based on $G$ to those based on $G/H$. See Mackey [97] for a more extensive account of these matters and of various applications of the theory of induced representations.

The results of §6.6 appeared in Mackey [88], [98], and they were generalized by Mackey [93] to groups $G$ with a closed normal subgroup $N$ that need not be Abelian or a factor in a semi-direct product. We shall now briefly sketch how the general theory works. (A further generalization can be found in Fell and Doran [41].)

First, one can replace the hypothesis that $N$ is Abelian by the much weaker hypothesis that $N$ is type I (see §7.2). $\widehat{N}$ is taken to be the set of equivalence classes of irreducible representations of $N$, suitably topologized (see §7.2). $G$ acts on $\widehat{N}$ just as in the Abelian case — the action of $x \in G$ on a representation $\nu$ of $N$ yields the representation $(x\nu)(n) = \nu(x^{-1}nx)$ — and one must assume that this action is regular just as in §6.6. There is an analogue of Theorem 4.45 that gives a decomposition of any representation of $N$ as a direct integral of irreducibles (see §7.4), so as in Proposition 6.36 one can associate to each representation $\pi$ of $G$ a system of imprimitivity $(\pi, \widehat{N}, P)$. If $\pi$ is irreducible, the analogue of Proposition 6.37 holds, so as in Theorems 6.39 and 6.40 one obtains a correspondence between irreducible representations $\pi$ of $G$ and irreducible representations $\sigma$ of the stability groups $G_\nu$ ($\nu \in \widehat{N}$) such that $\sigma|N$ is a direct sum of copies of $\nu$.

One now wishes to analyze these representations of $G_\nu$ in terms of representations of $G_\nu/N$, and here is where the really new feature appears: one must consider not just ordinary representations of $G_\nu/N$ but **projective representations** (also called ray representations or multiplier representations), that is, continuous homomorphisms from $G_\nu/N$ into the unitary group on a Hilbert space modulo the scalar multiples of the identity. This complicates the picture considerably. The compensating virtue is that the whole theory of induced representations can be developed for projective representations, and it is then self-contained: one can analyze the projective representations of $G$ in terms of those of $N$ and those of subgroups of $G/N$.

If $G$ does not act regularly on $\widehat{N}$, Proposition 6.37 breaks down, and so does all of the subsequent analysis. To analyze the representations of $G$ in this case, one must consider not only transitive actions of $G$ on subsets of $\widehat{N}$ (the actions of $G$ on orbits) but ergodic ones — that is, actions of $G$ on subsets of $\widehat{N}$ equipped with ergodic quasi-invariant measures. This leads to a description of the irreducible representations of $G$ which, in the case of semi-direct products $G = N \ltimes H$ where the $G$-action reduces to an $H$-action, is similar to the one we gave for the discrete Heisenberg group in the preceding section. This description can be recast in a form that emphasizes the analogy with induced representations via the notion of "virtual groups," according to which an irreducible representation of $G$ is induced from a representation of a virtual subgroup rather than the honest subgroup $G_\nu$. These ideas were developed by Mackey in the early 1960s; he explicated the concepts related to virtual groups in [95] and gave a lucid sketch of the results for semi-direct products in [96]. However, the theory was worked out in full technical detail only later by Ramsay [115].

The irreducible representations of the $ax + b$ group were first determined by Gelfand and Naimark [49], and those of the Poincaré group by Wigner [150]. The first proof of the Stone-von Neumann theorem appeared in von Neumann [142]; it is reproduced in Folland [42].

A more extensive account of the theory of induced representations and its ramifications can be found in Kaniuth and Taylor [71]. For more about representations of the Mautner group, see Baggett [3], Kirillov [77], and Cowling [27], and for more about representations of the discrete Heisenberg group, see Baggett, Mitchell, and Ramsay [4] and the references given there.

# 7

## Further Topics in Representation Theory

This chapter is a survey of some general results about unitary representations of noncompact, non-Abelian groups, together with some discussion of concrete cases. The proofs of many of the theorems in this subject are lengthy and technical and involve ideas beyond the scope of this book. Hence, to a large extent we shall content ourselves with providing definitions and statements of the theorems, together with references to sources where a detailed treatment can be found. (In particular, the "notes and references" for this material are scattered throughout the chapter instead of being collected in a separate section at the end.)

Our principal object of concern is the set of equivalence classes of irreducible unitary representations of a locally compact group $G$. This set is called the (**unitary**) **dual space** of $G$. As in Chapters 4 and 5, we denote it by $\widehat{G}$, and we denote the equivalence class of an irreducible representation $\pi$ by $[\pi]$.

In this chapter most of the main results are subject to the hypothesis that $G$ is second countable. This has, in particular, the following consequences. First, there is a countable base for the topology of $G$ consisting of sets with compact closure. The linear combinations of their characteristic functions are dense in $L^p(G)$ for $p < \infty$, and hence $L^p(G)$ is separable. Second, $G$ itself is separable, and it follows that if $\pi$ is a cyclic representation of $G$, the Hilbert space $\mathcal{H}_\pi$ is separable. (In particular, $\mathcal{H}_\pi$ is separable whenever $\pi$ is irreducible.) In view of Proposition 3.3, to understand the representations of $G$ it is enough to consider those acting on separable Hilbert spaces.

## 7.1 The Group C* Algebra

Let $G$ be a locally compact group. In order to describe the structure of $\widehat{G}$, we need to digress to construct a modification of the group algebra

$L^1(G)$. If $f \in L^1(G)$, we define

$$\|f\|_* = \sup_{[\pi] \in \widehat{G}} \|\pi(f)\|.$$

Clearly $\| \cdot \|_*$ is a seminorm on $L^1(G)$ that satisfies $\|f\|_* \leq \|f\|_1$.

**7.1 Proposition.** *Let $r(G)$ be the set of equivalence classes of unitary representations of $G$ (irreducible or not); let $\mathcal{P}_1$ be the set of normalized functions of positive type on $G$ as in §3.3, and let $\mathcal{E}(\mathcal{P}_1)$ be the set of extreme points in $\mathcal{P}_1$. Then for any $f \in L^1(G)$,*

$$\|f\|_*^2 = \sup_{[\pi] \in r(G)} \|\pi(f)\|^2 = \sup_{\phi \in \mathcal{P}_1} \int (f^* * f)\phi = \sup_{\phi \in \mathcal{E}(\mathcal{P}_1)} \int (f^* * f)\phi.$$

*Proof.* Denote the four numbers whose equality is asserted, in the order given above, by $A_1, \ldots, A_4$. Trivially, $A_1 \leq A_2$. If $\pi$ is a unitary representation of $G$ and $u$ is a unit vector in $\mathcal{H}_\pi$, let $\phi(x) = \langle \pi(x)u, u \rangle$. Then $\phi \in \mathcal{P}_1$ and $\|\pi(f)u\|^2 = \int (f^* * f)\phi$ (Proposition 3.15), so $A_2 \leq A_3$. That $A_3 \leq A_4$ follows from Theorem 3.27, for if $\phi \in \mathcal{P}_1$, $\int (f^* * f)\phi$ is a limit of convex combinations of numbers of the form $\int (f^* * f)\psi$ with $\psi \in \mathcal{E}(\mathcal{P}_1)$. Finally, if $\phi \in \mathcal{E}(\mathcal{P}_1)$ then $\phi(x) = \langle \pi(x)u, u \rangle$ for some irreducible representation $\pi$ and some unit vector $u \in \mathcal{H}_\pi$ by Proposition 3.20 and Theorem 3.25, whence $\int (f^* * f)\phi = \|\pi(f)u\|^2$, so $A_4 \leq A_1$. $\square$

**7.2 Corollary.** $f \mapsto \|f\|_*$ *is a norm on $L^1(G)$.*

*Proof.* $\| \cdot \|_*$ is obviously a seminorm on $L^1(G)$. If $\|f\|_* = 0$ then $\pi(f) = 0$ for every unitary representation of $G$ by Proposition 7.1. But if we take $\pi$ to be the left regular representation, then $\pi(f)g = f * g$, and $f * g$ approximates $f$ when $g$ is an approximate identity; so $f = 0$. $\square$

We obviously have

$$\|f * g\|_* = \sup_{[\pi] \in \widehat{G}} \|\pi(f)\pi(g)\| \leq \|f\|_* \|g\|_*,$$

$$\|f^*\|_* = \sup_{[\pi] \in \widehat{G}} \|\pi(f)^*\| = \|f\|_*,$$

$$\|f^* * f\|_* = \sup_{[\pi] \in \widehat{G}} \|\pi(f)^*\pi(f)\| = \sup_{[\pi] \in \widehat{G}} \|\pi(f)\|^2 = \|f\|_*^2.$$

Hence the algebra operations and the involution on $L^1(G)$ extend continuously to the completion of $L^1(G)$ with respect to the norm $\| \cdot \|_*$ and make that completion into a C* algebra, called the **group C* algebra** of $G$ and denoted by $C^*(G)$. We shall denote the elements of $C^*(G)$ by letters such as $f$ even though they are not, in general, functions on $G$.

*Example.* Suppose $G$ is Abelian. Then $\|f\|_* = \|\widehat{f}\|_{\sup}$, so $C^*(G)$ is isometrically $*$-isomorphic to $C_0(\widehat{G})$.

Any $*$-representation of $L^1(G)$ extends uniquely to a $*$-representation of $C^*(G)$, so by Theorems 3.9 and 3.11 there is a one-to-one correspondence between unitary representations of $G$ and nondegenerate $*$-representations of $C^*(G)$. If $\pi$ is such a representation, its kernel

$$\ker(\pi) = \{f \in C^*(G) : \pi(f) = 0\}$$

is a closed two-sided ideal of $C^*(G)$. Ideals of the form $\ker(\pi)$ where $\pi$ is *irreducible* are called **primitive ideals** of $C^*(G)$. The space of all primitive ideals of $C^*(G)$ is denoted by $\mathrm{Prim}(G)$:

$$\mathrm{Prim}(G) = \{\ker(\pi) : [\pi] \in \widehat{G}\}.$$

If $U$ is a nonempty subset of $\mathrm{Prim}(G)$, we define $\overline{U} \subset \mathrm{Prim}(G)$ by

$$\overline{U} = \left\{ \mathcal{I} \in \mathrm{Prim}(G) : \mathcal{I} \supset \bigcap_{\mathcal{J} \in U} \mathcal{J} \right\}.$$

We also set $\overline{\varnothing} = \varnothing$.

**7.3 Proposition.** *For any* $U, V \subset \mathrm{Prim}(G)$,

$$\overline{U} \supset U, \qquad \overline{\overline{U}} = \overline{U}, \qquad \overline{U \cup V} = \overline{U} \cup \overline{V}.$$

*Proof.* It is obvious that $\overline{U} \supset U$ and that $\bigcap_{\mathcal{J} \in \overline{U}} \mathcal{J} = \bigcap_{\mathcal{J} \in U} \mathcal{J}$, so that $\overline{\overline{U}} = \overline{U}$. Also, since $U \subset U \cup V$ and $V \subset U \cup V$ we have $\overline{U} \subset \overline{U \cup V}$ and $\overline{V} \subset \overline{U \cup V}$; hence $\overline{U} \cup \overline{V} \subset \overline{U \cup V}$. To prove the reverse inclusion, suppose $\ker(\pi) \notin \overline{U} \cup \overline{V}$. Then there exist $f \in \bigcap_{\mathcal{J} \in U} \mathcal{J}$ and $g \in \bigcap_{\mathcal{J} \in V} \mathcal{J}$ such that $\pi(f) \neq 0$ and $\pi(g) \neq 0$. Pick a vector $u \in \mathcal{H}_\pi$ such that $\pi(f)u \neq 0$. Since $\pi(g) \neq 0$ and $\pi$ is irreducible, there exists $h \in C^*(G)$ such that $\pi(g)\pi(h)\pi(f)u \neq 0$. But $g * h * f \in \bigcap_{\mathcal{J} \in U \cup V} \mathcal{J}$, so $\ker(\pi) \not\supset \bigcap_{\mathcal{J} \in U \cup V} \mathcal{J}$ and hence $\ker(\pi) \notin \overline{U \cup V}$. $\square$

It now follows from a theorem of Kuratowski (Kuratowski [83, §§4–5], Folland [45, Exercise 4.12]) that there is a unique topology on $\mathrm{Prim}(G)$ with respect to which $\overline{U}$ is the closure of $U$, for any $U \subset \mathrm{Prim}(G)$. This topology is called the **hull-kernel topology** or the **Jacobson topology**. We observe that if $\mathcal{I}$ and $\mathcal{J}$ are distinct elements of $\mathrm{Prim}(G)$ then either $\mathcal{I} \not\subset \mathcal{J}$ or $\mathcal{J} \not\subset \mathcal{I}$, and hence either $\mathcal{I} \notin \overline{\{\mathcal{J}\}}$ or $\mathcal{J} \notin \overline{\{\mathcal{I}\}}$. In other words, the hull-kernel topology is always $T_0$. However, it may not satisfy any stronger separation properties, as we shall see below.

*Remark.* One can consider primitive ideals in $L^1(G)$ rather than $C^*(G)$. The space $\mathrm{Prim}_{L^1}(G)$ of all such ideals carries its own hull-kernel topology, and the map $\Psi(\mathcal{I}) = \mathcal{I} \cap L^1(G)$ is a continuous surjection from $\mathrm{Prim}(G)$ to $\mathrm{Prim}_{L^1}(G)$. However, it need not be a homeomorphism. It is known that $\Psi$ is a homeomorphism when $G$ has polynomial growth (i.e., for any compact neighborhood $V$ of the identity, the Haar measure of $VV\cdots V$ ($n$ factors) grows at most polynomially in $n$), but not, for example, when $G$ is a noncompact connected semisimple Lie group. See Boidol et al. [17].

Many of the results described below about representations of $G$ can be rephrased as results about nondegenerate *-representations of $C^*(G)$, and as such can be generalized to nondegenerate *-representations of arbitrary C* algebras. Indeed, this is the natural way of attacking many problems in representation theory; see Dixmier [32] or Pedersen [111].

## 7.2   The Structure of the Dual Space

Let $G$ be a locally compact group. If $\pi$ is an irreducible representation of $G$, the kernel $\ker(\pi) \in \mathrm{Prim}(G)$ clearly depends only on the equivalence class of $\pi$, and the map $[\pi] \mapsto \ker(\pi)$ is a surjection from $\widehat{G}$ onto $\mathrm{Prim}(G)$. We can therefore pull back the hull-kernel topology on $\mathrm{Prim}(G)$ to $\widehat{G}$. That is, we make $\widehat{G}$ into a topological space by declaring the open sets to be those of the form $\{[\pi] : \ker(\pi) \in U\}$ where $U$ is open in $\mathrm{Prim}(G)$. This topology is commonly called the **Fell topology** on $\widehat{G}$; it was introduced in Fell [37]. To understand it more clearly, let us see what it means in some familiar cases.

**7.4 Proposition.** *When $G$ is Abelian, the Fell topology on $\widehat{G}$ is the usual topology on $\widehat{G}$ as defined in Chapter 4. When $G$ is compact, the Fell topology on $\widehat{G}$ is the discrete topology.*

*Proof.* Suppose $G$ is Abelian. As we observed in §7.1, $C^*(G)$ is essentially $C_0(\widehat{G})$; the primitive ideals of $C_0(\widehat{G})$ are the maximal ideals $\{f : f(\xi_0) = 0\}$ for $\xi_0 \in \widehat{G}$; and the hull-kernel closure of a set $U \subset \widehat{G}$ is just $\nu(\iota(U))$ in the notation of §4.5. It therefore follows from Theorem 4.49 that the ordinary closed sets in $\widehat{G}$ are the hull-kernel closed sets.

Now suppose $G$ is compact. If $[\pi] \in \widehat{G}$, let $\chi_\pi(x) = \mathrm{tr}\,\pi(x)$ be the character of $\pi$. It follows easily from the Schur orthogonality relations (5.8) that $\pi(\chi_\pi)$ is $(\dim \mathcal{H}_\pi)^{-1}$ times the identity operator on $\mathcal{H}_\pi$, and that $\pi'(\chi_\pi) = 0$ if $[\pi'] \neq [\pi]$. Hence, $\chi_\pi$ belongs to $\bigcap_{[\pi'] \neq [\pi]} \ker(\pi')$ but not to $\ker(\pi)$. This means that $[\pi]$ is not in the closure of $\widehat{G} \setminus \{[\pi]\}$, and hence that $\{[\pi]\}$ is open. $\qquad\qquad\qquad\qquad\qquad\qquad\qquad\qquad\qquad\square$

When $G$ is neither Abelian nor compact, $\widehat{G}$ is usually non-Hausdorff. We shall see some examples below, but for the moment we give a heuristic reason why this phenomenon is to be expected. It often happens that one can construct a family of representations $\pi_t$ of $G$ that depend continuously (in some suitable sense) on a real parameter $t$, such that $\pi_t$ is irreducible for all $t \neq 0$ but $\pi_0$ is reducible. In such a situation, all the irreducible components of $\pi_0$ (or rather their equivalence classes) will be limits of $[\pi_t]$ as $t \to 0$; to put it another way, every neighborhood of any of those components will contain $[\pi_t]$ for $t$ sufficiently small. In any event, the components of $\pi_0$ will correspond to non-Hausdorff points of $\widehat{G}$.

Another description of the Fell topology on $\widehat{G}$ is available. Namely, let $\mathcal{E}(\mathcal{P}_1)$ be the set of extreme points of the set of normalized functions of positive type on $G$. $\mathcal{E}(\mathcal{P}_1)$ is a subset of $L^\infty(G)$ and as such is endowed with the weak* topology. On the other hand, for each $\phi \in \mathcal{E}(\mathcal{P}_1)$ one has the representation $\pi_\phi$ given by (3.19), and the map $\phi \mapsto [\pi_\phi]$ is a surjection from $\mathcal{E}(\mathcal{P}_1)$ to $\widehat{G}$ by Corollary 3.24 and Theorem 3.25. It can then be shown that the Fell topology on $\widehat{G}$ is the one generated by the map $\phi \mapsto [\pi_\phi]$; that is, $U \subset \widehat{G}$ is open if and only if its inverse image in $\mathcal{E}(\mathcal{P}_1)$ is open. See Dixmier [32, §3.4].

Further information about the Fell topology can be found in Dixmier [32], Fell and Doran [40], and Kaniuth and Taylor [71].

Another aspect of the structure of $\widehat{G}$ that is often more directly relevant than its topology is its structure as a measurable space. Of course one can always consider the $\sigma$-algebra of Borel sets defined by the Fell topology, but for second countable groups there is another natural $\sigma$-algebra on $\widehat{G}$ that is sometimes more suitable. To define it, we need some terminology.

For each positive integer $n$ let $\mathcal{H}_n$ be a fixed Hilbert space of dimension $n$, and let $\mathcal{H}_\infty$ be a fixed separable infinite-dimensional Hilbert space. (For example: take $\mathcal{H}_n = \mathbb{C}^n$ for $n < \infty$ and $\mathcal{H}^\infty = l^2(\mathbb{Z})$.) For $n = 1, 2, \ldots, \infty$ let $\mathrm{Irr}_n(G)$ be the set of irreducible representations of $G$ on $\mathcal{H}_n$ (where we do not identify equivalent but distinct representations), and let $\mathrm{Irr}(G)$ be the union of all the $\mathrm{Irr}_n(G)$. For each $n$ we define the $\sigma$-algebra $\mathcal{B}_n$ on $\mathrm{Irr}_n(G)$ to be the smallest $\sigma$-algebra with respect to which all the functions

$$(7.5) \qquad \pi \mapsto \langle \pi(x)u, v \rangle \qquad (x \in G, \ u, v \in \mathcal{H}_n)$$

are measurable, and we define the $\sigma$-algebra $\mathcal{B}$ on $\mathrm{Irr}(G)$ by requiring that $E \in \mathcal{B}$ if and only if $E \cap \mathrm{Irr}_n(G) \in \mathcal{B}_n$ for all $n$.

Now suppose $G$ is second countable. The map $\pi \mapsto [\pi]$ from $\mathrm{Irr}(G)$ to $\widehat{G}$ is then a surjection, and we can use it to define a quotient $\sigma$-algebra on $\widehat{G}$. Namely, the **Mackey Borel structure** on $\widehat{G}$ (introduced in Mackey [92], [98]) is the $\sigma$-algebra $\mathcal{M}$ on $\widehat{G}$ consisting of all $E \subset \widehat{G}$ such that $\{\pi \in \mathrm{Irr}(G) : [\pi] \in E\} \in \mathcal{B}$.

One could use the same procedure to define a topology on $\widehat{G}$. Namely, one imposes on $\mathrm{Irr}_n(G)$ the weakest topology that makes all the functions (7.5) continuous, on $\mathrm{Irr}(G)$ the topology for which $U \subset \mathrm{Irr}(G)$ is open precisely when $U \cap \mathrm{Irr}_n(G)$ is open for all $n$, and on $\widehat{G}$ the quotient topology induced by the map $\pi \mapsto [\pi]$. This topology is less natural than the Fell topology because it makes all the sets $\widehat{G}_n = \{[\pi] : \pi \in \mathrm{Irr}_n(G)\}$ closed, whereas (as we suggested above) one may wish to regard certain lower-dimensional representations as limits of higher-dimensional ones. However, it can be shown (Dixmier [32, §3.5]) that these two topologies induce the same relative topology on $\widehat{G}_n$ for every $n$. It follows easily that the Mackey Borel structure includes the family of Borel sets for the Fell topology, but in general it is strictly larger. In particular, the Mackey Borel structure distinguishes points — every singleton set in $\widehat{G}$ belongs to $\mathcal{M}$ (Mackey [92], or Dixmier [32, §3.8]) — but the Fell Borel sets do so only when the map $[\pi] \mapsto \ker(\pi)$ from $\widehat{G}$ to $\mathrm{Prim}(G)$ is injective.

It is of interest to know how well-behaved $\widehat{G}$ is as a topological space or as a measurable space. On the topological side, the main questions are whether $\widehat{G}$ satisfies one of the separation axioms $T_0$ or $T_1$. On the measure-theoretic side, the principal conditions to be considered are the following.

Let $(X, \mathcal{M})$ be a measurable space, i.e., a set equipped with a $\sigma$-algebra. $\mathcal{M}$ is **countably separated** if there is a countable family $\{E_j\}_1^\infty$ in $\mathcal{M}$ such that each $x \in X$ satisfies $\{x\} = \bigcap_{\{j : x \in E_j\}} E_j$. $(X, \mathcal{M})$ (or just $\mathcal{M}$) is **standard** if $(X, \mathcal{M})$ is measurably isomorphic to a Borel subset of a complete separable metric space. (Two measurable spaces $(X, \mathcal{M})$ and $(Y, \mathcal{N})$ are measurably isomorphic if there is a bijection $f : X \to Y$ such that $E \in \mathcal{N}$ if and only if $f^{-1}(E) \in \mathcal{M}$.) Standardness easily implies countable separation, and it is an even stronger condition than might at first appear. By a remarkable theorem of Kuratowski [83, §37.II] (see also Pedersen [111, Theorem 4.6.13] or the appendix in Takesaki [134, vol. I]), if $(X, \mathcal{M})$ is standard there are only two possibilities: either $X$ is countable and $\mathcal{M}$ is the $\sigma$-algebra of all subsets of $X$, or $X$ has the cardinality of the continuum and $(X, \mathcal{M})$ is measurably isomorphic to the unit interval $[0, 1]$ with its $\sigma$-algebra of Borel sets.

It turns out also that the topological or measure-theoretic structure of $\widehat{G}$ is closely related to certain types of behavior of the representations of $G$. Before coming to the main theorems, we therefore need some more terminology.

A unitary representation $\pi$ of $G$ is **primary** if the von Neumann algebra $\mathcal{A}(\pi)$ generated by $\pi$, or equivalently its commutant $\mathcal{C}(\pi)$, is a factor, i.e., its center consists of scalar multiples of $I$. (Primary representations are also known as *factor representations*.) By Schur's lemma, every irreducible representation is primary. More generally, if $\pi$ is a direct sum of irreducible representations, $\pi$ is primary if and only if all its irreducible subrepresentations are unitarily equivalent. (We shall prove this in the next section as Theorem 7.15.) The group $G$ is called **type I** if every primary representation of $G$ is a direct sum of copies of some irreducible representation, or equivalently, if $\mathcal{A}(\pi)$ and $\mathcal{C}(\pi)$ are factors of type I for every primary representation $\pi$. (This uninspired terminology has unfortunately become ubiquitous. In view of the results we are about to present below, the reader might prefer to follow Kirillov [77] in calling type I groups *tame* and non-type-I groups *wild*.)

*Example 1.* Every compact group is type I. This follows immediately from Theorem 5.2 and Theorem 7.15 below.

*Example 2.* Every Abelian group is type I. Indeed, if $\pi$ is a representation of an Abelian group $G$, $\mathcal{A}(\pi)$ is commutative. Hence, if $\pi$ is primary we must have $\pi(x) = \langle x, \xi \rangle I$ for some $\xi \in \widehat{G}$, and a choice of orthonormal basis for $\mathcal{H}_\pi$ then exhibits $\pi$ as a direct sum of copies of the one-dimensional representation $\xi$.

*Example 3.* The Heisenberg groups $H_n$ discussed in §6.7 are type I. Indeed, the center $Z$ of $H_n$ is the set of elements of the form $(0, 0, t)$, so if $\pi$ is a primary representation of $H_n$, the operators $\pi(0, 0, t)$ must be scalar multiples of the identity, so that $\pi(0, 0, t) = e^{2\pi i h t} I$ for some $h \in \mathbb{R}$. If $h \neq 0$, the Stone-von Neumann theorem says that $\pi$ is a multiple of the irreducible representation $\rho_h$. If $h = 0$, $\pi$ factors through $H_n/Z \cong \mathbb{R}^{2n}$ and hence (as in Example 2 above) must be a multiple of one of the representations $\pi_{b,\beta}$ of Corollary 6.52.

One more set of definitions. A C* algebra $\mathcal{A}$ is said to be **CCR** if whenever $\pi$ is an irreducible *-representation of $\mathcal{A}$, $\pi(f)$ is a compact operator for every $f \in \mathcal{A}$. ("CCR" stands for "completely continuous representations," "completely continuous operator" being a once-common synonym for "compact operator." CCR algebras are also sometimes called **liminaire** or **liminal**, the former being a French synonym for "CCR" invented by Dixmier.) A C* algebra $\mathcal{A}$ is said to be **GCR** (or **postliminaire** or **postliminal**) if every nonzero quotient C* algebra of $\mathcal{A}$ possesses a nonzero two-sided ideal that is CCR. (These notions are due to Kaplansky [72]; we refer to this paper or Dixmier [32] for a fuller explanation of the significance of the GCR condition.)

The group $G$ is called **CCR** (resp. **GCR**) if $C^*(G)$ is CCR (resp. GCR). Since $L^1(G)$ is dense in $C^*(G)$, $G$ is CCR if and only if $\pi(f)$ is compact whenever $\pi$ is irreducible and $f \in L^1(G)$. It is obvious that Abelian groups and compact groups are CCR, simply because their irreducible representations are all finite-dimensional.

Finally we are ready to state the main theorems that give the equivalence of various "nice" conditions on the structure of $\widehat{G}$ and the behavior of the representations of $G$.

**7.6 Theorem.** *If $G$ is a second countable locally compact group, the following are equivalent:*

 i. *$G$ is type I.*
 ii. *The Fell topology on $\widehat{G}$ is $T_0$.*
 iii. *The map $[\pi] \mapsto \ker(\pi)$ from $\widehat{G}$ to $\mathrm{Prim}(G)$ is injective.*
 iv. *The Mackey Borel structure on $\widehat{G}$ is countably separated.*
 v. *The Mackey Borel structure on $\widehat{G}$ is standard.*
 vi. *The Mackey Borel structure on $\widehat{G}$ coincides with the $\sigma$-algebra of Borel sets for the Fell topology.*
 vii. *If $[\pi] \in \widehat{G}$, $\pi[C^*(G)]$ contains all compact operators on $\mathcal{H}_\pi$.*
 viii. *$G$ is GCR.*

**7.7 Theorem.** *A locally compact group $G$ is CCR if and only if the Fell topology on $\widehat{G}$ is $T_1$. In particular, every CCR group is type I.*

Theorems 7.6 and 7.7 are proved in a remarkable paper of Glimm [51]; some parts of them were proved independently by Dixmier and Kaplansky. The equivalence of (ii) and (iii) is obvious since $\mathrm{Prim}(G)$ is always $T_0$, and the implication (v) $\implies$ (iv) is easy; the other implications are all more or less difficult. The proofs can also be found in Dixmier [32].

Measurable spaces that are not countably separated are pathological — the classic Lebesgue nonmeasurable subsets of $\mathbb{R}$ are typical — and the equivalence of (i) and (iv) is strong evidence that the irreducible representations of non-type-I groups are essentially impossible to classify in any reasonable way. The situation is even worse than that: even if one knows the irreducible representations of a group $G$, only when $G$ is type I is there a reasonable way to classify arbitrary representations (up to equivalence) in terms of irreducible ones. We shall explore this point further in §7.4.

It remains to address the question of which groups are type I or CCR. The following general results are known; see also Theorem 7.10 for the case of solvable Lie groups.

**7.8 Theorem.**
a. *Every connected semisimple Lie group is CCR.*
b. *Every connected nilpotent Lie group is CCR.*
c. *Every connected real algebraic group is type I.*
d. *A discrete group is type I if and only if it possesses an Abelian normal subgroup of finite index.*

(a) is due to Harish-Chandra [61]; see also Harish-Chandra [60] for a direct proof that such groups are type I. (b) is due to Dixmier [29] and Kirillov [76]; see also Corwin and Greenleaf [26]. (c) is due to Dixmier [28], and (d) is due to Thoma [136].

> *Example 4.* Let us consider the discrete Heisenberg group H and its central quotients $H_q$ ($q \in \mathbb{Z}^+$) discussed in §6.8. For each $q$, $\{(j,k,l) : q|j\}$ is an Abelian normal subgroup of finite index of $H_q$, so $H_q$ is type I. On the other hand, H is not type I, and it provides a good illustration of the failure of the conditions on the structure of the dual space $\widehat{H}$ in Theorem 7.6. We have $\widehat{H} = \bigcup_{\omega \in \mathbb{R}/\mathbb{Z}} \widehat{H}^\omega$, where $\widehat{H}^\omega$ is the set of equivalence classes of irreducible representations with central character $\chi_\omega$, and we know from the results in §6.8 that $\widehat{H}^\omega$ is a 2-torus when $\omega$ is rational and an enormous mess when $\omega$ is irrational. In the latter case we exhibited small pieces of $\widehat{H}^\omega$ that are Borel isomorphic to classic Lebesgue nonmeasurable sets, which shows that the Mackey Borel structure on $\widehat{H}$ is not standard or countably separated. Moreover, when $\omega$ is rational, the map $[\pi] \mapsto \ker(\pi)$ is injective on $\widehat{H}^\omega$, but when $\omega$ is irrational, *every* representation of H (irreducible or not) with central character $\chi_\omega$ has the same kernel in $C^*(G)$, so the Fell topology on $\widehat{H}$ fails quite dramatically to be $T_0$. In particular, this common kernel is the kernel of $\mathrm{ind}_{\mathbb{Z}}^{H} \chi_\omega$, whose intersection with $l^1(H)$ is easily seen to be simply the set of all $f \in l^1(H)$ such that $\sum_l f(j,k,l) e^{2\pi i \omega l} = 0$ for all $j,k$. This result follows from Proposition 5.65 of Kaniuth and Taylor [71], where one can also find an explicit description of the hull-kernel topology on Prim(H).
>
> Finally, as far as the actual "type I" condition goes, pick an irrational $\omega \in \mathbb{R}$ and let
>
> $$[\pi(j,k,l)f](t,m) = e^{2\pi i(kt+\lambda\omega)} f(t + j\omega, m + j),$$
>
> acting on $L^2(\mu)$ where $\mu$ is the product of Lebesgue measure and counting measure on $(\mathbb{R}/\mathbb{Z}) \times \mathbb{Z}$. Then $\pi$ is a representation of H, and the von Neumann algebra it generates is a classic example of a factor of type II. (See Dixmier [31, Chapter I.9].)

The determination of $\widehat{G}$ for various specific types of groups $G$ is an ongoing endeavor that has occupied the attention of many mathematicians over the past forty years. We briefly describe the situation for connected noncompact, non-Abelian Lie groups. (For compact ones, see §5.5. The only connected Abelian Lie groups are products of $\mathbb{R}^n$ and $\mathbb{T}^m$, $n, m \geq 0$.)

For simply connected nilpotent Lie groups there is a beautiful and simple description of the dual space that we now give. First, a little background; see Helgason [64] for fuller explanations. Suppose $G$ is a Lie group with Lie algebra $\mathfrak{g}$. First, there is a natural linear action of $G$ on $\mathfrak{g}$ called the **adjoint action**: $\mathrm{Ad}(x)Y$ is the velocity vector to the curve $t \mapsto x[\exp tY]x^{-1}$ at $t = 0$. This yields a linear action of $G$ on the (real) dual space $\mathfrak{g}^*$ of $\mathfrak{g}$ called the **coadjoint action**, given by $\mathrm{Ad}^*(x) = [\mathrm{Ad}(x^{-1})]^*$. Second, if $\lambda \in \mathfrak{g}^*$ and $\mathfrak{h}$ is a subalgebra of $\mathfrak{g}$ such that $\lambda = 0$ on $[\mathfrak{h}, \mathfrak{h}]$, then $\lambda|\mathfrak{h}$ is an algebra homomorphism from $\mathfrak{h}$ to $\mathbb{R}$. If $H$ is a Lie subgroup of $G$ with Lie algebra $\mathfrak{h}$, we shall call a one-dimensional representation $\sigma$ of $H$ such that $\sigma(\exp X) = e^{2\pi i \lambda(X)}$ for $X \in \mathfrak{h}$ a **lifting** of $\lambda$ to $H$. If $G$ is simply connected and nilpotent, there is a unique connected $H$ whose Lie algebra is $\mathfrak{h}$, and every $\lambda \in \mathfrak{g}^*$ such that $\lambda([\mathfrak{h}, \mathfrak{h}]) = 0$ has a unique lifting $\sigma_\lambda$ to this $H$: namely, $H = \exp \mathfrak{h}$ and $\sigma_\lambda(\exp X) = e^{2\pi i \lambda(X)}$.

**7.9 Theorem.** *Let $G$ be a simply connected nilpotent Lie group. Given $\lambda \in \mathfrak{g}^*$, let $\mathfrak{h}$ be a maximal subalgebra of $\mathfrak{g}$ such that $\lambda = 0$ on $[\mathfrak{h}, \mathfrak{h}]$, and let $H$ and $\sigma_\lambda$ be as above. Then $\mathrm{ind}_H^G(\sigma_\lambda)$ is irreducible, and its equivalence class depends only on the orbit of $\lambda$ under the coadjoint action. The map $\mathcal{O}_\lambda \mapsto [\mathrm{ind}_H^G(\sigma_\lambda)]$ is a bijection from the set of coadjoint orbits to $\widehat{G}$ which is a homeomorphism with respect to the natural quotient topology on the set of orbits and the Fell topology on $\widehat{G}$.*

This theorem is due to Kirillov [76] except for the fact that the map $[\mathrm{ind}_H^G(\sigma_\lambda)] \mapsto \mathcal{O}_\lambda$ is continuous, which was proved by Brown [21]. See also Moore [106] and Corwin and Greenleaf [26] for expositions of the Kirillov theory.

The analysis underlying Theorem 7.9 also applies, with certain modifications, to solvable Lie groups. Indeed, for a solvable group $G$ such that the exponential map $\exp : \mathfrak{g} \to G$ is a diffeomorphism (such groups are called **exponential solvable** groups), the construction in Theorem 7.9 carries over with almost no change — except that one must be a little more careful in the choice of the subalgebra $\mathfrak{h}$ — to give a one-to-one correspondence between $\widehat{G}$ and the space of coadjoint orbits in $\mathfrak{g}^*$. For more general connected solvable groups $G$ the situation is considerably more complicated, but suitable extensions of the Kirillov construction, involving an extensive use of the Mackey machine, yield necessary and sufficient conditions for $G$ to be type I and a complete parametrization of its dual space when it is. (The Mautner group discussed in §6.8 is the simplest example of a connected solvable Lie group that is not type I.) In particular, we have the following complement to Theorem 7.8.

**7.10 Theorem.** *A simply connected solvable Lie group $G$ is type I if and only if the space of coadjoint orbits is countably separated and every $\lambda \in \mathfrak{g}^*$ has a lifting to the group $G_\lambda = \{x \in G : \mathrm{Ad}^*(x)\lambda = \lambda\}$. Every exponential solvable group is type I.*

This theorem, as well as the description of $\widehat{G}$ alluded to above, is due to Takenouchi [133] and Bernat [9] in the exponential solvable case, and to Auslander and Kostant [2] in the general case. See Moore [106] for a very readable survey of these results.

The study of representations of connected semisimple (or, more generally, reductive) Lie groups is an enormous undertaking that involves techniques far outside the scope of this book; we refer to Knapp [79] for a good exposition of (many parts of) the theory. The first major result was the determination of the dual space of $SL(2, \mathbb{R})$ by Bargmann [8] in 1947; we shall describe his result in §7.6. The next three decades saw the discovery of large families of irreducible representations for general semisimple Lie groups (enough to decompose the regular representation; see §7.5), but the full dual space remained unknown except in a few cases. Only in the 1980s was the dual space determined for large classes of reductive groups, including groups of real rank one ($SO(n, 1)$, $SU(n, 1)$, $Sp(n, 1)$, and $F_{4,-20}$), the complex classical groups ($SL(n, \mathbb{C})$, $SO(n, \mathbb{C})$, and $Sp(n, \mathbb{C})$), and $GL(n, F)$ where $F$ is $\mathbb{R}$, $\mathbb{C}$, or $\mathbb{H}$. These results are due to Baldoni Silva and Barbasch [6], Barbasch [7], and Vogan [139]. The research of the last thirty years in this field has proceeded in some perhaps surprising directions that reveal deep connections with algebraic geometry. Vogan's engaging expository paper [141] gives a sketch of these developments as well as an account of a computational *tour de force* relating to the representations of one of the real noncompact simple groups of type $E_8$.

Finally, we mention that the correspondence between irreducible representations and orbits of the coadjoint action, which works so beautifully in the nilpotent case and to a large extent in the solvable case, has also been a very useful principle in the study of representations of other Lie groups. This principle was first elaborated by Kostant [82] in his theory of "geometric quantization," and it has reappeared with various modifications in much subsequent work. See Kirillov [77] for an exposition of this philosophy, and Vogan [140] for a nice account of its role in the representation theory of reductive groups.

## 7.3    Tensor Products of Representations

In this section we discuss (outer) tensor products of representations. The main results are a characterization of the type I primary representations of a group and a recipe for constructing the irreducible representations of a product group out of representations of the factors. The necessary background on tensor products of Hilbert spaces and tensor products of operators is in Appendix 3.

Suppose $G_1$ and $G_2$ are locally compact groups, and $\pi_1$ and $\pi_2$ are representations of $G_1$ and $G_2$ on $\mathcal{H}_1$ and $\mathcal{H}_2$, respectively. By Theorem A.19(e), we can define a unitary representation $\pi_1 \otimes \pi_2$ of $G_1 \times G_2$ on $\mathcal{H}_1 \otimes \mathcal{H}_2$ by

$$(\pi_1 \otimes \pi_2)(x, y) = \pi_1(x) \otimes \pi_2(y).$$

$\pi_1 \otimes \pi_2$ is called the **Kronecker product** or **outer tensor product** of $\pi_1$ and $\pi_2$. When $G_1$ and $G_2$ are the same group $G$, the **inner tensor product** of $\pi_1$ and $\pi_2$ is the representation of $G$ obtained by restricting $\pi_1 \otimes \pi_2$ to the diagonal subgroup of $G \times G$:

$$x \mapsto (\pi_1 \otimes \pi_2)(x, x) = \pi_1(x) \otimes \pi_2(x).$$

This representation is also commonly denoted by $\pi_1 \otimes \pi_2$. We have encountered inner tensor products at a couple of points in Chapter 5, but they will play no further role in this book, so we shall not trouble to invent a separate notation for them.

The key to our results is the following theorem, which was already sketched informally in §1.6.

**7.11 Theorem.** *Suppose $\mathcal{H}_1$ and $\mathcal{H}_2$ are Hilbert spaces, and $\mathcal{S}$ is a subset of $\mathcal{L}(\mathcal{H}_1)$ such that the only bounded operators on $\mathcal{H}_1$ that commute with every $S \in \mathcal{S}$ are scalar multiples of $I$. Then the bounded operators on $\mathcal{H}_1 \otimes \mathcal{H}_2$ that commute with $S \otimes I$ for every $S \in \mathcal{S}$ are precisely those of the form $I \otimes T$ where $T \in \mathcal{L}(\mathcal{H}_2)$.*

*Proof.* Clearly $(S \otimes I)(I \otimes T) = S \otimes T = (I \otimes T)(S \otimes I)$ for any $S$ and $T$. On the other hand, suppose $L \in \mathcal{L}(\mathcal{H}_1 \otimes \mathcal{H}_2)$. A choice of orthonormal basis $\{v_j\}_{j \in J}$ for $\mathcal{H}_2$ identifies $\mathcal{H}_2$ with $l^2(J)$ and hence identifies $\mathcal{H}_1 \otimes \mathcal{H}_2$ with the direct sum $\mathcal{H}_1^J = \bigoplus_J \mathcal{H}_1$ as in Proposition A.18 (Appendix 3). As discussed at the beginning of §1.6, $L$ is then represented as a matrix $[L_{ij}]$ of operators on $\mathcal{H}_1$; specifically, $L(u \otimes v_j) = \sum_i (L_{ij}u) \otimes v_i$. For $S \in \mathcal{L}(\mathcal{H}_1)$, since $L(S \otimes I)(u \otimes v_j) = L(Su \otimes v_j) = \sum_i (L_{ij}Su) \otimes v_i$ and $(S \otimes I)L(u \otimes v_j) = \sum_i (SL_{ij}u) \otimes v_i$, $L$ commutes with $S \otimes I$ if and only if all $L_{ij}$ commute with $S$, and this is the case for all $S \in \mathcal{S}$ if and only if $L_{ij} = t_{ij}I$ for some scalars $t_{ij}$. But then $L(u \otimes v_j) = u \otimes (\sum_i t_{ij}v_i)$, so $L = I \otimes T$ where $T$ is the operator on $\mathcal{H}_2$ given by $T(\sum_j c_j v_j) = \sum_{ij} t_{ij}c_j v_i$. $T$ is bounded because if $u \in \mathcal{H}_1$ is a unit vector and $v \in \mathcal{H}_2$,

$$\|Tv\| = \|u \otimes Tv\| = \|L(u \otimes v)\| \le \|L\| \, \|u \otimes v\| = \|L\| \, \|v\|. \qquad \square$$

We now return to the discussion of representations $\pi_1$ and $\pi_2$ of locally compact groups $G_1$ and $G_2$ and their Kronecker product $\pi_1 \otimes \pi_2$.

**7.12 Theorem.** $\pi_1 \otimes \pi_2$ *is irreducible if and only if $\pi_1$ and $\pi_2$ are both irreducible.*

*Proof.* If $\mathcal{M} \subset \mathcal{H}_2$ is an invariant subspace for $\pi_2$, one easily sees that $\mathcal{H}_1 \otimes \mathcal{M}$ is an invariant subspace for $\pi_1 \otimes \pi_2$; similarly for $\pi_1$. Hence $\pi_1 \otimes \pi_2$ is reducible if either factor is. On the other hand, suppose $\pi_1$ and $\pi_2$ are irreducible and $L \in \mathcal{C}(\pi_1 \otimes \pi_2)$. Restricting attention to the group $G_1 \times \{1\}$, we see that $L$ commutes with $\pi_1(x) \otimes I$ for every $x \in G_1$. By Schur's lemma and Theorem 7.11, therefore, $L = I \otimes T$ for some $T \in \mathcal{L}(\mathcal{H}_2)$. But the same reasoning applied to $\{1\} \times G_2$ shows that $L = S \otimes I$ for some $S \in \mathcal{L}(\mathcal{H}_1)$. These conditions can only hold simultaneously if $S$ and $T$, and hence $L$, are multiples of the identity. Hence $\pi_1 \otimes \pi_2$ is irreducible by Schur's lemma. $\qquad \square$

If $\mathcal{H}_1$ and $\mathcal{H}_2$ are Hilbert spaces and $\pi$ is a representation of $G$ on $\mathcal{H}_1$, we denote by $\pi \otimes I$ the representation of $G$ on $\mathcal{H}_1 \otimes \mathcal{H}_2$ given by

$$(\pi \otimes I)(x) = \pi(x) \otimes I.$$

(The Hilbert space $\mathcal{H}_2$ on which the second factor $I$ acts is understood from the context. This construction is really a special case of the tensor product of two representations, in which $G_1 = G$ and $G_2$ is the trivial group.)

**7.13 Proposition.** *Let $\mathcal{H}_1$ and $\mathcal{H}_2$ be Hilbert spaces, and suppose $\pi$ and $\pi'$ are irreducible representations of $G$ on $\mathcal{H}_1$.*
 a. *If $\mathcal{M} \subset \mathcal{H}_1 \otimes \mathcal{H}_2$ is an irreducible subspace for $\pi \otimes I$, then $(\pi \otimes I)^{\mathcal{M}}$ is equivalent to $\pi$.*
 b. *If $\pi \otimes I$ and $\pi' \otimes I$ are equivalent, so are $\pi$ and $\pi'$.*

*Proof.* For (a), choose $A_0 \in \mathcal{M}$ and $v_0 \in \mathcal{H}_2$ such that $A_0 v_0 \neq 0$, and define $V : \mathcal{H}_1 \otimes \mathcal{H}_2 \to \mathcal{H}_1$ by $VA = Av_0$. (Recall that $A$ is an antilinear operator from $\mathcal{H}_2$ to $\mathcal{H}_1$.) Then

$$V(\pi(x) \otimes I)A = \big[(\pi(x) \otimes I)A\big](v_0) = \pi(x)A(v_0) = \pi(x)VA,$$

so $V$ intertwines $\pi \otimes I$ and $\pi$. In particular, $V|\mathcal{M}$ intertwines $(\pi \otimes I)^{\mathcal{M}}$

and $\pi$, and it is nonzero since $VA_0 \neq 0$. Hence $V|\mathcal{M}$ is an equivalence by Schur's lemma.

(b) is an easy consequence of (a). Pick a nonzero $v_0 \in \mathcal{H}_2$ and let $\mathcal{M} = \mathcal{H}_1 \otimes v_0$. Clearly $\mathcal{M}$ is invariant under $\pi \otimes I$ and $(\pi \otimes I)^{\mathcal{M}}$ is equivalent to $\pi$. If $U$ is an equivalence of $\pi \otimes I$ and $\pi' \otimes I$, $U|\mathcal{M}$ is an equivalence of $(\pi \otimes I)^{\mathcal{M}}$ and $(\pi' \otimes I)^{U(\mathcal{M})}$. The former is equivalent to $\pi$ and the latter, by part (a), is equivalent to $\pi'$. $\qquad\square$

**7.14 Corollary.** *Suppose $\pi_1, \pi_1'$ and $\pi_2, \pi_2'$ are irreducible representations of $G_1$ and $G_2$, respectively. Then $\pi_1 \otimes \pi_2$ is unitarily equivalent to $\pi_1' \otimes \pi_2'$ if and only if $\pi_1$ and $\pi_2$ are equivalent to $\pi_1'$ and $\pi_2'$, respectively.*

*Proof.* The "if" implication is obvious, and the "only if" implication follows easily from Proposition 7.13(b) by considering $\pi_1 \otimes I$ and $I \otimes \pi_2$ as the restrictions of $\pi_1 \otimes \pi_2$ to $G_1 \times \{1\}$ and $\{1\} \times G_2$. $\qquad\square$

A choice of an orthonormal basis $\{v_\alpha\}_{\alpha \in A}$ for $\mathcal{H}_2$ gives an isomorphism of $\mathcal{H}_2$ with $l^2(A)$. In view of Proposition A.18, it is then easy to see that $\pi \otimes I$ is equivalent to a direct sum of copies of $\pi$, one for each $\alpha \in A$. Hence, the condition that the group $G$ be type I may be rephrased as saying that every primary representation of $G$ is equivalent to a representation of the form $\pi \otimes I$ where $\pi$ is irreducible. In this connection, we can now give the promised characterization of completely decomposable primary representations.

**7.15 Theorem.** *Suppose $\{\pi_\alpha : \alpha \in A\}$ is a collection of irreducible representations of $G$. Then $\bigoplus_{\alpha \in A} \pi_\alpha$ is primary if and only if the $\pi_\alpha$'s are all equivalent.*

*Proof.* If each $\pi_\alpha$ is equivalent to some representation $\pi$ then $\bigoplus \pi_\alpha$ is equivalent to $\pi \otimes I$ acting on $\mathcal{H}_\pi \otimes l^2(A)$. By Schur's lemma and Theorem 7.11, the map $T \mapsto I \otimes T$ is an isomorphism from $\mathcal{L}(l^2(A))$ to $\mathcal{C}(\pi \otimes I)$, and $\mathcal{L}(l^2(A))$ has trivial center, so $\pi \otimes I$ is primary.

On the other hand, if the $\pi_\alpha$'s are not all equivalent, let $\Sigma$ denote the set of their equivalence classes. If $\mathcal{H}_\alpha$ is the space on which $\pi_\alpha$ acts, we then have $\bigoplus_{\alpha \in A} \mathcal{H}_\alpha = \bigoplus_{\sigma \in \Sigma} \mathcal{M}_\sigma$ where $\mathcal{M}_\sigma = \bigoplus_{\pi_\alpha \in \sigma} \mathcal{H}_\alpha$. Let $P_\sigma$ be the orthogonal projection onto $\mathcal{M}_\sigma$. Clearly $P_\sigma \in \mathcal{C}(\bigoplus \pi_\alpha)$. Moreover, if $T \in \mathcal{C}(\bigoplus \pi_\alpha)$ and $\pi_\alpha \in \sigma$, $P_\tau T|\mathcal{H}_\alpha = 0$ for all $\tau \neq \sigma \in \Sigma$ by Proposition 7.13(a) and Schur's lemma. It follows that $T$ maps each subspace $\mathcal{M}_\sigma$ into itself and hence that $TP_\sigma = P_\sigma T$ for all $\sigma$. But this says that each $P_\sigma$ is in the center of $\mathcal{C}(\bigoplus \pi_\alpha)$, so $\bigoplus \pi_\alpha$ is not primary. $\qquad\square$

It follows from Theorem 7.12 and Corollary 7.14 that the map

$$(7.16) \qquad\qquad ([\pi_1], [\pi_2]) \mapsto [\pi_1 \otimes \pi_2]$$

is a well-defined injection from $\widehat{G}_1 \times \widehat{G}_2$ into $(G_1 \times G_2)\widehat{\phantom{.}}$. If either $G_1$ or $G_2$ is type I, it is a bijection:

**7.17 Theorem.** *If either $G_1$ or $G_2$ is type I, then every irreducible representation $\pi$ of $G_1 \times G_2$ is equivalent to a representation of the form $\pi_1 \otimes \pi_2$, and hence the map (7.16) is a bijection from $\widehat{G}_1 \times \widehat{G}_2$ to $(G_1 \times G_2)\widehat{\phantom{.}}$.*

*Proof.* Suppose $G_1$ is type I (the argument is the same with $G_1$ and $G_2$ switched). Let $\pi^1(x) = \pi(x,1)$ and $\pi^2(y) = \pi(1,y)$. $\pi^1$ and $\pi^2$ are representations of $G_1$ and $G_2$ on $\mathcal{H}_\pi$, and since $(x,1)(1,y) = (x,y) = (1,y)(x,1)$, $\pi^2(y) \in \mathcal{C}(\pi^1)$ for every $y \in G_2$. Thus if $T$ is in the center of $\mathcal{C}(\pi^1)$ then $T \in \mathcal{C}(\pi^2)$ and hence $T \in \mathcal{C}(\pi)$. But then $T = cI$ by Schur's lemma, so $\pi^1$ is primary. Therefore, after performing a unitary transformation we may assume that $\mathcal{H}_\pi = \mathcal{H}_1 \otimes \mathcal{H}_2$ and $\pi^1 = \pi_1 \otimes I$ for some irreducible representation $\pi_1$ of $G_1$. Then, by Schur's lemma and Theorem 7.11, $\pi^2(y) = I \otimes \pi_2(y)$ for some $\pi_2(y) \in \mathcal{L}(\mathcal{H}_2)$. It is easily checked that $\pi_2$ is a unitary representation of $G_2$ (necessarily irreducible by Theorem 7.12) and that $\pi = \pi_1 \otimes \pi_2$. $\square$

When $G_1$ and $G_2$ are not type I, the conclusion of Theorem 7.17 may fail; see Mackey [86]. We shall present a concrete example of this phenomenon in §7.6.

The Kronecker product and the inducing construction both give ways of manufacturing representations of a group $G$ from representations of its subgroups. As ways of producing *irreducible* representations of $G$ they apply in quite different situations; nonetheless, they are related. In fact, we have the following proposition, whose proof we leave as an instructive exercise for the reader.

**7.18 Proposition.** *Let $G = G_1 \times G_2$, and identify $G_1$ with the subgroup $G_1 \times \{1\}$ of $G$. If $\pi$ is a representation of $G_1$, $\mathrm{ind}_{G_1}^G(\pi)$ is canonically equivalent to $\pi \otimes \lambda$, where $\lambda$ is the left regular representation of $G_2$.*

## 7.4 Direct Integral Decompositions

In this section we sketch the theory of direct integrals, to which we have alluded in a number of places. This theory is originally due to von Neumann [145], and another version of it was developed by Godement [54]. Our treatment more or less follows Dixmier [31], to which we refer for a fuller discussion. The theory works well only under certain countability assumptions; accordingly, in this section (except for Theorem 7.28) we consider only representations of second countable groups on separable Hilbert spaces.

First we must define the direct integral of a family $\{\mathcal{H}_\alpha\}_{\alpha \in A}$ of Hilbert spaces with respect to a measure $\mu$ on the parameter space $A$. Roughly speaking, this space should consist of functions $f$ on $A$ such that $f(\alpha) \in \mathcal{H}_\alpha$ for each $\alpha$ and $\int \|f(\alpha)\|_{\mathcal{H}_\alpha}^2 \, d\mu(\alpha) < \infty$. If the $\mathcal{H}_\alpha$'s are all copies of a fixed Hilbert space $\mathcal{H}$, there is no difficulty in this. From an abstract point of view there is no real harm in the assumption that $\mathcal{H}_\alpha = \mathcal{H}$ for all $\alpha$ (cf. Proposition 7.21 below), but it is unnatural and inconvenient for concrete applications. On the other hand, if the $\mathcal{H}_\alpha$'s are all different, it is not clear what one should mean by measurability of such functions $f$, and there are certain pitfalls to be avoided. To come up with a workable definition will require a certain amount of toil, which we now undertake.

Throughout this section, $(A, \mathcal{M})$ will denote a measurable space, i.e., a set equipped with a $\sigma$-algebra. A family $\{\mathcal{H}_\alpha\}_{\alpha \in A}$ of nonzero separable Hilbert spaces indexed by $A$ will be called a **field** of Hilbert spaces over $A$, and an element of $\prod_{\alpha \in A} \mathcal{H}_\alpha$ — that is, a map $f$ on $A$ such that $f(\alpha) \in \mathcal{H}_\alpha$ for each $\alpha$ — will be called a **vector field** on $A$. We denote the inner product and norm on $\mathcal{H}_\alpha$ by $\langle \cdot, \cdot \rangle_\alpha$ and $\| \cdot \|_\alpha$. A **measurable field of Hilbert spaces** over $A$ is a field of Hilbert spaces $\{\mathcal{H}_\alpha\}$ together with a countable set $\{e_j\}_1^\infty$ of vector fields with the following properties:

i. the functions $\alpha \mapsto \langle e_j(\alpha), e_k(\alpha) \rangle_\alpha$ are measurable for all $j, k$,

ii. the linear span of $\{e_j(\alpha)\}_1^\infty$ is dense in $\mathcal{H}_\alpha$ for each $\alpha$.

Before proceeding further, let us examine some examples.

> *Example 1.* Let $\mathcal{H}$ be a separable Hilbert space with orthonormal basis $\{e_j\}$. If we set $\mathcal{H}_\alpha = \mathcal{H}$ and $e_j(\alpha) = e_j$ for all $\alpha$, we obtain a measurable field of Hilbert spaces over $A$, called a **constant field**.

> *Example 2.* Suppose $A$ is discrete (i.e., $\mathcal{M}$ consists of all subsets of $A$) and $\{\mathcal{H}_\alpha\}$ is an arbitrary field of Hilbert spaces over $A$. For each $\alpha$ let $d(\alpha) = \dim \mathcal{H}_\alpha$ and let $\{e_j(\alpha)\}_1^{d(\alpha)}$ be an orthonormal basis for $\mathcal{H}_\alpha$. If we set $e_j(\alpha) = 0$ when $j > d(\alpha)$, the vector fields $e_j$ make $\{\mathcal{H}_\alpha\}$ into a measurable field.

> *Example 3.* Let $A$ be a second countable manifold and $\mathcal{V} = \{\mathcal{H}_\alpha\}$ a vector bundle over $A$ whose fibers $\mathcal{H}_\alpha$ are Hilbert spaces. By the local triviality of $\mathcal{V}$, a partition of unity argument gives the existence of a countable family $\{e_j\}$ of continuous sections of $\mathcal{V}$ whose linear span is dense in the fiber over every point. $\mathcal{V}$ then becomes a measurable field of Hilbert spaces with respect to the Borel $\sigma$-algebra on $A$.

*Example 4.* Let $G$ be a second countable locally compact group, and let $A = \mathcal{P}_0 \setminus \{0\}$ be the set of functions $\phi$ of positive type on $G$ with $0 < \|\phi\|_\infty \leq 1$. $\mathcal{P}_0$ is a compact Hausdorff space with the weak* topology, so $A$ is a locally compact Hausdorff space. For each $\phi \in A$ let $\mathcal{H}_\phi$ be the Hilbert space obtained by completing $L^1(G)$ with respect to the semi-inner product $\langle f, g \rangle_\phi = \int (g^* * f)\phi$ as in §3.3. If $f \in L^1(G)$, let $\widetilde{f}(\phi)$ be the image of $f$ in $\mathcal{H}_\phi$; then $\langle \widetilde{f}(\phi), \widetilde{g}(\phi) \rangle_\phi = \int (f^* * g)\phi$ is a continuous function of $\phi$ for every $f, g \in L^1(G)$. If $\{f_j\}$ is any countable dense subset of $L^1(G)$, the vector fields $\{\widetilde{f}_j\}$ then make $\{\mathcal{H}_\phi\}$ into a measurable field of Hilbert spaces over $A$ with respect to the Borel $\sigma$-algebra on $A$.

The following proposition provides the crucial information about the structure of measurable fields of Hilbert spaces.

**7.19 Proposition.** *Let $\{\mathcal{H}_\alpha\}$, $\{e_j\}$ be a measurable field of Hilbert spaces over $A$, with $\dim \mathcal{H}_\alpha = d(\alpha) \in [1, \infty]$. Then $\{\alpha \in A : d(\alpha) = m\}$ is measurable for $m = 1, 2, \ldots, \infty$. Moreover, there is a sequence $\{u_k\}_1^\infty$ of vector fields with the following properties:*

*i. for each $\alpha$, $\{u_k(\alpha)\}_1^{d(\alpha)}$ is an orthonormal basis for $\mathcal{H}_\alpha$, and $u_k(\alpha) = 0$ for $k > d(\alpha)$ (if $d(\alpha) < \infty$);*

*ii. for each $k$ there is a measurable partition of $A$, $A = \bigcup_{l=1}^\infty A_l^k$, such that on each $A_l^k$, $u_k(\alpha)$ is a finite linear combination of the $e_j(\alpha)$'s with coefficients depending measurably on $\alpha$.*

*Proof.* First, define a sequence $\{f_j\}$ of vector fields inductively as follows: $f_1(\alpha)$ is the first of the vectors $e_1(\alpha), e_2(\alpha), \ldots$ that is nonzero; for $j > 1$, $f_j(\alpha)$ is the first of the vectors $e_1(\alpha), e_2(\alpha), \ldots$ that is not in the linear span of $f_1(\alpha), \ldots, f_{j-1}(\alpha)$ if such a vector exists, and otherwise $f_j(\alpha) = 0$. Clearly $\{f_j(\alpha)\}$ and $\{e_j(\alpha)\}$ have the same linear span for each $\alpha$.

*Claim:* Suppose $k$ is a positive integer. There is a measurable partition $\{A_l^k\}_{l=1}^\infty$ of $A$ such that for all $j \leq k$ and all $l$, either $f_j(\alpha) = 0$ for all $\alpha \in A_l^k$ or $f_j(\alpha) = e_{m(j)}(\alpha) \neq 0$ for all $\alpha \in A_l^k$ where $m(j)$ is independent of $\alpha$.

We establish the claim by induction on $k$. For $k = 1$, let $N_l = \{\alpha : e_l(\alpha) \neq 0\}$ and $A_l^1 = N_l \setminus \bigcup_1^{l-1} N_i$. Clearly $f_1(\alpha) = e_l(\alpha)$ for $\alpha \in A_l^1$. Since the $e_j(\alpha)$'s span $\mathcal{H}_\alpha$ for each $\alpha$ we have $\bigcup_l A_l^1 = A$. Also, $e_j(\alpha) = 0$ if and only if $\langle e_j(\alpha), e_j(\alpha) \rangle_\alpha = 0$, and since these inner products are measurable in $\alpha$, it follows that $A_l^1$ is measurable for all $l$. Hence the claim is proved for $k = 1$.

Now suppose that $k > 1$ and that the sets $A_l^j$ have been constructed for $j < k$. It suffices to prove the claim with $A$ replaced by $B = A_l^{k-1}$ for an arbitrary integer $l$, as one then obtains the required partition of $A$ by combining the resulting partitions of the $A_l^{k-1}$'s. Now, on $B$, either

$f_{k-1} = 0$, in which case $f_k = 0$ and we can take the trivial partition of $B$, or $f_j = e_{m(j)}$ for $j < k$ and $f_1, \ldots, f_{k-1}$ are pointwise linearly independent. In the latter case, for each $m$ the set $B_m = \{\alpha \in B : f_k(\alpha) = e_m(\alpha)\}$ is given by

$$B_m = \{\alpha : D[e_{m(1)}(\alpha), \ldots, e_{m(k-1)}(\alpha), e_l(\alpha)] = 0 \text{ for } l < m,$$
$$D[e_{m(1)}(\alpha), \ldots, e_{m(k-1)}(\alpha), e_m(\alpha)] \neq 0\},$$

where
$$D[v_1, \ldots v_k] = \det(\langle v_i, v_j \rangle)_{i,j=1}^k.$$

(We are using the fact that $v_1, \ldots, v_k$ are linearly dependent if and only if $D[v_1, \ldots, v_k] = 0$.) It follows that the $B_m$'s are measurable, and they together with $B_0 = \{\alpha \in B : f_k(\alpha) = 0\}$ make up the required partition of $B$. The claim is established.

It follows immediately from the claim that $\{\alpha : f_k(\alpha) \neq 0\}$ is measurable for each $k$. Since

$$\{\alpha : d(\alpha) = m\} = \{\alpha : f_m(\alpha) \neq 0\} \setminus \{\alpha : f_{m+1}(\alpha) \neq 0\},$$

the first assertion of the proposition is proved. To prove the second one, simply apply the Gram-Schmidt process pointwise to the sequence $\{f_k\}$ to obtain the sequence $\{u_k\}$. Then $\{u_k\}$ has property (i) by construction. Moreover, on each $A_l^k$, $u_k$ is a linear combination of $e_{m(1)}, \ldots, e_{m(k)}$ with coefficients that are continuous functions of the inner products $\langle e_{m(i)}, e_{m(j)} \rangle$ and hence are measurable, so (ii) holds too.          □

Given a measurable field of Hilbert spaces $\{\mathcal{H}_\alpha\}$, $\{e_j\}$ on $A$, a vector field $f$ on $A$ will be called **measurable** if $\langle f(\alpha), e_j(\alpha) \rangle_\alpha$ is a measurable function on $A$ for each $j$.

**7.20 Proposition.** *Let $\{u_k\}$ be as in Proposition 7.19. A vector field $f$ on $A$ is measurable if and only if $\langle f(\alpha), u_k(\alpha) \rangle_\alpha$ is a measurable function on $A$ for each $k$. If $f$ and $g$ are measurable vector fields, $\langle f(\alpha), g(\alpha) \rangle_\alpha$ is a measurable function.*

*Proof.* If $f$ is measurable, it follows from property (ii) of $\{u_k\}$ that $\langle f, u_k \rangle$ is measurable for each $k$. On the other hand, by property (i) of $\{u_k\}$ we have
$$\langle f(\alpha), e_j(\alpha) \rangle_\alpha = \sum_k \langle f(\alpha), u_k(\alpha) \rangle_\alpha \langle u_k(\alpha), e_j(\alpha) \rangle_\alpha.$$

$\langle u_k, e_j \rangle$ is measurable by (ii) again, so if $\langle f, u_k \rangle$ is measurable for all $k$, so is $\langle f, e_j \rangle$. Likewise, if $f$ and $g$ are measurable, $\langle f, g \rangle = \sum_k \langle f, u_k \rangle \langle u_k, g \rangle$ is measurable.          □

Finally we are ready to define direct integrals. Suppose $\{\mathcal{H}_\alpha\}, \{e_j\}$ is a measurable field of Hilbert spaces over $A$, and suppose $\mu$ is a measure on $A$. The **direct integral** of the spaces $\mathcal{H}_\alpha$ with respect to $\mu$, denoted by

$$\int^\oplus \mathcal{H}_\alpha \, d\mu(\alpha),$$

is the space of measurable vector fields $f$ on $A$ such that

$$\|f\|^2 = \int \|f(\alpha)\|_\alpha^2 \, d\mu(\alpha) < \infty.$$

(The integrand is measurable by Proposition 7.20.) An easy modification of the usual proof that $L^2(\mu)$ is complete shows that $\int^\oplus \mathcal{H}_\alpha \, d\mu(\alpha)$ is a Hilbert space with inner product

$$\langle f, g \rangle = \int \langle f(\alpha), g(\alpha) \rangle_\alpha \, d\mu(\alpha).$$

Let us see how this works for the first three examples of measurable fields of Hilbert spaces discussed above.

1. In the case of a constant field, $\mathcal{H}_\alpha = \mathcal{H}$ for all $\alpha$, $\int^\oplus \mathcal{H}_\alpha \, d\mu(\alpha)$ is just the space of measurable functions from $A$ to $\mathcal{H}$ that are square-integrable with respect to $\mu$. We denote this space by $L^2(A, \mu, \mathcal{H})$.

2. If $A$ is discrete and $\mu$ is counting measure on $A$, then $\int^\oplus \mathcal{H}_\alpha \, d\mu(\alpha)$ is nothing but $\bigoplus_{\alpha \in A} \mathcal{H}_\alpha$.

3. If the $\mathcal{H}_\alpha$'s are the fibers of a vector bundle $\mathcal{V}$, $\int^\oplus \mathcal{H}_\alpha \, d\mu(\alpha)$ is the space of sections of $\mathcal{V}$ that are square-integrable with respect to $\mu$.

We now make a few simple remarks about direct integrals. First, $\int^\oplus \mathcal{H}_\alpha \, d\mu(\alpha)$ really depends only on the equivalence class of $\mu$. More precisely, if $\mu'$ is another measure on $A$ such that $\mu$ and $\mu'$ are mutually absolutely continuous, it is easily verified that the map $f \mapsto f\sqrt{d\mu/d\mu'}$ defines a unitary isomorphism from $\int^\oplus \mathcal{H}_\alpha \, d\mu(\alpha)$ to $\int^\oplus \mathcal{H}_\alpha \, d\mu'(\alpha)$.

Second, if $\{A_m\}_1^\infty$ is a measurable partition of $A$, there is an obvious isomorphism

$$\int_A^\oplus \mathcal{H}_\alpha \, d\mu(\alpha) \cong \bigoplus_{m=1}^\infty \int_{A_m}^\oplus \mathcal{H}_\alpha \, d\mu(\alpha).$$

Third, suppose $\dim \mathcal{H}_\alpha = d$ is independent of $\alpha$. Let $\{u_j\}$ be as in Proposition 7.19; these vector fields furnish an orthonormal basis for $\mathcal{H}_\alpha$ for each $\alpha$ and hence an identification of $\mathcal{H}_\alpha$ and $\mathbb{C}^d$ (where $\mathbb{C}^\infty = l^2$). Once this identification is made, the field $\{\mathcal{H}_\alpha\}$ becomes a constant field, and by Proposition 7.20, the vector fields that are measurable with respect to the original measurable field become the vector fields that are measurable with respect to this constant field. In short, the choice of $\{u_j\}$ yields an isomorphism between $\int^{\oplus} \mathcal{H}_\alpha \, d\mu(\alpha)$ and $L^2(A, \mu, \mathbb{C}^d)$.

If we combine the last two remarks and Proposition 7.19, we obtain the following result.

**7.21 Proposition.** *Let* $\{\mathcal{H}_\alpha\}$, $\{e_j\}$ *be a measurable field of Hilbert spaces over* $A$, *and let* $\mu$ *be a measure on* $A$. *For* $m = 1, 2, \ldots, \infty$, *let* $A_m = \{\alpha \in A : \dim \mathcal{H}_\alpha = m\}$. *Then a choice of vector fields* $\{u_j\}$ *as in Proposition 7.19 defines a unitary isomorphism*

$$\int^{\oplus} \mathcal{H}_\alpha \, d\mu(\alpha) \cong L^2(A_\infty, \mu, l^2) \oplus \bigoplus_1^{\infty} L^2(A_m, \mu, \mathbb{C}^m).$$

At this point we can answer a question that must have been in the reader's mind, namely, the extent to which $\int^{\oplus} \mathcal{H}_\alpha \, d\mu(\alpha)$ depends on the choice of $\{e_j\}$ in the definition of measurable field of Hilbert spaces. Evidently there is some dependence, for $\{e_j\}$ is used to specify the measurable vector fields out of which $\int^{\oplus} \mathcal{H}_\alpha \, d\mu(\alpha)$ is made up.

Indeed, suppose $\{\epsilon_k\}$ is another sequence of vector fields that define a measurable field structure on $\{\mathcal{H}_\alpha\}$. On the one hand, if the inner products $\langle e_j(\alpha), \epsilon_k(\alpha) \rangle_\alpha$ are all measurable functions of $\alpha$, the sets of $\{e_j\}$-measurable vector fields and $\{\epsilon_k\}$-measurable vector fields coincide, and hence so do the direct integrals fashioned out of them. To see this, let $\{u_j\}$ be as in Proposition 7.19. If $f$ is $\{e_j\}$-measurable, the inner products $\langle f(\alpha), u_j(\alpha) \rangle_\alpha$ and $\langle \epsilon_k(\alpha), u_j(\alpha) \rangle_\alpha$ are measurable in $\alpha$; hence so is

$$\langle f(\alpha), \epsilon_k(\alpha) \rangle_\alpha = \sum_j \langle f(\alpha), u_j(\alpha) \rangle_\alpha \langle u_j(\alpha), \epsilon_k(\alpha) \rangle_\alpha,$$

so $f$ is $\{\epsilon_k\}$-measurable.

On the other hand, the inner products $\langle e_j(\alpha), \epsilon_k(\alpha) \rangle_\alpha$ need not be measurable, and in this case the sets of $\{e_j\}$-measurable vector fields and $\{\epsilon_k\}$-measurable vector fields will be quite different. For example, let $E$ be a nonmeasurable set in $A$, let $\phi = \chi_E - \chi_{A \setminus E}$, and let $\epsilon_j(\alpha) = \phi(\alpha) e_j(\alpha)$. Then the pointwise inner products of the $\epsilon_j$'s are the same as those of the $e_j$'s since $|\phi| \equiv 1$, so $\{\epsilon_j\}$ defines a structure of measurable field on $\{\mathcal{H}_\alpha\}$ just as $\{e_j\}$ does — but not the same one. More generally, one can obtain a new structure of measurable field by replacing $e_j(\alpha)$ by $U(\alpha) e_j(\alpha)$ where $U(\alpha)$ is a unitary operator on $\mathcal{H}_\alpha$ depending completely arbitrarily on $\alpha$.

However, Proposition 7.21 shows that up to an isomorphism that respects the direct integral structure (although not the identity of the individual vectors in the spaces $\mathcal{H}_\alpha$), $\int^{\oplus} \mathcal{H}_\alpha \, d\mu(\alpha)$ does not depend on the choice of $\{e_j\}$: the spaces $L^2(A_m, \mu, \mathbb{C}^m)$ and $L^2(A_\infty, \mu, l^2)$ are quite canonical. Hence, we shall sometimes omit mentioning $\{e_j\}$ in referring to direct integrals of Hilbert spaces.

We now consider direct integrals of operators. Let $\{\mathcal{H}_\alpha\}$, $\{e_j\}$ be a measurable field of Hilbert spaces over $A$. A **field of operators** over $A$ is an element $T$ of $\prod_{\alpha \in A} \mathcal{L}(\mathcal{H}_\alpha)$; $T$ is **measurable** if $\alpha \mapsto T(\alpha)f(\alpha)$ is a measurable vector field whenever $f$ is a measurable vector field.

**7.22 Proposition.** *Let $T$ be a field of operators over $A$. Then $T$ is measurable if and only if $\langle T(\alpha)e_j(\alpha), e_k(\alpha)\rangle_\alpha$ is a measurable function on $A$ for all $j, k$.*

*Proof.* "Only if" is obvious. If

$$\langle T(\alpha)e_j(\alpha), e_k(\alpha)\rangle_\alpha = \langle e_j(\alpha), T(\alpha)^* e_k(\alpha)\rangle_\alpha$$

is measurable for all $j, k$ then $T^*(\alpha)e_k(\alpha)$ is measurable for all $k$; but then if $f$ is a measurable vector field,

$$\langle T(\alpha)f(\alpha), e_k(\alpha)\rangle_\alpha = \langle f(\alpha), T(\alpha)^* e_k(\alpha)\rangle_\alpha$$

is measurable for all $k$ by Proposition 7.20, so $T(\alpha)f(\alpha)$ is measurable. $\square$

Suppose $\mu$ is a measure on $A$ and $T$ is a measurable field of operators such that

(7.23) $$\|T\|_\infty = \operatorname*{ess\,sup}_{\alpha \in A} \|T(\alpha)\| < \infty.$$

Then $\|T(\alpha)f(\alpha)\|_\alpha \le \|T\|_\infty \|f(\alpha)\|_\alpha$ a.e., so $T$ defines a bounded operator on $\int^\oplus \mathcal{H}_\alpha \, d\mu(\alpha)$ which we denote by $\int^\oplus T(\alpha) \, d\mu(\alpha)$, or by $\int^\oplus T$ for short:

$$\left[ \left( \int^\oplus T(\alpha) \, d\mu(\alpha) \right) f \right](\alpha) = \left[ (\textstyle\int^\oplus T)f \right](\alpha) = T(\alpha)f(\alpha).$$

$\int^\oplus T$ is called the **direct integral** of the field $T$. It has the obvious elementary algebraic properties:

(7.24) $$\int^\oplus T + \int^\oplus S = \int^\oplus (T + S),$$
$$\int^\oplus T \int^\oplus S = \int^\oplus TS, \qquad (\textstyle\int^\oplus T)^* = \int^\oplus T^*.$$

(*Note:* If $\mu$ is not semi-finite — a case that could perfectly well be excluded from the beginning — the usual modification of the essential supremum is called for in (7.23): "$\|T\|_\infty \le \lambda$" means that $\|T(\alpha)\| \le \lambda$ for all $\alpha$ except in a locally null set. We shall neglect this point in what follows; the reader may repair the resulting solecisms.)

**7.25 Proposition.** *If $T$ is an essentially bounded measurable field of operators, then $\|\int^{\oplus} T\| = \|T\|_{\infty}$.*

*Proof.* Obviously $\|\int^{\oplus} T\| \leq \|T\|_{\infty}$. If $f \in \int^{\oplus} \mathcal{H}_{\alpha} \, d\mu(\alpha)$ and $\phi \in L^{\infty}(\mu)$,

$$\int |\phi(\alpha)|^2 \|T(\alpha)f(\alpha)\|_{\alpha}^2 \, d\mu(\alpha) = \|\int^{\oplus} T(\phi f)\|^2$$

$$\leq \|\int^{\oplus} T\|^2 \|\phi f\|^2 = \|\int^{\oplus} T\|^2 \int |\phi(\alpha)|^2 \|f(\alpha)\|_{\alpha}^2 \, d\mu(\alpha),$$

from which it follows that

(7.26) $$\|T(\alpha)f(\alpha)\|_{\alpha} \leq \|\int^{\oplus} T\| \|f(\alpha)\|_{\alpha} \quad \text{a.e..}$$

In fact, (7.26) holds for any measurable vector field $f$, as one sees by applying (7.26) with $f(\alpha)$ replaced by $\|f(\alpha)\|_{\alpha}^{-1} f(\alpha) \chi_E(\alpha)$ where $E$ is an arbitrary subset of $A$ of finite measure. Now, by taking linear combinations of the $e_j$'s with coefficients in a countable dense subset of $\mathbb{C}$ we can make a sequence $\{f_j\}$ of measurable vector fields such that $\{f_j(\alpha)\}$ is dense in $\mathcal{H}_{\alpha}$ for each $\alpha$. On applying (7.26) to the $f_j$'s, we see that $\|T(\alpha)\| \leq \|\int^{\oplus} T\|$ for almost every $\alpha$, i.e., $\|T\|_{\infty} \leq \|\int^{\oplus} T\|$. □

**7.27 Corollary.** *If $\int^{\oplus} T(\alpha) \, d\mu(\alpha) = \int^{\oplus} S(\alpha) \, d\mu(\alpha)$ then $T(\alpha) = S(\alpha)$ a.e.*

*Proof.* If $\int^{\oplus}(T - S) = 0$ then $\|T - S\|_{\infty} = 0$. □

A simple but important special case of direct integrals of operators arises when the operators $T(\alpha)$ are all scalar multiples of the identity, $T(\alpha) = \phi(\alpha)I$ with $\phi \in L^{\infty}(\mu)$, in which case $\int^{\oplus} T$ is just the operator $f \mapsto \phi f$. We call such operators on $\int^{\oplus} \mathcal{H}_{\alpha} \, d\mu(\alpha)$ **diagonal** operators.

Now suppose that $G$ is a locally compact group and that $\pi_{\alpha}$ is a unitary representation of $G$ on $\mathcal{H}_{\alpha}$ for each $\alpha \in A$, such that $\alpha \mapsto \pi_{\alpha}(x)$ is a measurable field of operators for each $x \in G$; we shall call $\{\pi_{\alpha}\}_{\alpha \in A}$ a **measurable field of representations** of $G$. Since $\|\pi_{\alpha}(x)\| \equiv 1$, we can form the direct integral

$$\pi(x) = \int^{\oplus} \pi_{\alpha}(x) \, d\mu(\alpha).$$

$\pi$ is a unitary representation of $G$ on $\int^{\oplus} \mathcal{H}_{\alpha} \, d\mu(\alpha)$. Indeed, that $\pi(x)$ is unitary for each $x$ and that $\pi(xy) = \pi(x)\pi(y)$ follows from (7.24), and the strong continuity of $\pi$ follows from the strong continuity of the $\pi_{\alpha}$'s together with the dominated convergence theorem. $\pi$ is called the **direct integral** of the representations $\pi_{\alpha}$.

*Example.* Let $A = \mathcal{P}_0 \setminus \{0\}$ and $\{\mathcal{H}_\phi\}_{\phi \in A}$ be as in the fourth example at the beginning of this section. For each $\phi \in A$ let $\pi_\phi$ be the canonical representation of $G$ on $\mathcal{H}_\phi$. Then

$$\langle \pi_\phi(x) \widetilde{f}(\phi), \widetilde{g}(\phi) \rangle_\phi = \int (g^* * L_x f) \phi$$

is a continuous function of $\phi$ for all $f, g \in L^1(G)$, so $\{\pi_\phi\}$ is a measurable field of representations, and we can form $\int^\oplus \pi_\phi \, d\mu(\phi)$ for any Radon measure $\mu$ on $A$.

We now consider the problem of expressing an arbitrary unitary representation of a group $G$ as a direct integral of irreducible representations. When $G$ is compact, the problem is solved by Theorem 5.2, and when $G$ is Abelian it is essentially solved by Theorem 4.45. Indeed, we have:

**7.28 Theorem.** *Let $G$ be a locally compact Abelian group, and let $\pi$ be a unitary representation of $G$ on $\mathcal{H}$. Then $\pi$ is equivalent to a direct integral of irreducible representations.*

*Proof.* By combining Theorem 4.45 and the Spectral Theorem 1.47 (or the arguments leading to them), one sees that there is a measure space $(\Omega, \mu)$, where $\Omega$ is a disjoint union of copies of $\widehat{G}$, and a unitary map $U : \mathcal{H} \to L^2(\mu)$ such that $U\pi(x)U^{-1}$ is multiplication by the function $\xi \mapsto \langle x, \xi \rangle$ on each copy of $\widehat{G}$. In other words, each $\omega \in \Omega$ determines a character $\xi_\omega \in \widehat{G}$ such that $U\pi(x)U^{-1}f(\omega) = \langle x, \xi_\omega \rangle f(\omega)$ for $f \in L^2(\mu)$, and the map $\omega \mapsto \xi_\omega$ (the identity map on each copy of $\widehat{G}$) is obviously measurable. But this says that $U\pi(\cdot)U^{-1}$ is the direct integral $\int^\oplus \xi_\omega \, d\mu(\omega)$ acting on $\int^\oplus \mathbb{C} \, d\mu = L^2(\mu)$. $\square$

We now return to the general (noncompact, non-Abelian) case and present the main theorems on direct integral decompositions. We motivate the fundamental existence theorem with the following considerations. If $\pi = \int^\oplus \pi_\alpha \, d\mu(\alpha)$, the diagonal operators $f \mapsto \phi f$ ($\phi \in L^\infty(\mu)$) all belong to $\mathcal{C}(\pi)$. In particular, the operators $f \mapsto \chi_E f$, where $E$ is a measurable subset of $A$, are projections whose ranges are invariant subspaces for $\pi$. The following theorem effectively asserts the converse, that for any projection-valued measure $P$ with values in $\mathcal{C}(\pi)$ there is a direct integral decomposition of $\pi$ with respect to which the $P(E)$'s are precisely the projections that are diagonal. However, if one has a commuting family of projections in $\mathcal{C}(\pi)$, the von Neumann algebra they generate will be a commutative subalgebra of $\mathcal{C}(\pi)$, and the theorem is phrased in terms of such subalgebras.

**7.29 Theorem.** *Suppose $G$ is a second countable locally compact group, $\pi$ is a unitary representation of $G$ on a separable Hilbert space $\mathcal{H}$, and $\mathcal{B}$ is a commutative von Neumann subalgebra of $\mathcal{C}(\pi)$. Then there is a standard measure space $(A, \mathcal{M}, \mu)$, a measurable field $\{\mathcal{H}_\alpha\}$ of Hilbert spaces on $A$, a measurable field $\{\pi_\alpha\}$ of representations of $G$, and a unitary map $U : \mathcal{H} \to \int^\oplus \mathcal{H}_\alpha \, d\mu(\alpha)$, such that:*

*i. $U\pi(x)U^{-1} = \int^\oplus \pi_\alpha(x) \, d\mu(\alpha)$ for $x \in G$;*

*ii. $U\pi(f)U^{-1} = \int^\oplus \pi_\alpha(f) \, d\mu(\alpha)$ for $f \in L^1(G)$;*

*iii. $U\mathcal{B}U^{-1}$ is the algebra of diagonal operators on $\int^\oplus \mathcal{H}_\alpha \, d\mu(\alpha)$.*

Theorem 7.29 is essentially due to von Neumann [145]; other proofs have been given by Godement [54], Segal [127], and Mackey [98]. The proofs of von Neumann and Mackey are highly measure-theoretic, while those of Godement and Segal are more functional-analytic. The complete proof of Theorem 7.29 is too long to give here, but it is impossible to resist the opportunity to sketch the ideas of the Godement-Segal argument, as they involve a beautiful interplay of techniques developed earlier in this book. Here, then, is an outline of the proof *d'après* Godement [54].

*Proof.* First, since $\mathcal{H}$ is separable, $\pi$ is the direct sum of countably many cyclic representations, and it suffices to prove the theorem for each of these; hence we shall assume that $\pi$ is cyclic. Let $z$ be a unit cyclic vector for $\pi$; let $\Delta$ be the spectrum of $\mathcal{B}$; let $\mu_{u,v}$ be the measure on $\Delta$ associated to $u, v \in \mathcal{H}$ as in the spectral theorem ($\langle Tu, v \rangle = \int \widehat{T} \, d\mu_{u,v}$ for $T \in \mathcal{B}$); and let $\mu_0 = \mu_{z,z}$. Suppose $S \in \mathcal{L}(\mathcal{H})$ commutes with every operator in $\mathcal{B}$. If $T$ is any positive operator in $\mathcal{B}$ we have

$$\int \widehat{T} \, d\mu_{Sz,z} = \langle TSz, z \rangle = \langle ST^{1/2}z, T^{1/2}z \rangle \le \|S\| \, \|T^{1/2}z\| = \|S\| \int \widehat{T} \, d\mu_0.$$

It follows that $\mu_{Sz,z}$ is absolutely continuous with respect to $\mu_0$ with Radon-Nikodym derivative $\delta_S \in L^\infty(\mu_0)$ satisfying $\|\delta_S\|_\infty \le \|S\|$. Moreover, since $\delta_{S^*S} \, d\mu_0 = d\mu_{Sz,Sz}$ is a positive measure, we have $\delta_{S^*S} \ge 0$ for all $S$.

By taking $S = \pi(x)^*\pi(y)$ with $x, y \in G$ and using the cyclicity of $z$, one easily sees that all the spectral measures $\mu_{u,v}$ are absolutely continuous with respect to $\mu$ and that $\text{supp}(\mu) = \Delta$. Hence, for every $S$ as above, the spectral functional calculus yields an operator $T_S$ such that $\langle T_S u, v \rangle = \int \delta_S \, d\mu_{u,v}$ for $u, v \in \mathcal{H}$. (The point is that $\delta_S$ is only defined $\mu_0$-a.e., but it is then defined $\mu_{u,v}$-a.e. for all $u, v$.) $T_S$ commutes with every operator that commutes with $\mathcal{B}$, so the von Neumann density theorem 1.56 implies that $T_S$ is actually in $\mathcal{B}$. But this means that after modification on a $\mu_0$-null set, $\delta_S$ equals the continuous function $\widehat{T}_S$. In short, to each $S \in \mathcal{L}(\mathcal{H})$ that commutes with $\mathcal{B}$ we have associated a unique continuous function $\delta_S$ on $\Delta$ such that $\|\delta_S\|_{\sup} \leq \|S\|$ and $\delta_{S^*S} \geq 0$.

Now take $S = \pi(f)$ for $f \in L^1(G)$. For each $h \in \Delta$, the map $f \mapsto \delta_{\pi(f)}(h)$ is a linear functional on $L^1(G)$ such that $|\delta_{\pi(f)}(h)| \leq \|f\|_1$ and $\delta_{\pi(f^**f)}(h) \geq 0$, so it is given by integration against a function of positive type $\phi_h \in \mathcal{P}_0$. The map $h \mapsto \phi_h$ (call it $\Phi$) is continuous from $\Delta$ to the closed unit ball of $L^\infty(G)$ with the weak* topology, and the latter is second countable since $L^1(G)$ is separable. Let $A = \Phi(\Delta) \setminus \{0\}$, and let $\mu$ be the push-forward of $\mu_0$ to $A$, $\mu(E) = \mu_0(\Phi^{-1}(E))$ for $E$ a Borel subset of $A$. Then $(A, \mu)$ is a standard measure space since $\Phi(\Delta)$ is compact and metrizable.

For $\phi \in A$, let $\mathcal{H}_\phi$ and $\pi_\phi$ be the Hilbert space and representation of $G$ canonically associated to $\phi$ as in §3.3. The correspondence

$$\pi(f)z \mapsto \tilde{f}(\phi) = \text{image of } f \text{ in } \mathcal{H}_\phi \qquad (f \in L^1(G))$$

sets up a unitary map from $\mathcal{H}$ to $\int^\oplus \mathcal{H}_\phi \, d\mu(\phi)$ that intertwines $\pi$ and $\int^\oplus \pi_\phi \, d\mu(\phi)$. Finally, if $T \in \mathcal{B}$, let $\nu_T$ be the push-forward of the measure $\hat{T} \, d\mu_0$ to $A$. One checks that $|\nu_T| \leq \|T\|\mu$, so that $d\nu_T = \gamma_T \, d\mu$ for some $\gamma_T \in L^\infty(\mu)$; then $UTU^{-1}$ is the diagonal operator $g \mapsto \gamma_T g$. $\qquad\square$

Particular choices of the algebra $\mathcal{B}$ in Theorem 7.29 will lead to particularly interesting decompositions of the representation $\pi$. The following theorem gives the two most important cases. (The reader may find it an instructive exercise to prove this theorem for the special case where $\pi$ is a direct sum of irreducible representations; Theorem 7.15 and its proof may be useful.)

**7.30 Theorem.** *Let $G$, $\pi$, $\mathcal{B}$, $A$, $\mu$, and $\pi_\alpha$ be as in Theorem 7.29.*
  a. *If $\mathcal{B}$ is the center of $\mathcal{C}(\pi)$, $\pi_\alpha$ is primary for $\mu$-almost every $\alpha$.*
  b. *$\pi_\alpha$ is irreducible for $\mu$-almost every $\alpha$ if and only if $\mathcal{B}$ is a maximal commutative $*$-subalgebra of $\mathcal{C}(\pi)$.*

Part (a) is due (in essence) to von Neumann [145] and part (b) to Mautner [102]. Other proofs of part (b) have been given by Godement [54] and Segal [127]. Both parts of the theorem are proved in Mackey [98]and Dixmier [32].

Theorems 7.29 and 7.30(b) yield the existence of an irreducible decomposition of an arbitrary unitary representation of a second countable group.

*Remark:* There is a temptation to try to give a simple proof of this result as follows. Let $\pi$ be a representation of $G$; by Proposition 3.3 we may assume that $\pi$ has a unit cyclic vector $u$. Let $\psi(x) = \langle\pi(x)u, u\rangle$ be the associated function of positive type. By Choquet's theorem there is a measure $\mu$ on $\mathcal{E}(\mathcal{P}_1)$ such that $\psi = \int \phi \, d\mu(\phi)$. We form the direct integral $\int^{\oplus} \pi_\phi \, d\mu(\phi)$ as in the example preceding Theorem 7.28; each $\pi_\phi$ is irreducible by Theorem 3.25. Moreover, the correspondence $\pi(f)u \mapsto \tilde{f}$ is easily seen to define an isometry $V$ from $\mathcal{H}_\pi$ into $\int^{\oplus} \mathcal{H}_\phi \, d\mu(\phi)$ that intertwines $\pi$ and $\int^{\oplus} \pi_\phi \, d\mu(\phi)$. The trouble is that the measure $\mu$ is not unique, and unless it is chosen carefully, the Hilbert space $\mathcal{H}_\mu = \int^{\oplus} \mathcal{H}_\phi \, d\mu(\phi)$ will be too big, so that $V$ is not surjective. (It can happen, for example, that $\dim \mathcal{H}_\pi < \infty$ but $\dim \mathcal{H}_\mu = \infty$.)

The next question is the extent to which direct integral decompositions are unique. The answer requires a little thought. We have already observed that the measure $\mu$ can be replaced by an equivalent measure without changing anything, and in the preceding theorems there is no uniqueness in the measure space $A$, much less in the measure on it. Indeed, consider a simple example: let $\pi$ be the trivial representation of $G$ on an infinite-dimensional Hilbert space $\mathcal{H}$ ($\pi(x) = I$ for all $x$). We can think of $\mathcal{H}$ as $l^2 = \int^{\oplus} \mathbb{C} \, d\mu$ where $\mu$ is counting measure on $\mathbb{Z}$ or as $L^2([0,1]) = \int^{\oplus} \mathbb{C} \, d\lambda$ where $\lambda$ is Lebesgue measure on $[0,1]$; thus $\pi = \int^{\oplus} \pi_0 \, d\mu = \int^{\oplus} \pi_0 \, d\lambda$ where $\pi_0$ is the trivial representation of $G$ on $\mathbb{C}$, and $\mu$ and $\lambda$ bear no resemblance to one another!

To see what form a uniqueness theorem should take, let us consider the case where $\pi$ is a direct sum of irreducibles, $\pi = \bigoplus \pi_j$. Here the answer is clear: the equivalence classes of the $\pi_j$'s and the multiplicity with which each class occurs are uniquely determined by $\pi$. This suggests that we should consider direct integral decompositions whose parameter space is $\widehat{G}$. As one might suspect, this works well precisely when $\widehat{G}$ is measure-theoretically "decent," that is, in view of Theorem 7.6, when $G$ is type I.

Let us consider $\widehat{G}$ as a measurable space with the Mackey Borel structure. For $n = 1, 2, \ldots, \infty$, let $\widehat{G}_n \subset \widehat{G}$ be the set of equivalence classes of $n$-dimensional irreducible representations of $G$. If one fixes an $n$-dimensional Hilbert space $\mathcal{H}_n$ for each $n$, there is a canonical measurable field of Hilbert spaces $\{\mathcal{H}_p\}_{p \in \widehat{G}}$ over $\widehat{G}$ such that $\mathcal{H}_p = \mathcal{H}_n$ for $p \in \widehat{G}_n$, obtained by forming the appropriate constant field over each $\widehat{G}_n$ and putting them together.

**7.31 Lemma.** *Let $A$ be a subset of $\widehat{G}$ on which the Mackey Borel structure is standard. There is a measurable field of representations $\{\pi_p\}$ over $\widehat{G}$, acting on the canonical field of Hilbert spaces over $\widehat{G}$, such that $\pi_p \in p$ for each $p \in A$.*

This result is due to Mackey [93, Theorem 10.2]; another proof for the case where $G$ is type I can be found in Dixmier [32, §4.6]. (The only point is the measurable selection of $\pi_p \in p$ for $p \in A$; for $p \notin A$ one can take $\pi_p$ to be the trivial representation on $\mathcal{H}_p$.)

A measure $\mu$ on $\widehat{G}$ will be called **standard** if there is a set $A \subset \widehat{G}$ on which the Mackey Borel structure is standard such that $\mu(\widehat{G} \setminus A) = 0$. In this case, we may form the direct integral $\int^{\oplus} \pi_p \, d\mu(p)$ where $\pi_p$ is as in Lemma 7.31. This representation depends, up to unitary equivalence, only on the equivalence class of $\mu$ and not on the particular field $\{\pi_p\}$. If $G$ is type I, by Theorem 7.6 we may take $A = \widehat{G}$ in Lemma 7.31, and every measure on $\widehat{G}$ is standard.

**7.32 Theorem.** *Suppose $G$ is second countable and type I, and $\pi$ is a unitary representation of $G$ on a separable Hilbert space. Let $\{\pi_p\}_{p \in \widehat{G}}$ be as in Lemma 7.31 with $A = \widehat{G}$. There exist finite measures $\mu_1, \mu_2, \ldots, \mu_\infty$ on $\widehat{G}$, uniquely determined to within equivalence, such that*
   i. *$\mu_j \perp \mu_k$ for $j \neq k$,*
   ii. *$\pi$ is equivalent to $\rho_1 \oplus 2\rho_2 \oplus \cdots \oplus \infty\rho_\infty$, where $\rho_n = \int^{\oplus} \pi_p \, d\mu_n(p)$ and $n\rho_n$ denotes the direct sum of $n$ copies of $\rho_n$.*

For the proof, see Dixmier [32, §8.6], or Mackey [98, Theorems 1.21 and 2.15].

When $G$ is Abelian, Theorem 7.32 follows from Theorem 4.45 and the spectral multiplicity theory for commutative C* algebras (for which, see Halmos [58] or Nelson [109]), and it is valid (when properly rephrased) without any countability assumptions on $G$ or $\mathcal{H}$.

It remains to ask what happens when $G$ is not type I. The answer is: *terrible things*. The uniqueness theorem 7.32 breaks down completely; more specifically, if $\sigma$ is a primary representation of $G$ that is not of the form $n\pi$ (in the notation of Theorem 7.40) where $\pi$ is irreducible, there is generally no uniqueness in its direct integral decomposition. Such a representation can have two decompositions,

$$\int^{\oplus} \pi_\alpha \, d\mu(\alpha) \cong \sigma \cong \int^{\oplus} \rho_\beta \, d\nu(\beta),$$

such that the $\pi_\alpha$'s and the $\rho_\beta$'s are all irreducible and *no* $\pi_\alpha$ is equivalent to *any* $\rho_\beta$. In fact, one has the following theorem.

**7.33 Theorem.** *If $G$ is not type I, there exist mutually singular standard measures $\mu$ and $\nu$ on $\widehat{G}$ such that $\int^{\oplus} \pi_p \, d\mu(p)$ and $\int^{\oplus} \pi_p \, d\nu(p)$ (defined as in the remarks following Lemma 7.31) are equivalent primary representations.*

This result is due to Dixmier [30]. We shall present a concrete example of this phenomenon in §7.6; others can be found in Mackey [89], [98, §3.5], and Kirillov [77, §19].

When $G$ is not type I, one can do a little better by considering primary decompositions instead of irreducible decompositions. Indeed, Theorem 7.30(a) shows that there is a natural decomposition of any representation as a direct integral of primary representations. The only non-canonical thing here is the measure space $(A, \mu)$ on which the integral is based (the various proofs of Theorem 7.29 give quite different $A$'s), and this defect can be remedied.

Some terminology: two representations $\pi$ and $\rho$ of $G$ are called **quasi-equivalent** if there is no subrepresentation $\pi'$ of $\pi$ such that $\mathcal{C}(\pi', \rho) = \{0\}$, and no subrepresentation $\rho'$ of $\rho$ such that $\mathcal{C}(\pi, \rho') = \{0\}$. Unitary equivalence obviously implies quasi-equivalence, and the two notions coincide for irreducible representations. More generally, if $\pi$ and $\rho$ are direct sums of irreducible representations, $\pi$ and $\rho$ are quasi-equivalent if and only if exactly the same irreducible equivalence classes occur in these direct sums, perhaps with different multiplicities.

Let $\check{G}$ be the set of quasi-equivalence classes of primary representations of $G$; $\check{G}$ is called the **quasi-dual** of $G$. There is a natural map from $\widehat{G}$ to $\check{G}$ taking each equivalence class of irreducible representations to its quasi-equivalence class. When $G$ is type I, every primary representation is quasi-equivalent to the irreducible representation of which it is a multiple, so this map is a bijection. However, when $G$ is not type I, $\check{G}$ is larger than $\widehat{G}$.

$\check{G}$ is endowed with a $\sigma$-algebra just like the Mackey Borel structure on $\widehat{G}$ (see Ernest [36] or Dixmier [32]), and one defines standard measures on $\check{G}$ just as on $\widehat{G}$. The refined form of Theorems 7.29 and 7.30(a) is then the following.

**7.34 Theorem.** *Suppose $G$ is second countable and $\pi$ is a representation of $G$ on a separable Hilbert space. There is a standard measure $\mu$ on $\check{G}$, a measurable field of Hilbert spaces $\{\mathcal{H}_p\}$ over $\check{G}$, and a measurable field of representations $\{\pi_p\}$ on $\{\mathcal{H}_p\}$ such that $\pi_p \in p$ for $\mu$-almost every $p \in \check{G}$; there is a unitary equivalence $U$ between $\pi$ and $\int^{\oplus} \pi_p \, d\mu(p)$; and the center of $U\mathcal{C}(\pi)U^{-1}$ is the algebra of diagonal operators on $\int^{\oplus} \mathcal{H}_p \, d\mu(p)$. If $\mu'$ and $\{\pi'_p\}$ also have these properties, then $\mu$ is equivalent to $\mu'$ and $\pi_p$ is equivalent to $\pi'_p$ for $\mu$-almost every $p$.*

This result is due to Ernest [36]. The proof can be found in Dixmier [32, §8.4]; see also Pedersen [111, §4.12] for the C* algebra version. The decomposition $\pi \cong \int^{\oplus} \pi_p \, d\mu(p)$ in this theorem is called the **central decomposition** of $\pi$.

## 7.5   The Plancherel Theorem

The subject of this section is, roughly speaking, the explicit decomposition of the regular representation of a locally compact group $G$ as a direct integral of irreducible representations. More precisely, let $G$ be a unimodular locally compact group. (We shall consider the nonunimodular case at the end of this section.) We have the right and left regular representations of $G$ on $L^2(G)$,

$$\rho(x)f(y) = R_x f(y) = f(yx), \qquad \lambda(x)f(y) = L_x f(y) = f(x^{-1}y).$$

Since $R_x L_y = L_y R_x$ for all $x, y \in G$ (this is just the associative law), the representations $\rho$ and $\lambda$ can be combined to give a representation $\tau$ of $G \times G$ on $L^2(G)$:

$$\tau(x,y) = R_x L_y = L_y R_x; \qquad \tau(x,y)f(z) = f(y^{-1}zx).$$

$\tau$ is called the **two-sided regular representation** of $G$ (although it is actually a representation of $G \times G$). Assuming that $G$ is second countable, we shall be interested in obtaining the decomposition of $\tau$ into irreducible representations $\sigma$ and the corresponding decomposition of $L^2(G)$ as a direct integral $\int^{\oplus} \mathcal{H}_\sigma \, d\mu(\sigma)$. When $G$ is type I, this will yield the decomposition of $R$ and $L$ into primary representations that are explicitly of the form $\pi \otimes I$ with $\pi$ irreducible, and the isomorphism between $L^2(G)$ and $\int^{\oplus} \mathcal{H}_\sigma \, d\mu(\sigma)$ will take the form of a "Fourier inversion theorem" and a "Parseval formula." This is the Plancherel theorem.

To begin with, let us see how this works when $G$ is Abelian or compact. In both these cases, we have already proved the essential results, and it is just a matter of interpreting them properly.

When $G$ is Abelian, the Fourier transform $\mathcal{F}$ is a unitary map from $L^2(G)$ to $L^2(\widehat{G}) = \int_{\widehat{G}}^{\oplus} \mathbb{C} \, d\xi$ (where $d\xi$ denotes Haar measure on $\widehat{G}$) such that $\mathcal{F}\rho(x)\mathcal{F}^{-1}g(\xi) = \langle x, \xi\rangle g(\xi)$ and $\mathcal{F}\lambda(x)\mathcal{F}^{-1}g(\xi) = \overline{\langle x, \xi\rangle}g(\xi)$. In other words, when we regard elements $\xi$ of $\widehat{G}$ as representations of $G$, we have $\mathcal{F}\rho\mathcal{F}^{-1} = \int^{\oplus} \xi \, d\xi$ and $\mathcal{F}\lambda\mathcal{F}^{-1} = \int^{\oplus} \overline{\xi} \, d\xi$, and hence $\mathcal{F}\tau\mathcal{F}^{-1} = \int^{\oplus} \xi \otimes \overline{\xi} \, d\xi$. This gives the decomposition of $\rho$, $\lambda$, and $\tau$ into irreducibles.

When $G$ is compact, we have the Peter-Weyl decomposition $L^2(G) = \bigoplus_{[\pi] \in \widehat{G}} \mathcal{E}_\pi$, and Theorem 5.9 says precisely that the restriction of the two-sided regular representation to $\mathcal{E}_\pi$ is equivalent to $\pi \otimes \overline{\pi}$, so $\tau \cong \bigoplus_{[\pi] \in \widehat{G}} \pi \otimes \overline{\pi}$. We can be more explicit: if $f \in L^2(G)$ and $[\pi] \in \widehat{G}$, the Fourier transform $\widehat{f}(\pi)$ as defined in §5.3 is a linear operator on the finite-dimensional space $\mathcal{H}_\pi$. It can therefore be regarded as an *antilinear* map from the *dual* $\mathcal{H}_\pi^* = \mathcal{H}_{\overline{\pi}}$ of $\mathcal{H}_\pi$ into $\mathcal{H}_\pi$, that is, an element of $\mathcal{H}_\pi \otimes \mathcal{H}_{\overline{\pi}}$.

When this identification is made, 5.18 says that

$$[\tau(x,y)f]\widehat{\ }(\pi) = \pi(x)\widehat{f}(\pi)\overline{\pi}(y)^* = [\pi(x) \otimes \overline{\pi}(y)]\widehat{f}(\pi).$$

In other words, the map $f \mapsto \widehat{f}(\pi)$ intertwines $\tau$ with $\pi \otimes \overline{\pi}$. The Fourier transform on $G$ can therefore be regarded as the explicit realization of the unitary map $U : L^2(G) \to \bigoplus \mathcal{H}_\pi \otimes \mathcal{H}_{\overline{\pi}}$ that turns $\tau$ into $\bigoplus \pi \otimes \overline{\pi}$. (The normalization factors $d_\pi$ are incorporated into the norm on the space $\bigoplus \mathcal{H}_\pi \otimes \mathcal{H}_{\overline{\pi}}$, which should be regarded as $\int^\oplus \mathcal{H}_\pi \otimes \mathcal{H}_{\overline{\pi}} \, d\mu([\pi])$) where $\mu$ is $d_\pi$ times counting measure on $\widehat{G}$.) The inverse of $U$ is given by the inversion formula (5.15), and the unitarity of $U$ is expressed by the Parseval formula (5.16).

With this in mind, it is easy to conjecture the correct generalization to non-compact groups. Indeed, suppose $G$ is second countable, unimodular, and type I. By Theorem 7.6 and Lemma 7.31, there is a measurable field of irreducible representations over $\widehat{G}$ such that the representation at the point $p \in \widehat{G}$ belongs to the equivalence class $p$. We shall assume that such a field has been fixed once and for all and identify the points of $\widehat{G}$ with the representations in this field. If $f \in L^1(G)$, then, we define the **Fourier transform** of $f$ (as in the compact case) to be the measurable field of operators over $\widehat{G}$ given by

$$(7.35) \qquad\qquad \widehat{f}(\pi) = \int f(x)\pi(x^{-1}) \, dx.$$

The basic properties (5.17) and (5.18) of the Fourier transform remain valid in this general situation. We wish to think of $\widehat{f}(\pi)$ as an element of $\mathcal{H}_\pi \otimes \mathcal{H}_{\overline{\pi}}$. However, when $\pi$ is infinite-dimensional, $\mathcal{H}_\pi \otimes \mathcal{H}_{\overline{\pi}}$ can be identified not with the full space of bounded operators on $\mathcal{H}_\pi$ but with the space of Hilbert-Schmidt operators (see Appendix 2). One of the things that must be proved, therefore, is that $\widehat{f}(\pi)$ is Hilbert-Schmidt for a suitably large class of $f$'s and $\pi$'s. However, this turns out to be the case, and here is the resulting theorem.

Let

$$\mathcal{J}^1 = L^1(G) \cap L^2(G), \quad \mathcal{J}^2 = \text{linear span of } \{f * g : f,g \in \mathcal{J}^1\}.$$

The elements of $\mathcal{J}^2$ are *finite* linear combinations of convolutions of elements of $\mathcal{J}^1$. They are continuous functions since $L^2 * L^2 \subset C_0$. $\mathcal{J}^2$ will play the role here that the space $\mathcal{B}^1(G)$ played in §4.3, and indeed $\mathcal{J}^2 \subset \mathcal{B}^1(G)$ when $G$ is Abelian.

**7.36 Theorem** (The Plancherel Theorem). *Suppose $G$ is a second countable, unimodular, type I group. There is a measure $\mu$ on $\widehat{G}$, uniquely* *determined once the Haar measure on $G$ is fixed, with the following properties. The Fourier transform $f \mapsto \hat{f}$ maps $\mathfrak{I}^1$ into $\int^{\oplus} \mathcal{H}_\pi \otimes \mathcal{H}_{\bar{\pi}} \, d\mu(\pi)$, and it extends to a unitary map from $L^2(G)$ onto $\int^{\oplus} \mathcal{H}_\pi \otimes \mathcal{H}_{\bar{\pi}} \, d\mu(\pi)$ that intertwines the two-sided regular representation $\tau$ with $\int^{\oplus} \pi \otimes \bar{\pi} \, d\mu(\pi)$. For $f, g \in \mathfrak{I}^1$ one has the Parseval formula*

$$(7.37) \qquad \int f(x)\overline{g(x)} \, dx = \int \mathrm{tr}\big[\hat{f}(\pi)\hat{g}(\pi)^*\big] \, d\mu(\pi),$$

*and for $h \in \mathfrak{I}^2$ one has the Fourier inversion formula*

$$(7.38) \qquad h(x) = \int \mathrm{tr}\big[\hat{\pi}(x)\hat{h}(\pi)\big] \, d\mu(\pi) \qquad (x \in G).$$

This theorem is due (in a slightly different form) to Segal [126], [127] and Mautner [103]; the proof may be found in Dixmier [32, §18.8]. Let us amplify the statement a bit.

First, the fact that the Fourier transform maps $\mathfrak{I}^1$ into $\int^{\oplus} \mathcal{H}_\pi \otimes \mathcal{H}_{\bar{\pi}} \, d\mu(\pi)$ means, in particular, that when $f \in \mathfrak{I}^1$, $\hat{f}(\pi)$ is Hilbert-Schmidt for $\mu$-almost every $\pi$ and its Hilbert-Schmidt norm is square-integrable on $\hat{G}$. This being the case, if $f, g \in \mathfrak{I}^1$ then $(f * g)\hat{}(\pi) = \hat{g}(\pi)\hat{f}(\pi)$ is trace-class for $\mu$-almost every $\pi$ and its trace is integrable on $\hat{G}$. (See Appendix 2.) It follows that the integral in (7.38) is well-defined for all $h \in \mathfrak{I}^2$.

Second, (7.37) (which is just a restatement of the fact that the Fourier transform is unitary) is formally equivalent to (7.38). Indeed, (7.37) follows from (7.38) by taking $h = g^* * f$ and $x = 1$. On the other hand, (7.38) for $x = 1$ follows at least formally from (7.37) by taking $f = h$ and $g = $ an approximate identity, and (7.38) for general $x$ follows from (7.38) for $x = 1$ by replacing $h$ with $R_x h$.

Third, if we restrict the two-sided regular representation $\tau$ to the groups $G \times \{1\}$ and $\{1\} \times G$, we obtain the decompositions of the right and left regular representations into type I primary representations,

$$\rho \cong \int^{\oplus} \pi \otimes I \, d\mu(\pi), \qquad \lambda \cong \int^{\oplus} I \otimes \bar{\pi} \, d\mu(\pi),$$

where the equivalences are given by the Fourier transform. (These are actually the central decompositions of $\rho$ and $\lambda$, as follows from Theorem 7.39 below.) In conjunction with Proposition 7.21, they immediately yield the canonical decompositions of $\rho$ and $\lambda$ into irreducibles as in Theorem 7.32. Indeed, let $\hat{G}_n = \{\pi \in \hat{G} : \dim \mathcal{H}_\pi = n\}$. Then the measures $\mu_n$ of Theorem 7.32 are given by $\mu_n(E) = \mu(E \cap \hat{G}_n)$ in the case of $\rho$ and $\mu_n(E) = \mu(\overline{E} \cap G_n)$ in the case of $\lambda$, where $\overline{E} = \{\bar{\pi} : \pi \in E\}$.

The measure $\mu$ in the Plancherel theorem is called the **Plancherel measure** on $\widehat{G}$. When $G$ is Abelian, Plancherel measure is Haar measure, and when $G$ is compact, it is given by $\mu(E) = \sum_{\pi \in E} d_\pi$. In both these cases, the support of the Plancherel measure is all of $\widehat{G}$; that is, there is no nonempty open $U \subset \widehat{G}$ such that $\mu(U) = 0$. It may come as something of a surprise that this is not the case in general. Indeed, the condition that $\operatorname{supp} \mu = \widehat{G}$ is equivalent to the amenability of the group $G$, a condition that has many other characterizations; see Dixmier [32] and Greenleaf [56]. For connected Lie groups, amenability is equivalent to having a solvable normal subgroup with compact quotient; in particular, noncompact semisimple groups are not amenable.

The determination of the Plancherel measure for non-compact, non-Abelian groups is a difficult problem. The Plancherel measure has been explicitly determined for all connected semisimple Lie groups by Harish-Chandra [62], [63]; see also Knapp [79]. (This was done despite the lack of understanding of the full dual space for many semisimple groups! The "unknown" representations form a set of Plancherel measure zero.) For connected nilpotent Lie groups, one has a fair amount of information about the Plancherel measure in terms of the Kirillov theory, and it has been explicitly determined in some special cases; see Corwin and Greenleaf [26]. Kleppner and Lipsman [78] have shown how to study Plancherel measure in terms of the Mackey machine; in many cases they are able to compute the Plancherel measure of $G$ in terms of the Plancherel measures of a normal subgroup $N$ and the various little groups in $G/N$.

Some further insight into the Plancherel theorem is afforded by the following theorem of Segal [125]:

**7.39 Theorem.** *Let $G$ be a unimodular locally compact group with regular representations $\rho$ and $\lambda$, and let $\mathcal{R}$ and $\mathcal{L}$ be the von Neumann algebras generated by $\{\rho(x) : x \in G\}$ and $\{\lambda(x) : x \in G\}$, respectively.*
  a. *$\mathcal{R} = \mathcal{C}(\lambda)$ and $\mathcal{L} = \mathcal{C}(\rho)$.*
  b. *An operator $T \in \mathcal{L}(L^2(G))$ commutes with every element of $\mathcal{C}(\rho)$ if and only if $T \in \mathcal{C}(\lambda)$, and vice versa.*
  c. *$\mathcal{C}(\rho) \cap \mathcal{C}(\lambda)$ is the common center of $\mathcal{C}(\rho)$ and $\mathcal{C}(\lambda)$.*

*Proof.* (b) follows from (a) and the von Neumann density theorem (1.56), and (c) is an immediate corollary of (b). Hence, since the situation is symmetric in $\rho$ and $\lambda$, it suffices to show that $\mathcal{L} = \mathcal{C}(\rho)$.

On the one hand, since left translations commute with right translations it is clear that $\lambda(x) \in \mathcal{C}(\rho)$ for all $x \in G$ and hence that $\mathcal{L} \subset \mathcal{C}(\rho)$. To prove the reverse inclusion, suppose $T \in \mathcal{C}(\rho)$. If $f, g \in L^1 \cap L^2$ we have

$$\lambda(f)g = \int f(y)[\lambda(y)g]\,dy = f * g = \int [\rho(y)f]g(y^{-1})\,dy = \rho(\widetilde{g})f,$$

where $\widetilde{g}(x) = g(x^{-1})$. The operator $\rho(\widetilde{g})$ belongs to $\mathcal{R}$ by Theorem 3.12, so it commutes with $T$. Hence,

$$T\lambda(f)g = T(f*g) = T\rho(\widetilde{g})f = \rho(\widetilde{g})Tf = Tf*g.$$

Suppose $Tf \in L^1$: we then have $T\lambda(f) = \lambda(Tf)$, and hence $T\lambda(f) \in \mathcal{L}$ by Theorem 3.12 again. An additional approximation argument, for which we refer the reader to Segal [125], shows that the same conclusion holds without assuming $Tf \in L^1$. But $\lambda(f) \to I$ strongly as $f$ runs through an approximate identity, so $T \in \mathcal{L}$. $\qquad\square$

The Plancherel theorem admits various partial generalizations to groups that fail to be second countable, unimodular, or type I. We now discuss the most important of these.

Suppose $G$ is unimodular and second countable but not type I. There is little hope of obtaining a canonical irreducible decomposition for the left or right regular representation, but — a pleasant surprise — there is such a decomposition for the two-sided regular representation. Indeed, let $\tau = \int^{\oplus} \tau_\alpha \, d\mu(\alpha)$ be the central decomposition of $\tau$ as in Theorem 7.34. Since $\mathcal{C}(\tau) = \mathcal{C}(\rho) \cap \mathcal{C}(\lambda)$ is Abelian by Theorem 7.39, this is actually a decomposition into irreducible representations by Theorem 7.30(b). It follows easily that the representations $\rho_\alpha(x) = \tau_\alpha(x, 1)$ and $\lambda_\alpha(x) = \tau_\alpha(1, x)$ of $G$ are primary, and we have $\rho = \int^{\oplus} \rho_\alpha \, d\mu(\alpha)$, $\lambda = \int^{\oplus} \lambda_\alpha \, d\mu(\alpha)$. These are the central decompositions of $\rho$ and $\lambda$, by Theorem 7.39. The difference between this and the type I case is that $\tau_\alpha$ is usually not of the form $\pi \otimes \overline{\pi}$ for an irreducible representation $\pi$ of $G$, and $\rho_\alpha$ and $\lambda_\alpha$ are usually not multiples of irreducible representations.

There is an analogue of the Parseval formula (7.37) here. Namely, one defines the operators $\widehat{f}(\lambda_\alpha)$ and $\widehat{f}(\rho_\alpha)$ as in (7.35), and then

$$\int f(x)\overline{g(x)} \, dx = \int \mathrm{tr}_{\rho_\alpha}\left[ \widehat{f}(\rho_\alpha)\widehat{g}(\rho_\alpha)^* \right] d\mu(\alpha)$$
$$= \int \mathrm{tr}_{\lambda_\alpha}\left[ \widehat{f}(\lambda_\alpha)\widehat{g}(\lambda_\alpha)^* \right] d\mu(\alpha),$$

where $\mathrm{tr}_{\rho_\alpha}$ and $\mathrm{tr}_{\lambda_\alpha}$ are the generalized traces associated to the von Neumann algebras generated by the representations $\rho_\alpha$ and $\lambda_\alpha$. This result is due to Segal [126] and Mautner [103]; we refer the reader to Dixmier [31], [32] for a proof and an explanation of the generalized traces. (An important part of the theorem is that the only von Neumann algebras that arise in this situation are ones for which the appropriate traces exist; i.e., factors of type III do not occur.) The Plancherel formula for the Mautner group is worked out in a concrete fashion in Cowling [27].

One word of caution: for some groups the preceding results are essentially vacuous because $\rho$ and $\lambda$ are primary to begin with, and hence (by Theorem 7.39) $\tau$ is irreducible. Examples include the rational $ax+b$ group (defined like the usual $ax+b$ group but with $a$ and $b$ rational and with the discrete topology) and free groups on more than one generator. We shall discuss the latter example in §7.6; see Mackey [98, §3.5] for the former.

We now turn to the case of nonunimodular groups. Suppose $G$ is second countable and type I but not unimodular, and let $\Delta$ be the modular function of $G$. The definitions of $\rho$ and $\tau$ must be modified to make them unitary, as follows:

$$\rho(x) = \Delta(x)^{1/2} R_x, \qquad \tau(x,y) = \Delta(x)^{1/2} R_x L_y.$$

With this change, $\tau$ is still unitarily equivalent to $\int^\oplus \pi \otimes \overline{\pi}\, d\mu(\pi)$ for a suitable measure $\mu$ on $\widehat{G}$, but the equivalence is *not* given by the Fourier transform (7.35). This can easily be seen by observing the effect on the Parseval formula (7.37) when $f$ and $g$ are replaced by $R_x f$ and $R_x g$. The left side changes by a factor of $\Delta(x)^{-1}$; but since the trace is a unitary invariant and

$$(R_x f)\widehat{\phantom{f}}(\pi) = \int f(yx)\pi(y^{-1})\, dy = \int f(y)\pi(xy^{-1})\Delta(x)^{-1}\, dy$$
$$= \Delta(x)^{-1}\pi(x)\widehat{f}(\pi),$$

the right side changes by a factor of $\Delta(x)^{-2}$. Hence (7.37) cannot be valid, i.e., the Fourier transform (7.35) cannot be unitary.

The solution to this problem is to modify the Fourier transform. First, as a matter of convenience and adherence to convention, we shall replace $\pi(x^{-1})$ by $\pi(x)$ in (7.35), i.e., we consider the operators $\pi(f)$. This change has no effect on the difficulty just mentioned. (It could also be made in the unimodular case, where it would merely result in the relabeling of various things in the Plancherel theorem.) But suppose we can find, for each $\pi \in \widehat{G}$, a (probably unbounded) self-adjoint operator $D_\pi$ on $\mathcal{H}_\pi$ with the property that

$$D_\pi \pi(x) = \Delta(x)^{1/2}\pi(x)D_\pi \qquad (x \in G).$$

If we then redefine the Fourier transform to be

(7.40) $$\widehat{f}(\pi) = \pi(f)D_\pi = \int f(y)\pi(y)D_\pi\, dy,$$

we see that

$$(R_x f)\widehat{\phantom{f}}(\pi) = \int f(y)\pi(yx^{-1})D_\pi\Delta(x)^{-1}\, dy$$
$$= \int f(y)\pi(y)D_\pi\pi(x^{-1})\Delta(x)^{-1/2}\, dy = \Delta(x)^{-1/2}\widehat{f}(\pi)\pi(x)^{-1}.$$

Hence, if we use *this* $\widehat{f}$ in (7.37), substitution of $R_x f$ for $f$ changes both sides by $\Delta(x)^{-1}$, so there is some hope that the formula is correct.

How can one construct such operators $D_\pi$? When $\pi$ is induced from a subgroup $H$ on which the modular function is trivial, this is easy. The Hilbert space $\mathcal{H}_\pi$ is then a space of vector-valued functions on $G$, and one can set

$$(7.41) \qquad\qquad (D_\pi f)(x) = \Delta(x)^{1/2} f(x).$$

(Since $\Delta(\xi) = 1$ for $\xi \in H$, $D_\pi$ does not affect the condition (6.3) in the definition of $\mathcal{H}_\pi$.) The domain of $D_\pi$ is the set of all $f \in \mathcal{H}_\pi$ such that $D_\pi f \in \mathcal{H}_\pi$.

We now have the ingredients for a nonunimodular Plancherel theorem that works for all groups satisfying a couple of technical conditions.

**7.42 Theorem.** *Suppose $G$ is second countable. Let $H = \ker(\Delta)$, and suppose that $H$ is type I and that $G$ acts regularly on $\widehat{H}$ (in the sense described in §6.6). There is a unique measure $\mu$ on $\widehat{G}$ with the following properties. First, $\mu$-almost every $\pi \in \widehat{G}$ is induced from a representation of $H$. Second, if for all such $\pi$ one defines $D_\pi$ by (7.41), the map $f \mapsto \widehat{f}$ defined by (7.40) gives a unitary isomorphism from $L^2(G)$ to $\int^{\oplus} \mathcal{H}_\pi \otimes \mathcal{H}_{\overline{\pi}}\, d\mu(\pi)$ that intertwines $\tau$ and $\int^{\oplus} \pi \otimes \overline{\pi}\, d\mu(\pi)$. In particular, although $G$ need not be type I, $\mu$-almost all of the primary representations in the central decomposition of the regular representations are type I.*

This theorem is due to Tatsuuma [135] (whose notation is a little different since he uses right Haar measure instead of left Haar measure). Tatsuuma also obtains a more general form of the theorem that applies when $H$ is not type I. Another extension of the theorem, including a clarification of the role of its hypotheses, has been obtained by Duflo and Moore [33]. For more about the Plancherel theorem for solvable Lie groups, see Pukanszky [114] and Moore [106].

## 7.6 Examples

In this section we present four examples to illustrate the ideas of this chapter.

**1. The Heisenberg groups.** The Heisenberg groups $H_n$ defined in §6.7 are simply connected nilpotent Lie groups, so the Kirillov theory (Theorem 7.9) applies to them. We can identify both $H_n$ and its Lie algebra with $\mathbb{R}^n \times \mathbb{R}^n \times \mathbb{R}$, with group multiplication given by (6.47) and Lie bracket given by

$$[(x, \xi, t), (x', \xi', t')] = (0, 0, x \cdot \xi' - \xi \cdot x').$$

The exponential map is then merely the identity. A simple calculation shows that the adjoint action is given by

$$\big[\mathrm{Ad}(x, \xi, t)\big](y, \eta, s) = (y, \eta, s + x \cdot \eta - \xi \cdot y)$$

and hence that the coadjoint action is given by

$$\big[\mathrm{Ad}^*(x, \xi, t)\big](b, \beta, r) = (b + rx, \beta - r\xi, r).$$

From this it is clear that the coadjoint orbits are the hyperplanes $r = h$, where $h$ is a nonzero real constant, and the singleton sets $\{(b, \beta, 0)\}$. Theorem 7.9 together with Corollary 6.51 therefore yields the following results. (See also Fell [35].)

$\widehat{H}_n$ can be identified with the quotient of $\mathbb{R}^n \times \mathbb{R}^n \times \mathbb{R}$ by the coadjoint action. The hyperplane $r = h$ corresponds to the representation $\rho_h$ of Corollary 6.52, while the singleton $\{(b, \beta, 0)\}$ corresponds to the representation $\pi_{b,\beta}$. $\widehat{H}_n$ is $T_1$ since the orbits are all closed, but it is not Hausdorff because the singleton orbits have no disjoint neighborhoods. See Figure 7.1.

We now compute the Fourier transform on the Heisenberg group. The Euclidean Fourier transform enters into this calculation in a decisive way, and we shall denote the Euclidean Fourier transform of a function $f$ on $\mathbb{R}^n \times \mathbb{R}^n \times \mathbb{R}$ with respect to its first, second, and third arguments by $\mathcal{F}_1 f$, $\mathcal{F}_2 f$, and $\mathcal{F}_3 f$ respectively:

$$\mathcal{F}_1 f(b, \xi, t) = \int e^{-2\pi i b \cdot x} f(x, \xi, t) \, dx, \text{ etc.}$$

We shall also use the easily verified fact that Lebesgue measure is a left and right Haar measure on $H_n$.

Suppose $f \in L^1(H_n) \cap L^2(H_n)$. For the one-dimensional representations $\pi_{b,\beta}$, we have

$$\widehat{f}(\pi_{b,\beta}) = \iiint f(x, \xi, t) e^{-2\pi i (b \cdot x + \beta \cdot \xi)} \, dx \, d\xi \, dt = \mathcal{F}_1 \mathcal{F}_2 \mathcal{F}_3 f(b, \beta, 0).$$

For the representation $\rho_h$ on $L^2(\mathbb{R}^n)$, $\widehat{f}(\rho_h)$ is the operator on $L^2(\mathbb{R}^n)$

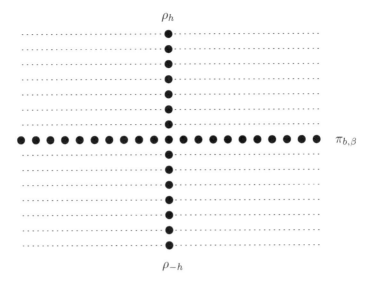

$\rho_h$

$\pi_{b,\beta}$

$\rho_{-h}$

**FIGURE 7.1**
The dual space of $H_n$.

given by

$$[\widehat{f}(\rho_h)\phi](y) = \iiint f(x,\xi,t)\left[\rho_h(-x,-\xi,-t)\phi\right](y)\,dx\,d\xi\,dt$$

$$= \iiint f(x,\xi,t)e^{-2\pi i h[t-\xi\cdot y-(\xi\cdot x/2)]}\phi(y+x)\,dx\,d\xi\,dt$$

$$= \iiint f(x-y,\xi,t)e^{-2\pi i h[t-\xi\cdot(x+y)/2]}\phi(x)\,dx\,d\xi\,dt.$$

That is, $\widehat{f}(\rho_h)$ is the integral operator on $L^2(\mathbb{R}^n)$ whose kernel is

(7.43)
$$K_h^f(y,x) = \iint f(x-y,\xi,t)e^{-2\pi i h[t-\xi\cdot(x+y)/2]}\,d\xi\,dt$$

$$= \mathcal{F}_2\mathcal{F}_3 f(x-y,\ -\tfrac{1}{2}h(x+y),\ h).$$

By a slight modification of Theorem A.17 (Appendix 3), the square of the Hilbert-Schmidt norm of this operator is

$$\||\widehat{f}(\rho_h)\||^2 = \iint |K_h^f(y,x)|^2\,dy\,dx.$$

The substitution $u = x - y$, $v = -\frac{1}{2}h(x+y)$ together with the Euclidean Parseval formula gives

$$\||\widehat{f}(\rho_h)\||^2 = \iint |\mathcal{F}_2\mathcal{F}_3 f(u, v, h)|^2 |h|^{-n} \, du \, dv$$

$$= |h|^{-n} \iint |\mathcal{F}_3 f(u, w, h)|^2 \, du \, dw.$$

Therefore, by another application of Parseval,

$$\int \||\widehat{f}(\rho_h)\||^2 |h|^n \, dh = \iiint |\mathcal{F}_3 f(u, w, h)|^2 \, du \, dw \, dh$$

$$= \iiint |f(u, w, t)|^2 \, du \, dw \, dt.$$

But this is the Parseval formula for the Heisenberg group! In other words, Plancherel measure $\mu$ on $\widehat{H}^n$ is given by

$$d\mu(\rho_h) = |h|^n \, dh, \qquad d\mu(\pi_{b,\beta}) = 0,$$

and the Fourier transform on $H_n$ is a unitary isomorphism from $L^2(H_n)$ to

$$\int^\oplus \mathcal{H}_{\rho_h} \otimes \mathcal{H}_{\overline{\rho}_h} |h|^n \, dh = \int^\oplus L^2(\mathbb{R}^n) \otimes L^2(\mathbb{R}^n)^* |h|^n \, dh.$$

If we identify $L^2(\mathbb{R}^n) \otimes L^2(\mathbb{R}^n)^*$ with $L^2(\mathbb{R}^n \times \mathbb{R}^n)$ (see Appendix 3), this space becomes

$$\int^\oplus L^2(\mathbb{R}^n \times \mathbb{R}^n) |h|^n \, dh = L^2(\mathbb{R}^n \times \mathbb{R}^n \times \mathbb{R}, \, |h|^n \, dy \, dx \, dh),$$

and the Fourier transform, regarded as a map from $L^2(H_n)$ to the latter space, is just the map

$$f \mapsto \widetilde{f}(y, x, h) = \mathcal{F}_2\mathcal{F}_3 f(x - y, -\tfrac{1}{2}h(x+y), h)$$

of (7.43).

These considerations also give a direct proof that $H_n$ is CCR. Indeed, the preceding arguments show that $\widehat{f}(\rho_h) = \rho_h(\overline{f})^*$ is Hilbert-Schmidt for every $h$ when $f$ belongs (say) to the Schwartz class of rapidly decaying smooth functions; since these functions are dense in $L^1$ and the norm limit of compact operators is compact, $\rho_h(f)$ is compact for all $f \in L^1$. (Of course, $\pi_{b,\beta}(f)$ is compact for trivial reasons.)

**2. The ax + b group.** The Lie algebra of the $ax + b$ group $G$ is $\mathbb{R}^2$ with Lie bracket $[(s, t), (s', t')] = (0, st' - s't)$, and the exponential map is $\exp(s, t) = (e^s, s^{-1}(e^s - 1)t)$. (The easy way to see this is to think of

$G$ as the group of $2 \times 2$ matrices of the form $\left(\begin{smallmatrix} a & b \\ 0 & 1 \end{smallmatrix}\right)$ and its Lie algebra as the algebra of $2 \times 2$ matrices of the form $\left(\begin{smallmatrix} s & t \\ 0 & 0 \end{smallmatrix}\right)$.) A simple calculation shows that the adjoint action is given by

$$\big[\mathrm{Ad}(a,b)\big](s,t) = (s,\, at - bs),$$

and hence the coadjoint action is given by

$$\big[\mathrm{Ad}^*(a,b)\big](\sigma,\tau) = (\sigma + a^{-1}b\tau,\, a^{-1}\tau).$$

Thus the orbits of the coadjoint action are the half-plane $\tau > 0$, the half plane $\tau < 0$, and the singleton sets $\{(\sigma, 0)\}$.

$\quad$ $G$ is an exponential solvable group, so the Kirillov picture applies to it: $\widehat{G}$ can be identified with the quotient of $\mathbb{R}^2$ by the coadjoint action. (See Fell [38] for a proof that this identification is a homeomorphism.) The half-planes $\tau > 0$ and $\tau < 0$ correspond to the representations $\pi^+$ and $\pi^-$ constructed in §6.7, and the singleton orbit $\{(\sigma, 0)\}$ corresponds to the one-dimensional representation $\pi_\sigma^0(a,b) = a^{i\sigma}$. $\widehat{G}$ is $T_0$ but not $T_1$, for the orbits $\tau > 0$ and $\tau < 0$ are open but not closed. See Figure 7.2.

$\quad$ We now consider the Plancherel theorem for $G$. We recall from §2.2

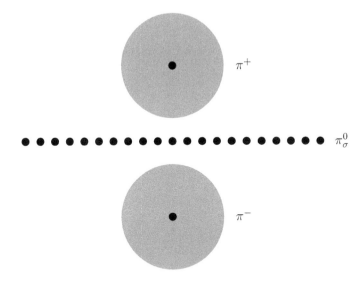

**FIGURE 7.2**
The dual space of the $ax + b$ group.

that left Haar measure on $G$ is $da\,db/a^2$ and right Haar measure is $da\,db/a$, so we must use the nonunimodular version of the Plancherel theorem, Theorem 7.42. As in the case of the Heisenberg group, it turns out that the one-dimensional representations have Plancherel measure zero, so we concern ourselves only with the representations $\pi^+$ and $\pi^-$, which (we recall) are the subrepresentations of

$$\pi(a,b)g(s) = a^{1/2}e^{2\pi ibs}g(as) \qquad (g \in L^2(\mathbb{R}))$$

on $\mathcal{H}_+ = L^2(0,\infty)$ and $\mathcal{H}_- = L^2(-\infty,0)$, respectively.

According to the discussion preceding Theorem 7.42, we should look for operators $D_+$ and $D_-$ on $\mathcal{H}_+$ and $\mathcal{H}_-$ such that

$$D_\pm \pi_\pm(a,b) = a^{-1/2}\pi^\pm(a,b)D_\pm.$$

(Here $a^{-1/2}$ is the square root of the modular function on $G$.) In view of (7.41), one easily sees that the operators

$$D_\pm\phi(s) = |s|^{1/2}\phi(s)$$

will work. We therefore define the modified Fourier transform according to (7.40):

$$\widehat{f}(\pi^\pm) = \pi^\pm(f)D_\pm.$$

Explicitly, for $\phi \in \mathcal{H}_+$,

$$\left[\widehat{f}(\pi^+)\phi\right](s) = \int_{-\infty}^{\infty}\int_0^\infty f(a,b)e^{2\pi ibs}a^{1/2}(as)^{1/2}\phi(as)\,\frac{da\,db}{a^2}$$

$$= \int_{-\infty}^{\infty}\int_0^\infty f(s^{-1}t,b)e^{2\pi ibs}s^{1/2}\phi(t)\,\frac{dt\,db}{t},$$

so $\widehat{f}(\pi^+)$ is the integral operator on $\mathcal{H}_+$ with kernel

$$K_+(s,t) = \frac{s^{1/2}}{t}\int_{-\infty}^{\infty}f(s^{-1}t,b)e^{2\pi ibs}\,db = \frac{s^{1/2}}{t}\mathcal{F}_2f(s^{-1}t,-s).$$

(Here, as above, $\mathcal{F}_2f$ is the Euclidean Fourier transform of $f$ in its second argument.) Likewise, $\widehat{f}(\pi^-)$ is the integral operator on $\mathcal{H}_-$ with kernel

$$K_-(s,t) = \frac{|s|^{1/2}}{|t|}\mathcal{F}_2f(s^{-1}t,-s).$$

The Hilbert-Schmidt norms of $\widehat{f}(\pi^\pm)$ (see Appendix 3) are therefore

$$\|\|\widehat{f}(\pi^+)\|\|^2 = \int_0^\infty\int_0^\infty \frac{s}{t^2}|\mathcal{F}_2f(s^{-1}t,-s)|^2\,dt\,ds$$

$$= \int_0^\infty\int_0^\infty |\mathcal{F}_2f(a,-s)|^2\,\frac{da\,ds}{a^2}$$

and

$$\||\widehat{f}(\pi^-)\||^2 = \int_{-\infty}^0 \int_{-\infty}^0 \frac{|s|}{t^2} |\mathcal{F}_2 f(s^{-1}t, -s)|^2 \, dt \, ds$$

$$= \int_{-\infty}^0 \int_0^\infty |\mathcal{F}_2 f(a, -s)|^2 \, \frac{da \, ds}{a^2}.$$

Therefore,

$$\||\widehat{f}(\pi^+)\||^2 + \||\widehat{f}(\pi^-)\||^2 = \int_{-\infty}^\infty \int_0^\infty |\mathcal{F}_2 f(a, -s)|^2 \, \frac{da \, ds}{a^2},$$

which, by the Euclidean Parseval formula, equals

$$\int_{-\infty}^\infty \int_0^\infty |f(a,b)|^2 \, \frac{da \, db}{a^2} = \|f\|_{L^2(G)}^2.$$

In short, the map $f \mapsto (\widehat{f}(\pi^+), \widehat{f}(\pi^-))$ is an isometry from $L^2(G)$ into

$$(\mathcal{H}_+ \otimes \mathcal{H}_+^*) \oplus (\mathcal{H}_- \otimes \mathcal{H}_-^*),$$

and an examination of these calculations shows without difficulty that it is actually unitary. Thus, Plancherel measure on $\widehat{G}$ is just counting measure on the two-element set $\{\pi^+, \pi^-\}$.

It is to be noted that the operators $\pi^\pm(f)$ are, in general, not Hilbert-Schmidt until they have been composed with the unbounded operators $D_\pm$. This accords with the fact that $G$ is not CCR. The latter assertion is true on general grounds (Theorem 7.7) since $\widehat{G}$ is not $T_1$, and it can also be seen in a more concrete fashion. Indeed, if $f \in L^1(G)$, the operators $\pi^\pm(f)$ are compact if and only if $\int_{-\infty}^\infty f(a,b) \, db = 0$ for almost every $a$. This is proved in Khalil [75], which also contains some other interesting results concerning the harmonic analysis of $G$.

**3. SL(2, R).** $SL(2, \mathbb{R})$ is the group of $2 \times 2$ real matrices of determinant one. Its representation theory has been well understood for a long time, but it would require far too much space to give a complete exposition of it here. We shall content ourselves with stating the results and giving some references. To begin with, we list some families of representations of $SL(2, \mathbb{R})$.

(i) *The trivial representation $\iota$*, acting on $\mathbb{C}$.

(ii) *The discrete series* $\{\delta_n^\pm : n \geq 2\}$. For $n \geq 2$, let $\mathcal{H}_n^+$ (resp. $\mathcal{H}_n^-$) be the space of holomorphic (resp. antiholomorphic) functions $f$ on the upper half plane $U = \{z = x + iy : y > 0\}$ such that

$$(7.44) \qquad \|f\|_{[n]}^2 = \iint_U |f(x+iy)|^2 y^{n-2} \, dx \, dy < \infty.$$

The representation $\delta_n^{\pm}$ of $SL(2,\mathbb{R})$ is defined on $\mathcal{H}_n^{\pm}$ by

(7.45)
$$\delta_n^+ \begin{pmatrix} a & b \\ c & d \end{pmatrix} f(z) = (-bz+d)^{-n} f\left(\frac{az-c}{-bz+d}\right),$$
$$\delta_n^- \begin{pmatrix} a & b \\ c & d \end{pmatrix} f(z) = (-b\bar{z}+d)^{-n} f\left(\frac{az-c}{-bz+d}\right).$$

The representations $\delta_n^{\pm}$ are unitary and irreducible, and they have another important property: their matrix coefficients $\langle \delta_n^{\pm}(x)f, g\rangle$ are square-integrable functions of $x \in SL(2,\mathbb{R})$. It follows, as in the proof of Theorem 5.9, that the representations $\delta_n^{\pm}$ occur as discrete summands in the decomposition of the regular representation into irreducibles; hence the name "discrete series."

(iii) *The mock discrete series* $\{\delta_1^+, \delta_1^-\}$. Let $\mathcal{H}_1^+$ be the space of holomorphic functions on the upper half plane such that

$$\|f\|_{[1]}^2 = \sup_{y>0} \int_{-\infty}^{\infty} |f(x+iy)|^2 \, dx < \infty,$$

and let $\mathcal{H}_1^-$ be the corresponding space of antiholomorphic functions. The representations $\delta_1^{\pm}$ of $SL(2,\mathbb{R})$ on $\mathcal{H}_1^{\pm}$ are given by (7.45). The matrix coefficients of $\delta_1^{\pm}$ are not square-integrable, but these representations bear a number of family relationships to the discrete series; hence the name "mock discrete series."

According to a well-known theorem of Paley and Wiener (see Dym and McKean [35, §3.4]) the Hilbert spaces $\mathcal{H}_1^{\pm}$ can be naturally identified with certain subspaces of $L^2(\mathbb{R})$, namely

$$\widetilde{\mathcal{H}}_1^+ = \{f \in L^2(\mathbb{R}) : \widehat{f}(\xi) = 0 \text{ for } \xi < 0\},$$
$$\widetilde{\mathcal{H}}_1^- = \{f \in L^2(\mathbb{R}) : \widehat{f}(\xi) = 0 \text{ for } \xi > 0\}.$$

The unitary map from $\mathcal{H}_1^{\pm}$ to $\widetilde{\mathcal{H}}_1^{\pm}$ simply takes a holomorphic or antiholomorphic function on the upper half plane to its boundary values on $\mathbb{R}$, and the inverse map is given by the Fourier inversion formula: if $f$ is in $\widetilde{\mathcal{H}}_1^+$ or $\widetilde{\mathcal{H}}_1^-$, the corresponding $F \in \mathcal{H}_1^+$ or $\mathcal{H}_1^-$ is given by

$$F(z) = \int e^{2\pi i \xi z} \widehat{f}(\xi) \, d\xi \quad \text{or} \quad F(z) = \int e^{2\pi i \xi \bar{z}} \widehat{f}(\xi) \, d\xi.$$

When these identifications are made, the representations $\delta_1^{\pm}$ are still given by (7.45), but with $x \in \mathbb{R}$ replacing $z \in U$.

(iv) *The principal series* $\{\pi_{it}^{\pm} : t \in \mathbb{R}\}$. These are the representations of $SL(2,\mathbb{R})$ induced from the one-dimensional representations of the upper triangular subgroup

$$P = \left\{ M_{a,b} = \begin{pmatrix} a & b \\ 0 & a^{-1} \end{pmatrix} : a \in \mathbb{R} \setminus \{0\}, \ b \in \mathbb{R} \right\}.$$

Any one-dimensional representation of $P$ must annihilate its commu-tator subgroup, namely $\{M_{a,b} : a = 1\}$, so it is easily seen that these representations are precisely

$$(7.46) \qquad \xi_{it}^+(M_{a,b}) = |a|^{it}, \quad \xi_{it}^-(M_{a,b}) = |a|^{it} \operatorname{sgn} a \qquad (t \in \mathbb{R}).$$

The principal series are then defined by

$$(7.47) \qquad \pi_{it}^+ = \operatorname{ind}_P^{SL(2,\mathbb{R})}(\xi_{it}^+), \qquad \pi_{it}^- = \operatorname{ind}_P^{SL(2,\mathbb{R})}(\xi_{it}^-).$$

$\pi_{it}^+$ and $\pi_{it}^-$ are known as the *spherical principal series* and *non-spherical principal series*, respectively.

The Hilbert spaces for these representations consist of complex-valued functions on $SL(2,\mathbb{R})$ satisfying certain covariance conditions on the cosets of $P$, and such functions are determined by their values on

$$\overline{N} = \left\{ \begin{pmatrix} 1 & 0 \\ t & 1 \end{pmatrix} : t \in \mathbb{R} \right\},$$

since $\overline{N}$ intersects each coset (except for a set of measure zero) in exactly one point. Since $\overline{N} \cong \mathbb{R}$, it is not hard to show that the map $f \mapsto f|\overline{N}$ sets up a unitary isomorphism from these Hilbert spaces to $L^2(\mathbb{R})$, and that the resulting realization of the representations $\pi_{it}^\pm$ on $L^2(\mathbb{R})$ is given by

$$\pi_{it}^\pm \begin{pmatrix} a & b \\ c & d \end{pmatrix} f(x) = m_\pm(-bx + d)| - bx + d|^{-1-it} f\left( \frac{ax - c}{-bx + d} \right),$$

where

$$m_+(y) = 1, \qquad m_-(y) = \operatorname{sgn} y.$$

The principal series representations are all irreducible except for $\pi_0^-$, which is the direct sum of the mock discrete series $\delta_1^+$ and $\delta_1^-$ (when all of these are realized on subspaces of $L^2(\mathbb{R})$). Moreover, $\pi_{-it}^+$ and $\pi_{-it}^-$ are equivalent respectively to $\pi_{it}^+$ and $\pi_{it}^-$, and otherwise these represen-tations are all inequivalent.

(v) *The complementary series* $\{\kappa_s : 0 < s < 1\}$. The Hilbert space for $\kappa_s$ is the set of all complex-valued functions $f$ on $\mathbb{R}$ such that

$$(7.48) \qquad \|f\|_{(s)}^2 = \frac{s}{2} \iint f(x)\overline{f(y)}|x - y|^{s-1} \, dx \, dy < \infty,$$

and the action of $SL(2, \mathbb{R})$ is like that of the spherical principal series:

$$\kappa_s \begin{pmatrix} a & b \\ c & d \end{pmatrix} f(x) = |-bx + d|^{-1-s} f\left(\frac{ax - c}{-bx + d}\right).$$

(The significance of the factor $s/2$ in (7.48) will be explained below.)

This completes our list of representations of $SL(2, \mathbb{R})$. These representations are all irreducible and inequivalent except that $\pi_{-it}^{\pm} \cong \pi_{it}^{\pm}$ and $\pi_0^- \cong \delta_1^+ \oplus \delta_1^-$ as mentioned in (iii), and every irreducible representation of $SL(2, \mathbb{R})$ is equivalent to one of them. Hence, we may parametrize the dual space $[SL(2, \mathbb{R})]\hat{}$ by identifying it with the following set of representations:

(7.49)
$$\{\iota\} \cup \{\delta_n^{\pm} : n \geq 1\} \cup \{\pi_{it}^+ : t \geq 0\} \cup \{\pi_{it}^- : t > 0\} \cup \{\kappa_s : 0 < s < 1\}.$$

These results are due to Bargmann [8], and an exposition of them can be found in a number of places, including Knapp [79] and Howe and Tan [68].

It remains to describe the Fell topology on $[SL(2, \mathbb{R})]\hat{}$. This is more or less what one would expect from the preceding discussion, although there is a bit of a surprise at the end of the complementary series. The situation can be most easily understood by drawing a picture: see Figure 7.3.

The meaning of Figure 7.3 is as follows. We identify $[SL(2, \mathbb{R})]\hat{}$ with

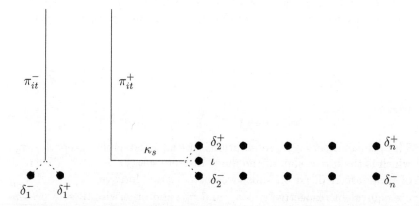

**FIGURE 7.3**
The dual space of $SL(2, \mathbb{R})$.

the set (7.49). If the five points $\delta_1^\pm$, $\delta_2^\pm$, and $\iota$ are omitted from (7.49), the resulting set can be embedded in the plane as a union of line segments and isolated points as indicated in Figure 7.3. The five exceptional points are non-Hausdorff points for the Fell topology. If $p$ is $\delta_1^+$ or $\delta_1^-$, the sets $\{p\} \cup \{\pi_{it}^- : 0 < t < t_0\}$ $(t_0 > 0)$ are a neighborhood base at $p$, and if $p$ is $\delta_2^+$, $\delta_2^-$, or $\iota$, the sets $\{p\} \cup \{\kappa_s : s_0 < s < 1\}$ $(0 < s_0 < 1)$ are a neighborhood base at $p$.

These facts seem to have first appeared explicitly in Miličić [105]. They follow from two pieces of information: first, an explicit knowledge of the characters of the representations in (7.49), which can be found in Knapp [79, §X.2], and two results of Fell ([37, Lemma 3.4 and Corollary 2 of Theorem 2.3]), which allow the topology to be read off from the behavior of the characters.

A few remarks are in order about the way the different series are joined together. First, the fact that $\pi_{it}^-$ converges to both $\delta_1^+$ and $\delta_1^-$ as $t \to 0$ is to be expected since $\pi_0^- = \delta_1^+ \oplus \delta_1^-$. Second, the norm (7.48) on the space for $\kappa_s$ converges to the norm of $L^2(\mathbb{R})$ as $s \to 0$, so that $\kappa_s \to \pi_0^+$. This can easily be shown by using the Fourier transform (e.g., see Folland [44, Exercise 4 of §10.2]) and is the reason for the factor of $s/2$ in (7.48).

Third, one can replace $it$ in (7.46) and (7.47) by an arbitrary complex number $s + it$ to obtain a family of (generally nonunitary) representations $\pi_{s+it}^\pm$ of $SL(2, \mathbb{R})$, the so-called nonunitary principal series. The complementary series $\kappa_s$ is obtained from $\pi_s^+$ $(0 < s < 1)$ by renorming the space on which it acts. On the other hand, $\pi_1^+$ contains subrepresentations equivalent (on the Lie algebra level, not unitarily) to $\delta_2^+$ and $\delta_2^-$, and its quotient by their direct sum is the trivial representation $\iota$. This accounts for the fact that $\kappa_s$ converges to $\delta_2^\pm$ and $\iota$ as $s \to 1$. (The other members of the discrete series, as well as the other irreducible [nonunitary] finite-dimensional representations of $SL(2, \mathbb{R})$, can also be obtained as subrepresentations and quotients of certain members of the nonunitary principal series; see Knapp [79].)

It should be emphasized that although the points $\delta_2^\pm$ and $\iota$ have no disjoint neighborhoods, they are attached to each other only through the complementary series: in the subset of $[SL(2, \mathbb{R})]\hat{\ }$ obtained by omitting the complementary series, these points are isolated. A similar statement holds for $\delta_1^\pm$.

Finally, we describe the Plancherel measure $\mu$ on $[SL(2, \mathbb{R})]\hat{\ }$, where $[SL(2, \mathbb{R})]\hat{\ }$ is parametrized by (7.49). The Plancherel measure of the complementary and mock discrete series and the trivial representation is zero, and on the principal and discrete series it is given by

$$d\mu(\pi_{it}^+) = \frac{t}{2} \tanh \frac{\pi t}{2}\, dt, \qquad d\mu(\pi_{it}^-) = \frac{t}{2} \coth \frac{\pi t}{2}\, dt,$$

$$\mu(\{\delta_n^+\}) = \mu(\{\delta_n^-\}) = n - 1.$$

(Of course this is determined only up to a constant factor, depending on the normalization of Haar measure on $SL(2,\mathbb{R})$.) For the proof, see Knapp [79].

Two significant features appear here that have not occurred in our other examples. First, the Plancherel measure has both a discrete and a continuous part. Second, the support of the Plancherel measure (the smallest closed set whose complement has measure zero) is not all of $[SL(2,\mathbb{R})]\hat{}\,$: it is the union of the principal, discrete, and mock discrete series.

**4. The free group on two generators.** Let $G$ be the free group on two generators $a$ and $b$ (with the discrete topology, and with counting measure as Haar measure), and let $\lambda$ be its left regular representation. We shall illustrate the bad phenomena that occur in non-type-I groups by showing that $\lambda$ is primary and that it admits two completely different decompositions into irreducible representations.

**7.50 Proposition.** $\lambda$ *is primary.*

*Proof.* Suppose $T$ belongs to the center of $\mathcal{C}(\lambda)$; thus $T$ commutes with all left translations (since $T \in \mathcal{C}(\lambda)$) and all right translations (since the latter belong to $\mathcal{C}(\lambda)$). Let $\delta \in L^2(G)$ be the function such that $\delta(1) = 1$ and $\delta(x) = 0$ for $x \neq 1$, and let $\phi = T\delta$. If $f \in L^1 \cap L^2$, we have $Tf = T(f * \delta) = f * \phi$ (since $T$ commutes with left translations) and $Tf = T(\delta * f) = \phi * f$ (since $T$ commutes with right translations). Thus $f * \phi = \phi * f$ for all $f$, and this easily implies that $\phi$ is constant on conjugacy classes (cf. the proof of Proposition 5.21). But every conjugacy class in $G$ except $\{1\}$ is infinite, so since $\phi \in L^2$ we must have $\phi = c\delta$. But then $T = cI$.     □

Next, to investigate the decomposition of $\lambda$ into irreducible representations, let us consider the cyclic subgroup $A$ of $G$ generated by $a$. $A$ is isomorphic to $\mathbb{Z}$, so each $\theta \in \mathbb{T}$ defines a character of $A$, whose value at $a^n$ is $e^{in\theta}$. (Here and in what follows, we identify $\mathbb{T}$ with $\mathbb{R}/2\pi\mathbb{Z}$.) Let $\pi_\theta$ be the representation of $G$ induced by this character. Thus, the Hilbert space $\mathcal{H}_\theta$ for $\pi_\theta$ consists of complex-valued functions $f$ on $G$ such that $f(xa^n) = e^{-in\theta}f(x)$, and its norm may be described as follows.

Each $x \neq 1$ in $G$ can be written uniquely as $x = x_1 x_2 \cdots x_n$ $(n \geq 1)$, where each $x_j$ is either $a$, $b$, $a^{-1}$, or $b^{-1}$, and $x_{j+1} \neq x_j^{-1}$ for all $j < n$. With this notation, let

$$\Omega = \{x = x_1 x_2 \cdots x_n : x_n = b \text{ or } b^{-1}\} \cup \{1\}.$$

Then $\Omega$ meets each left coset of $A$ in precisely one point, so the norm on $\mathcal{H}_\theta$ is given by

$$\|f\|^2 = \sum_{x \in \Omega} |f(x)|^2.$$

For each $z \in \Omega$, define

$$e_z(za^n) = e^{-in\theta}, \qquad e_z(x) = 0 \text{ for } x \notin zA;$$

then $\{e_z : z \in \Omega\}$ is an orthonormal basis for $\mathcal{H}_\theta$.

**7.51 Lemma.** *The operators $\pi_\theta(a^n)$ ($n \in \mathbb{Z}$) have a unique eigenvector up to scalar multiples, namely $e_1$, and the operators $\pi_\theta(b^n)$ have no eigenvectors.*

*Proof.* Evidently $[\pi_\theta(a^n)e_1](x) = e_1(a^{-n}x) = e^{in\theta}e_1(x)$. The action of $A$ on $G$ by left translations, $x \mapsto a^n x$, permutes the elements of $\Omega \setminus \{1\}$, and this action of $A$ on $\Omega \setminus \{1\}$ is faithful on every orbit of $A$. Hence the operators $\pi_\theta(a^n)$ act on $\{e_1\}^\perp$ by permuting the basis vectors $e_z$, and $\pi_\theta(a^n)$ is unitarily equivalent to a direct sum of copies (one for each orbit of $A$) of the shift operator $\{\alpha_k\} \mapsto \{\alpha_{k-n}\}$ on $l^2(\mathbb{Z})$. But the latter operator has no eigenvectors. (By Fourier analysis, it is unitarily equivalent to multiplication by $e^{in\theta}$ on $L^2(\mathbb{T})$.) For the same reason, the operators $\pi_\theta(b^n)$ have no eigenvectors, as they act on $\mathcal{H}_\theta$ by permutation of the basis vectors $e_z$, $z \in \Omega$. $\qquad \square$

**7.52 Lemma.** *The representations $\pi_\theta$ are irreducible and inequivalent.*

*Proof.* Suppose $T \in \mathcal{C}(\pi_\theta)$. By Lemma 7.51, $e_1$ is an eigenvector for $T$, say $Te_1 = ce_1$. But then if $z \in \Omega$, $Te_z = T\pi_\theta(z^{-1})e_1 = \pi_\theta(z^{-1})Te_1 = ce_z$. Thus $T = cI$, so $\pi_\theta$ is irreducible by Schur's lemma. If $\theta \neq \theta'$ (mod $2\pi$), then $\pi_\theta$ and $\pi_{\theta'}$ are inequivalent because $\pi_\theta(a^n)$ and $\pi_{\theta'}(a^n)$ have different eigenvalues on their unique eigenvectors. $\qquad \square$

If $f \in C_c(G)$ (i.e., $f$ is a function of finite support on $G$), define $\widetilde{f}_\theta : G \to \mathbb{C}$ by

$$\widetilde{f}_\theta(x) = \sum_{-\infty}^{\infty} f(xa^n)e^{in\theta}.$$

Then

$$\widetilde{f}_\theta(xa^k) = \sum f(xa^{k+n})e^{in\theta} = \sum f(xa^n)e^{i(n-k)\theta} = e^{-ik\theta}\widetilde{f}_\theta(x),$$

and the projection of the support of $\tilde{f}_\theta$ in $G/A$ is finite, so $\tilde{f}_\theta \in \mathcal{H}_\theta$. Moreover, by the Parseval formula for ordinary Fourier series,

$$\|f\|_{L^2(G)}^2 = \sum_{x \in G} |f(x)|^2 = \sum_{z \in \Omega} \sum_{-\infty}^{\infty} |f(za^n)|^2$$

(7.53)

$$= \sum_{z \in \Omega} \int_{\mathbb{T}} |\tilde{f}_\theta(z)|^2 \frac{d\theta}{2\pi} = \int_{\mathbb{T}} \|\tilde{f}_\theta\|_{\mathcal{H}_\theta}^2 \frac{d\theta}{2\pi}.$$

From this we can easily see that the map $f \mapsto \tilde{f}$ extends to a unitary isomorphism from $L^2(G)$ to $\int^\oplus \mathcal{H}_\theta \, d\mu(\theta)$, where $\mu$ is normalized Haar measure on $\mathbb{T}$. In more detail: for $z \in \Omega$ let $\delta^z$ be the function on $G$ whose value at $z$ is 1 and whose value elsewhere is 0. Then $(\widetilde{\delta^z})_\theta$ is the basis element $e_z$ of $\mathcal{H}_\theta$, so the vector fields $\widetilde{\delta^z}$ determine the structure of a measurable family of Hilbert spaces on $\{\mathcal{H}_\theta\}$, and we can form its direct integral $\int^\oplus \mathcal{H}_\theta \, d\mu(\theta)$. The formula (7.53) shows that the map $f \mapsto \tilde{f}$ is an isometry from $L^2(G)$ into this space; it is unitary since its inverse is given by $g \mapsto \check{g}$, where $\check{g}(x) = \int g(\theta, x) \, d\mu(\theta)$.

Moreover, since the action of $G$ in the representations $\lambda$ and $\pi_\theta$ is given in all cases by left translation, the map $f \mapsto \tilde{f}$ intertwines $\lambda$ with $\int^\oplus \pi_\theta \, d\mu(\theta)$. Thus we have obtained a direct integral decomposition of $\lambda$ into irreducible representations.

Now comes the crux of the matter: we can play the same game with $a$ and $b$ interchanged. That is, let $B$ be the cyclic subgroup generated by $b$. By inducing the characters of $B$ up to $G$, we obtain another family $\{\rho_\theta\}$ of irreducible, inequivalent representations of $G$ such that $\lambda \cong \int^\oplus \rho_\theta \, d\mu(\theta)$. But no $\rho_\theta$ is equivalent to any $\pi_{\theta'}$. Indeed, by Lemma 7.51, the restriction of $\pi_{\theta'}$ to $A$ has a one-dimensional invariant subspace, whereas by the analog of Lemma 7.51 with $a$ and $b$ switched, the restriction of $\rho_\theta$ to $A$ does not. In short, we have proved:

**7.54 Theorem.** *There exist irreducible representations $\pi_\theta$ and $\rho_\theta$ ($\theta \in \mathbb{T}$) of $G$, no two of which are equivalent, such that*

$$\lambda \cong \int_{\mathbb{T}}^\oplus \pi_\theta \, d\mu(\theta) \cong \int_{\mathbb{T}}^\oplus \rho_\theta \, d\mu(\theta),$$

*where $\mu$ is normalized Haar measure on $\mathbb{T}$.*

Theorem 7.54 and the arguments leading to it are due to Yoshizawa [151]. See also Mautner [102] for more about representations of discrete groups.

Finally, we remark that these results also give a counterexample to Theorem 7.17 when the groups are not type I. Indeed, let $\tau$ be the two-sided regular representation of $G$ (actually a representation of $G \times G$).

By Theorem 7.39 and Proposition 7.50, $\mathcal{C}(\tau) = \mathcal{C}(\lambda) \cap \mathcal{C}(\rho) = \mathbb{C}I$, so $\tau$ is irreducible. If $\tau$ were of the form $\pi_1 \otimes \pi_2$, the restriction of $\tau$ to $\{1\} \times G$ would be $I \otimes \pi_2$. But actually this restriction is $\lambda$, which is not a multiple of an irreducible representation.

# Appendices

## 1  A Hilbert Space Miscellany

### Sesquilinear Forms

Let $\mathcal{V}$ and $\mathcal{X}$ be complex vector spaces. A map $T : \mathcal{V} \to \mathcal{X}$ is **antilinear** (or **conjugate linear**) if $T(au + bv) = \bar{a}Tu + \bar{b}Tv$ for all $a, b \in \mathbb{C}$ and $u, v \in \mathcal{V}$. A map $B : \mathcal{V} \times \mathcal{V} \to \mathcal{X}$ is **sesquilinear** if $T(\cdot, v)$ is linear for each $v \in \mathcal{V}$ and $T(u, \cdot)$ is antilinear for each $u \in \mathcal{V}$. A sesquilinear map from $\mathcal{V} \times \mathcal{V}$ to $\mathbb{C}$ is called a **sesquilinear form** on $\mathcal{V}$. Sesquilinear maps are completely determined by their values on the diagonal, as follows.

**A.1 Theorem** (The Polarization Identity). *Suppose $B : \mathcal{V} \times \mathcal{V} \to \mathcal{X}$ is sesquilinear, and let $Q(v) = B(v, v)$. Then for all $u, v \in \mathcal{V}$,*

$$B(u, v) = \frac{1}{4}\big[Q(u + v) - Q(u - v) + iQ(u + iv) - iQ(u - iv)\big].$$

*Proof.* Simply expand the expression on the right ($Q(u + v) = Q(u) + Q(v) + B(u, v) + B(v, u)$, etc.) and collect terms. $\square$

A sesquilinear form $B$ on $\mathcal{V}$ is called **Hermitian** if $B(v, u) = \overline{B(u, v)}$ for all $u, v \in \mathcal{V}$ and **positive** if $B(u, u) \geq 0$ for all $u \in \mathcal{V}$.

**A.2 Corollary.** *A sesquilinear form $B$ is Hermitian if and only if $B(u, u) \in \mathbb{R}$ for all $u$. Every positive sesquilinear form is Hermitian.*

*Proof.* The first assertion follows easily from the polarization identity and the fact that $Q(au) = |a|^2 Q(u)$ for $a \in \mathbb{C}$, so that (for example) $Q(u - iv) = Q(i(u - iv)) = Q(v + iu)$. The second assertion follows immediately from the first one. $\square$

**A.3 Theorem** (The Schwarz and Minkowski Inequalities). *Let $B$ be a positive sesquilinear form on $\mathcal{V}$, and let $Q(u) = B(u, u)$. Then*

$$|B(u, v)|^2 \leq Q(u)Q(v), \qquad Q(u + v)^{1/2} \leq Q(u)^{1/2} + Q(v)^{1/2}.$$

*Proof.* The usual proofs of these inequalities for inner products do not depend on definiteness, so they apply to all positive forms. $\square$

273

## Direct Sums

Let $\{\mathcal{H}_\alpha\}_{\alpha \in A}$ be a family of Hilbert spaces. The **direct sum** $\bigoplus_{\alpha \in A} \mathcal{H}_\alpha$ is the set of all $v = (v_\alpha)_{\alpha \in A}$ in the Cartesian product $\prod_{\alpha \in A} \mathcal{H}_\alpha$ such that $\sum \|v_\alpha\|^2 < \infty$. (This condition implies, in particular, that $v_\alpha = 0$ for all but countably many $\alpha$.) $\bigoplus_{\alpha \in A} \mathcal{H}_\alpha$ is a Hilbert space with inner product

$$\langle u, v \rangle = \sum_{\alpha \in A} \langle u_\alpha, v_\alpha \rangle,$$

and the summands $\mathcal{H}_\alpha$ are embedded in it as mutually orthogonal closed subspaces.

If $\mathcal{H}$ is a Hilbert space and $\{\mathcal{M}_\alpha\}_{\alpha \in A}$ is a family of mutually orthogonal closed subspaces of $\mathcal{H}$ whose linear span is dense in $\mathcal{H}$, $\mathcal{H}$ can be identified with $\bigoplus_{\alpha \in A} \mathcal{M}_\alpha$. When we speak of direct sums of subspaces of a Hilbert space, we *always* assume that the subspaces are mutually orthogonal unless the contrary is explicitly stated.

## Isometries and Unitary Maps

Let $\mathcal{H}_1$ and $\mathcal{H}_2$ be Hilbert spaces and $T : \mathcal{H}_1 \to \mathcal{H}_2$ a bounded linear map. The **adjoint** of $T$ is the map $T^* : \mathcal{H}_2 \to \mathcal{H}_1$ defined by $\langle T^*v, u \rangle = \langle v, Tu \rangle$ for all $u \in \mathcal{H}_1$ and $v \in \mathcal{H}_2$. $T$ is an **isometry** if $\|Tu\| = \|u\|$ for all $u \in \mathcal{H}_1$. Since $\|Tu\|^2 = \langle Tu, Tu \rangle = \langle T^*Tu, u \rangle$ and $\|u\|^2 = \langle u, u \rangle$, the polarization identity implies that $T$ is an isometry if and only if $T^*T$ is the identity operator on $\mathcal{H}_1$. Isometries are injective but not necessarily surjective; a bijective isometry is called a **unitary** map. If $T$ is unitary then so is $T^{-1}$, so $T$ is unitary precisely when $T^*T$ and $TT^*$ are the identity operators on $\mathcal{H}_1$ and $\mathcal{H}_2$ respectively, i.e., when $T^* = T^{-1}$.

$T$ is called a **partial isometry** if $\|Tu\| = \|u\|$ whenever $u \perp \mathcal{N}(T)$. (Here and in what follows, $\mathcal{N}$ and $\mathcal{R}$ denote nullspace and range.) $T$ is a partial isometry precisely when $\langle T^*Tu, u \rangle = \langle u, u \rangle$ for $u \perp \mathcal{N}(T)$ and $\langle T^*Tu, u \rangle = 0$ for $u \in \mathcal{N}(T)$; by polarization, this means that $T^*T$ is the orthogonal projection onto $\mathcal{N}(T)^\perp$. Since $\mathcal{R}(T)^\perp = \mathcal{N}(T^*)$, it is an easy exercise to see that $T^*$ is also a partial isometry and that $TT^*$ is the orthogonal projection onto $\mathcal{R}(T)$.

## Decompositions of Operators

Suppose $\mathcal{H}$ is a Hilbert space and $T \in \mathcal{L}(\mathcal{H})$. $T$ is called **positive** if $\langle Tu, u \rangle \geq 0$ for all $u \in \mathcal{H}$. By Corollary A.2, every positive operator is self-adjoint, and it follows easily from the spectral theorem that a self-adjoint operator $T$ is positive if and only if its spectrum $\sigma(T)$ lies in $[0, \infty)$. In this case, one can apply the spectral functional calculus to the function $f(s) = \sqrt{s}$ on $\sigma(T)$ to obtain a positive operator whose square is $T$; we denote it by $\sqrt{T}$.

Now suppose $T$ is an arbitrary bounded operator on $\mathcal{H}$. $T^*T$ is always a positive operator, so we can define

(A.4) $$|T| = \sqrt{T^*T}.$$

Since $\||T|u\|^2 = \langle |T|^2 u, u \rangle = \langle T^*Tu, u \rangle = \|Tu\|^2$, the correspondence $|T|u \mapsto Tu$ extends uniquely to an isometry $V$ from the closure $\overline{\mathcal{R}(|T|)}$ of the range of $|T|$ to the closure of the range of $T$. But $\overline{\mathcal{R}(|T|)} = \mathcal{N}(|T|)^\perp$ since $|T|$ is self-adjoint, and $\mathcal{N}(|T|) = \mathcal{N}(T^*T) = \mathcal{N}(T)$. Hence, if we extend $V$ linearly to all of $\mathcal{H}$ by setting $V = 0$ on $\mathcal{N}(T)$, $V$ is a partial isometry such that $T = V|T|$. The factorization $T = V|T|$ is called the **polar decomposition** of $T$.

Since $V^*V$ is the orthogonal projection onto $\mathcal{N}(T)^\perp = \mathcal{R}(|T|)$, it follows easily that $V^*T = V^*V|T| = |T|$. Thus:

(A.5) $$T = V|T|, \qquad |T| = V^*T.$$

**A.6 Proposition.** *Every bounded operator on a Hilbert space is a linear combination of two self-adjoint operators and a linear combination of four unitary operators.*

*Proof.* If $T \in \mathcal{L}(\mathcal{H})$, then $A = (T + T^*)/2$ and $B = (T - T^*)/2i$ are self-adjoint and $T = A + iB$. This proves the first assertion; for the second, it is then enough to prove that every self-adjoint operator of norm $\leq 1$ is a linear combination of two unitary operators. But if $S = S^*$ and $\|S\| \leq 1$ then $\sigma(S) \subset [-1, 1]$, so $I - S^2$ is positive, and we have $S = \frac{1}{2}[f_+(S) + f_-(S)]$ where $f_\pm(s) = s \pm i\sqrt{1 - s^2}$. The operators $f_\pm(S)$ are unitary by the spectral functional calculus since $|f_\pm(s)| = 1$ for $s \in [-1, 1]$. $\qquad\square$

## Topologies on $\mathcal{L}(\mathcal{H})$

If $\mathcal{H}$ is a Hilbert space, the set $\mathcal{L}(\mathcal{H})$ of bounded linear operators on $\mathcal{H}$ carries three useful topologies:

- The **norm topology** is the topology induced by the operator norm $\|T\| = \sup_{\|u\|=1} \|Tu\|$.

- The **strong operator topology** is the topology induced by the semi-norms $T \mapsto \|Tu\|$, $u \in \mathcal{H}$. A net $\{T_\alpha\}$ in $\mathcal{L}(\mathcal{H})$ converges to $T$ strongly if and only if $\|T_\alpha u - Tu\| \to 0$ for every $u \in \mathcal{H}$.

- The **weak operator topology** is the topology induced by the semi-norms $T \mapsto |\langle Tu, v \rangle|$, $u, v \in \mathcal{H}$. A net $\{T_\alpha\}$ converges to $T$ weakly if and only if $\langle T_\alpha u, v \rangle \to \langle Tu, v \rangle$ for every $u, v \in \mathcal{H}$.

Thus, the norm topology is the topology of uniform convergence on bounded subsets of $\mathcal{H}$, the strong operator topology is the topology of pointwise convergence on $\mathcal{H}$, and the weak operator topology is the topology of weak pointwise convergence on $\mathcal{H}$.

*Examples*: Suppose $\{e_k\}_1^\infty$ is an orthonormal basis for $\mathcal{H}$. Define operators $T_n$ and $S_n$ on $\mathcal{H}$ for $n \geq 1$ by

$$T_n\left[\sum_1^\infty a_k e_k\right] = \sum_n^\infty a_k e_k, \qquad S_n\left[\sum_1^\infty a_k e_k\right] = \sum_1^\infty a_k e_{k+n}.$$

The sequences $\{T_n\}$ and $\{S_n\}$ are not convergent in norm, and $\{S_n\}$ is not even strongly convergent; indeed, if $n < m$, $\|(T_n - T_m)e_n\| = \|e_n\| = 1$ and $\|(S_n - S_m)e_1\| = \|e_n - e_m\| = \sqrt{2}$. However, $\{T_n\}$ converges strongly to 0, for if $u = \sum a_k e_k$, $\|T_n u\|^2 = \sum_n^\infty |a_k|^2 \to 0$; and $\{S_n\}$ converges weakly to 0, for if $u = \sum a_k e_k$ and $v = \sum b_k e_k$,

$$|\langle S_n u, v\rangle| = \left|\sum_{n+1}^\infty a_{k-n} b_k\right| \leq \|u\|\left[\sum_{n+1}^\infty |b_k|^2\right]^{1/2} \to 0.$$

## 2    Trace-Class and Hilbert-Schmidt Operators

Let $\mathcal{H}$ be a Hilbert space, which for convenience we assume to be separable. Suppose $T$ is a positive operator on $\mathcal{H}$. We say that $T$ is **trace-class** if $T$ has an orthonormal eigenbasis $\{e_n\}$ with eigenvalues $\{\lambda_n\}$ (necessarily nonnegative), and $\sum \lambda_n < \infty$. In this case we set $\operatorname{tr}(T) = \sum \lambda_n$. Note that every trace-class positive operator is compact, for $T$ is the norm limit of the finite-rank operators $T_N u = \sum_1^N \lambda_n \langle u, e_n\rangle e_n$.

**A.7 Proposition.** *If $T$ is positive and trace-class and $\{x_n\}$ is any orthonormal basis for $\mathcal{H}$, then $\sum \langle Tx_n, x_n\rangle = \operatorname{tr}(T)$.*

*Proof.* Let $\{e_j\}$ be an orthonormal eigenbasis for $T$ with eigenvalues $\{\lambda_j\}$. Since $x_n = \sum \langle x_n, e_j\rangle e_j$ and $\sum_n |\langle x_n, e_j\rangle|^2 = \|e_j\|^2 = 1$, we have

$$\sum_n \langle Tx_n, x_n\rangle = \sum_n \sum_j \langle x_n, e_j\rangle\langle Te_j, x_n\rangle = \sum_n \sum_j \lambda_j |\langle x_n, e_j\rangle|^2 = \sum_j \lambda_j.$$

Interchanging the sums is permissible since all terms are positive.    □

**A.8 Proposition.** *Suppose $T$ is positive and trace-class, $S \in \mathcal{L}(\mathcal{H})$, and $\{x_n\}$ is an orthonormal basis for $\mathcal{H}$. Then the sum $\sum \langle STx_n, x_n\rangle$ is absolutely convergent, and its value depends only on $S$ and $T$, not on $\{x_n\}$.*

*Proof.* Let $\{e_j\}$ be an orthonormal eigenbasis for $T$ with eigenvalues $\{\lambda_j\}$. Then

$$\langle STx_n, x_n \rangle = \sum_j \langle x_n, e_j \rangle \langle STe_j, x_n \rangle = \sum_j \lambda_j \langle x_n, e_j \rangle \langle Se_j, x_n \rangle.$$

Now,

$$\sum_n \sum_j \lambda_j |\langle x_n, e_j \rangle \langle Se_j, x_n \rangle|$$

$$\leq \sum_j \lambda_j \left[ \sum_n |\langle x_n, e_j \rangle|^2 \right]^{1/2} \left[ \sum_n |\langle Se_j, x_n \rangle|^2 \right]^{1/2}$$

$$= \sum_j \lambda_j \|e_j\| \, \|Se_j\| \leq \|S\| \sum_j \lambda_j < \infty.$$

This implies that $\sum \langle STx_n, x_n \rangle$ is absolutely convergent and that

$$\sum_n \langle STx_n, x_n \rangle = \sum_n \sum_j \langle \langle x_n, e_j \rangle STe_j, x_n \rangle$$

$$= \sum_j \sum_n \langle STe_j, \langle e_j, x_n \rangle x_n \rangle = \sum_j \langle STe_j, e_j \rangle.$$

$\square$

An operator $T \in \mathcal{L}(\mathcal{H})$ is called **trace-class** if the positive operator $|T|$, defined by (A.4), is trace-class.

**A.9 Proposition.** *Suppose $T$ is trace-class. Then $T$ is compact, and if $\{x_n\}$ is any orthonormal basis for $\mathcal{H}$, the sum $\sum \langle Tx_n, x_n \rangle$ is absolutely convergent and independent of $\{x_n\}$.*

*Proof.* Consider the polar decomposition $T = V|T|$ as in Appendix 1. $T$ is compact since $|T|$ is compact and $V$ is bounded, and the other assertion follows from Proposition A.8, with $T$ and $S$ replaced by $|T|$ and $V$. $\square$

If $T$ is trace-class, we set

$$\mathrm{tr}(T) = \sum \langle Tx_n, x_n \rangle,$$

where $\{x_n\}$ is any orthonormal basis for $\mathcal{H}$. This is well-defined by Proposition A.9.

**A.10 Proposition.** *The set of trace-class operators is a two-sided $*$-ideal in $\mathcal{L}(\mathcal{H})$. Moreover, if $T$ is trace-class and $S \in \mathcal{L}(\mathcal{H})$ then $\mathrm{tr}(ST) = \mathrm{tr}(TS)$.*

*Proof.* Clearly if $T$ is trace-class, so is $aT$ for any $a \in \mathbb{C}$. If $S$ and $T$ are trace-class, let $S = V|S|$, $T = W|T|$, and $S + T = X|S + T|$ be polar decompositions as in Appendix 1. Then $|S + T| = X^*(S + T)$ by (A.5). In particular, $|S + T|$ is compact, so it has an orthonormal eigenbasis $\{e_n\}$ by Theorem 1.52, and we have

$$\sum \langle |S + T|e_n, e_n \rangle = \sum \langle X^*(S + T)e_n, e_n \rangle$$
$$= \sum \langle X^*V|S|e_n, e_n \rangle + \sum \langle X^*W|T|e_n, e_n \rangle.$$

The sums on the right are absolutely convergent by Proposition A.8, so $S + T$ is trace-class. Next, suppose $T$ is trace-class and $U$ is unitary. Then $(UT)^*(UT) = T^*U^*UT = T^*T$, so $|UT| = |T|$ and hence $UT$ is trace-class. Also $(TU)^*TU = U^*T^*TU = U^{-1}(T^*T)U$, whence $|TU| = U^{-1}|T|U$. Thus $|TU|$ has the same eigenvalues as $|T|$, so $TU$ is trace-class. Moreover, if $\{x_n\}$ is an orthonormal basis,

$$\text{tr}(TU) = \sum \langle TUx_n, x_n \rangle = \sum \langle UTUx_n, Ux_n \rangle = \text{tr}(UT)$$

since $U$ is unitary and $\{Ux_n\}$ is again an orthonormal basis. Since every $S \in \mathcal{L}(\mathcal{H})$ is a linear combination of four unitary operators by Proposition A.6, it follows that $ST$ and $TS$ are trace-class and that $\text{tr}(ST) = \text{tr}(TS)$. Finally, if $T$ is trace-class and $T = V|T|$ is its polar decomposition, then $T^* = |T|V^*$ is trace-class since $|T|$ is. $\qquad\square$

An operator $T \in \mathcal{L}(\mathcal{H})$ is called **Hilbert-Schmidt** if $T^*T$ is trace-class. Since $T^*T$ is positive and $\langle T^*Tu, u \rangle = \|Tu\|^2$, it follows from Proposition A.7 that $T$ is Hilbert-Schmidt if and only if $\sum \|Tx_n\|^2 < \infty$ for some, and hence any, orthonormal basis $\{x_n\}$.

Every Hilbert-Schmidt operator is compact. Indeed, if $\{e_n\}$ is an eigenbasis for $T^*T$ with eigenvalues $\lambda_n$, it is also an eigenbasis for $|T|$ with eigenvalues $\sqrt{\lambda_n}$. $|T|$ is therefore the norm limit of operators of finite rank; hence $|T|$ is compact, and so is $T = V|T|$.

**A.11 Proposition.** *If $T$ is Hilbert-Schmidt, so is $T^*$. If $S$ and $T$ are Hilbert-Schmidt, then $ST$ is trace-class.*

*Proof.* If $\{x_n\}$ is an orthonormal basis for $\mathcal{H}$, we have

$$\sum_n \|Tx_n\|^2 = \sum_n \sum_m |\langle Tx_n, x_m \rangle|^2 = \sum_m \sum_n |\langle T^*x_m, x_n \rangle|^2$$
$$= \sum_m \|T^*x_m\|^2.$$

This proves the first assertion. For the second, let $ST = V|ST|$ be the polar decomposition of $ST$. $(ST)^*(ST)$ is compact and so has an orthonormal eigenbasis $\{e_n\}$ by Theorem 1.52. $\{e_n\}$ is also an eigenbasis for $|ST|$, and by (A.5) we have

$$\sum \langle |ST|e_n, e_n \rangle = \sum \langle V^* ST e_n, e_n \rangle = \sum \langle Te_n, S^* V e_n \rangle$$

$$\leq \left[ \sum \|Te_n\|^2 \right]^{1/2} \left[ \sum \|S^* V e_n\|^2 \right]^{1/2}.$$

But each $e_n$ belongs either to the nullspace of $V$ or its orthogonal complement, so the nonzero $Ve_n$'s are an orthonormal set. Since $S^*$ and $T$ are Hilbert-Schmidt, it follows that $\sum \|SVe_n\|^2 < \infty$ and $\sum \|Te_n\|^2 < \infty$, so $|ST|$ is trace-class. $\qquad\square$

If $\mathcal{H} = L^2(\mu)$ where $\mu$ is a $\sigma$-finite measure, the space of Hilbert-Schmidt operators on $\mathcal{H}$ can be identified with $L^2(\mu \times \mu)$, with $F \in L^2(\mu \times \mu)$ corresponding to the integral operator $T_F h(x) = \int F(x, y) h(y) \, d\mu(y)$. We shall explain this in more detail, using the language of tensor products, in Appendix 3.

---

## 3   Tensor Products of Hilbert Spaces

In the category of vector spaces over a fixed field, the tensor product of two vector spaces $V_1$ and $V_2$ is usually defined abstractly as a vector space $V_1 \otimes V_2$ such that any bilinear map from $V_1 \times V_2$ to another vector space factors uniquely through $V_1 \otimes V_2$. There are several ways of constructing concrete models of $V_1 \otimes V_2$; one is as the space of linear maps of finite rank from $V_2^*$ into $V_1$. We wish to modify this construction to obtain a tensor product in the category of Hilbert spaces.

Let $\mathcal{H}_1$ and $\mathcal{H}_2$ be Hilbert spaces, and consider bounded linear maps from $\mathcal{H}_2^*$ into $\mathcal{H}_1$, or equivalently, bounded *antilinear* maps from $\mathcal{H}_2$ into $\mathcal{H}_1$. The terminology and basic theory of bounded linear maps applies also to antilinear maps, with obvious modifications. For example, the composition of two antilinear maps is linear, and the composition of a linear map with an antilinear one is antilinear. The operator norm of an antilinear map $A$ is defined as in the linear case,

$$\|A\| = \sup_{\|x\|=1} \|Ax\|.$$

The adjoint of a bounded antilinear map is another such map, defined not by $\langle A^*u, v \rangle = \langle u, Av \rangle$ — which cannot be right since the left side is antilinear in both $u$ and $v$ whereas the right side is linear — but rather by

$$\langle A^*u, v \rangle = \langle Av, u \rangle.$$

Note that the map $A \mapsto A^*$ is linear rather than antilinear.

Suppose $A$ is an antilinear map from $\mathcal{H}_2$ to $\mathcal{H}_1$, and $\{u_\alpha\}$ and $\{v_\beta\}$ are orthonormal bases for $\mathcal{H}_1$ and $\mathcal{H}_2$, respectively. Then by the Parseval identity,

(A.12)
$$\begin{aligned} \sum_\beta \|Av_\beta\|^2 &= \sum_\beta \sum_\alpha |\langle Av_\beta, u_\alpha \rangle|^2 \\ &= \sum_\alpha \sum_\beta |\langle A^*u_\alpha, v_\beta \rangle|^2 = \sum_\alpha \|A^*u_\alpha\|^2. \end{aligned}$$

This shows that $\sum_\beta \|Av_\beta\|^2$ is independent of the choice of basis $\{v_\beta\}$, for if $\{w_\beta\}$ is another orthonormal basis for $\mathcal{H}_2$ we also have $\sum_\beta \|Aw_\beta\|^2 = \sum_\alpha \|A^*u_\alpha\|^2$. We define the **tensor product** of $\mathcal{H}_1$ and $\mathcal{H}_2$ to be the set $\mathcal{H}_1 \otimes \mathcal{H}_2$ of all antilinear $A : \mathcal{H}_2 \to \mathcal{H}_1$ such that $\sum_\beta \|Av_\beta\|^2 < \infty$ for some, and hence every, orthonormal basis $\{v_\beta\}$ for $\mathcal{H}_2$, and we set

$$\|\|A\|\|^2 = \sum_\beta \|Av_\beta\|^2.$$

**A.13 Theorem.** $\mathcal{H}_1 \otimes \mathcal{H}_2$ *is a Hilbert space with the norm* $\|\| \cdot \|\|$ *and associated inner product*

$$\langle A, B \rangle = \sum_\beta \langle Av_\beta, Bv_\beta \rangle,$$

*where* $\{v_\beta\}$ *is any orthonormal basis of* $\mathcal{H}_2$.

*Proof.* We first observe that if $A, B \in \mathcal{H}_1 \otimes \mathcal{H}_2$,

$$\sum |\langle Av_\beta, Bv_\beta \rangle| \le \sum \|Av_\beta\| \, \|Bv_\beta\| \le \|\|A\|\| \, \|\|B\|\| < \infty$$

by two applications of the Schwarz inequality. Thus the series $\sum \langle Av_\beta, Bv_\beta \rangle$ is absolutely convergent, and an application of the Parseval identity as in (A.12) then shows that $\sum \langle Av_\beta, Bv_\beta \rangle = \sum \langle A^*u_\alpha, B^*u_\alpha \rangle$ for any orthonormal bases $\{u_\alpha\}$ and $\{v_\beta\}$ of $\mathcal{H}_1$ and $\mathcal{H}_2$. Thus $\sum \langle Av_\beta, Bv_\beta \rangle$ is independent of the choice of basis $\{v_\beta\}$ and defines an inner product on $\mathcal{H}_1 \otimes \mathcal{H}_2$ whose associated norm is $\|\| \cdot \|\|$.

It remains to prove completeness. If $A \in \mathcal{H}_1 \otimes \mathcal{H}_2$ and $\epsilon > 0$, choose $\{v_\beta\}$ to be an orthonormal basis for $\mathcal{H}_2$ of which one element $v_0$ satisfies $\|Av_0\| > \|A\| - \epsilon$. It follows that $\|\|A\|\| > \|A\| - \epsilon$, and hence $\|\|A\|\| \ge \|A\|$. Therefore, if $\{A_n\}$ is Cauchy in $\mathcal{H}_1 \otimes \mathcal{H}_2$, it is Cauchy in the operator norm topology, and one only needs to show that its limit $A$ in the latter topology is also its limit in the topology of $\mathcal{H}_1 \otimes \mathcal{H}_2$. This we leave as an exercise for the reader. $\qquad\square$

The preceding constructions are a slight generalization of the ones in Appendix 2 pertaining to Hilbert-Schmidt operators. Indeed, for any Hilbert space $\mathcal{H}$ the space $\mathcal{H} \otimes \mathcal{H}^*$ (whose elements are antilinear maps from $\mathcal{H}^*$ to $\mathcal{H}$, or equivalently, linear maps from $\mathcal{H}$ to itself) is precisely the space of Hilbert-Schmidt operators on $\mathcal{H}$, and the inner product and norm on $\mathcal{H} \otimes \mathcal{H}^*$ are given by

$$\langle A, B \rangle = \mathrm{tr}(B^*A), \qquad \|\|A\|\| = \sqrt{\mathrm{tr}(A^*A)}.$$

$\|\|A\|\|$ is called the **Hilbert-Schmidt norm** of $A$.

If $u \in \mathcal{H}_1$ and $v \in \mathcal{H}_2$, the map $w \mapsto \langle v, w \rangle u$ ($w \in \mathcal{H}_2$) belongs to $\mathcal{H}_1 \otimes \mathcal{H}_2$; we denote it by $u \otimes v$:

$$(u \otimes v)(w) = \langle v, w \rangle u.$$

By picking an orthonormal basis for $\mathcal{H}_2$ of which $v/\|v\|$ is a member, we see that

$$\|\|u \otimes v\|\| = \frac{\|(u \otimes v)v\|}{\|v\|} = \frac{\langle v, v \rangle \|u\|}{\|v\|} = \|u\| \, \|v\|.$$

Similarly, for any $u, u' \in \mathcal{H}_1$ and $v, v' \in \mathcal{H}_2$,

(A.14) $$\langle u \otimes v, \, u' \otimes v' \rangle = \langle u, u' \rangle \langle v, v' \rangle$$

Next, if $A$ is a bounded antilinear map of finite rank from $\mathcal{H}_2$ to $\mathcal{H}_1$, let $u_1, \ldots u_n$ be an orthonormal basis for the range of $A$. Then $Aw = \sum_1^n \langle Aw, u_j \rangle u_j$. For each $j$, $w \mapsto \langle Aw, u_j \rangle$ is a bounded antilinear functional on $\mathcal{H}_2$, so there exists $v_j \in \mathcal{H}_2$ such that $\langle Aw, u_j \rangle = \langle v_j, w \rangle$. But this says that $A = \sum_1^n u_j \otimes v_j$. In other words, the space of antilinear maps of finite rank from $\mathcal{H}_2$ to $\mathcal{H}_1$ is contained in $\mathcal{H}_1 \otimes \mathcal{H}_2$ and consists of the linear span of the elements $u \otimes v$ for $u \in \mathcal{H}_1$ and $v \in \mathcal{H}_2$.

**A.15 Proposition.** *If $\{u_\alpha\}$, $\{v_\beta\}$ are orthonormal bases for $\mathcal{H}_1$, $\mathcal{H}_2$. Then $\{u_\alpha \otimes v_\beta\}$ is an orthonormal basis for $\mathcal{H}_1 \otimes \mathcal{H}_2$.*

*Proof.* It follows from (A.14) that $\{u_\alpha \otimes v_\beta\}$ is an orthonormal set in $\mathcal{H}_1 \otimes \mathcal{H}_2$. If $A \in \mathcal{H}_1 \otimes \mathcal{H}_2$ we have

$$\langle A, \, u_\alpha \otimes v_\beta \rangle = \sum_{\beta'} \langle Av_{\beta'}, (u_\alpha \otimes v_\beta)v_{\beta'} \rangle = \langle Av_\beta, u_\alpha \rangle,$$

so

$$\sum_{\alpha,\beta} |\langle A, \, u_\alpha \otimes v_\beta \rangle|^2 = \sum_{\alpha,\beta} |\langle Av_\beta, u_\alpha \rangle|^2 = \sum_\beta \|Av_\beta\|^2 = \|\|A\|\|^2.$$

Thus the Parseval identity holds, so $\{u_\alpha \otimes v_\beta\}$ is a basis. $\qquad\square$

We point out a couple of simple and useful isomorphisms. First, by (A.12), the map $A \mapsto A^*$ is a unitary isomorphism from $\mathcal{H}_1 \otimes \mathcal{H}_2$ to $\mathcal{H}_2 \otimes \mathcal{H}_1$, and $(u \otimes v)^* = v \otimes u$. Second, tensor products distribute across direct sums in the obvious way:

$$\mathcal{H}_1 \cong \mathcal{H}_1' \oplus \mathcal{H}_1'' \implies \mathcal{H}_1 \otimes \mathcal{H}_2 \cong (\mathcal{H}_1' \otimes \mathcal{H}_2) \oplus (\mathcal{H}_1'' \otimes \mathcal{H}_2).$$

We leave it as an exercise for the reader to make this precise.

Another interpretation of the tensor product is available when the Hilbert spaces in question are $L^2$ spaces. Suppose $(X, \mu)$ and $(Y, \nu)$ are $\sigma$-finite measure spaces. If $f \in L^2(\mu)$ and $g, h \in L^2(\nu)$, we have

$$(f \otimes g)(h) = \langle g, h \rangle f = \int f(\cdot) g(y) \overline{h(y)} \, d\nu(y),$$

so $f \otimes g$ is the antilinear integral operator whose kernel is the function

(A.16)        $$(f \otimes g)(x, y) = f(x) g(y)$$

on $X \times Y$. Similarly, the operator $\sum_1^n f_j \otimes g_j$ has the kernel $\sum_1^n f_j(x) g_j(y)$. We are therefore led to the following result.

**A.17 Theorem.** *Suppose $(X, \mu)$ and $(Y, \nu)$ are $\sigma$-finite measure spaces. The identification of $f \otimes g \in L^2(\mu) \otimes L^2(\nu)$ with the function (A.16) extends uniquely to an isometric isomorphism of $L^2(\mu) \otimes L^2(\nu)$ with $L^2(\mu \times \nu)$, whose inverse identifies $F \in L^2(\mu \times \nu)$ with the operator $h \mapsto \int F(\cdot, y) \overline{h(y)} \, d\nu(y)$.*

*Proof.* By (A.14),

$$\left\| \sum_1^n f_j \otimes g_j \right\|^2 = \sum_{j,k=1}^n \langle f_j, f_k \rangle \langle g_j, g_k \rangle$$

$$= \iint \left| \sum_1^n f_j(x) g_j(y) \right|^2 d\mu(x) \, d\nu(y).$$

Hence the stated identification extends by linearity and continuity to an isometric embedding of $L^2(\mu) \otimes L^2(\nu)$ into $L^2(\mu \times \nu)$. To complete the proof, by Proposition A.15 it is enough to show that if $\{f_\alpha\}$ and $\{g_\beta\}$ are orthonormal bases for $L^2(\mu)$ and $L^2(\nu)$ then $\{f_\alpha \otimes g_\beta\}$ is a basis for $L^2(\mu \times \nu)$. If $F \in L^2(\mu \times \nu)$, there is a set $N \subset Y$ with $\nu(N) = 0$ and $\int |F(x, y)|^2 \, d\mu(x) < \infty$ for all $y \notin N$. For every $\alpha$ the function

$$\phi_\alpha(y) = \langle F(\cdot, y), f_\alpha \rangle = \int F(x, y) \overline{f_\alpha(x)} \, d\mu(x)$$

is well-defined for $y \notin N$, and

$$\int |\phi_\alpha(y)|^2 \, d\nu(y) \le \int \|F(\cdot, y)\|_2^2 \, d\nu(y) = \|F\|_2^2 < \infty,$$

so $\phi_\alpha \in L^2(\nu)$. Moreover,

$$\langle \phi_\alpha, g_\beta \rangle = \iint F(x, y) \overline{f_\alpha(x) g_\beta(y)} \, d\mu(x) \, d\nu(y) = \langle F, f_\alpha \otimes g_\beta \rangle.$$

Now, $F(\cdot, y) = \sum \phi_\alpha(y) f_\alpha$ (where the series converges in $L^2(\mu)$) for all $y \notin N$, so by the Parseval identities on $L^2(\mu)$ and $L^2(\nu)$,

$$\|F\|_2^2 = \int \|F(\cdot, y)\|_2^2 \, d\nu(y) = \int \sum_\alpha |\phi_\alpha(y)|^2 \, d\nu(y)$$

$$= \sum_{\alpha, \beta} |\langle \phi_\alpha, g_\beta \rangle|^2 = \sum_{\alpha, \beta} |\langle F, f_\alpha \otimes g_\beta \rangle|^2.$$

Thus the Parseval identity holds for the orthonormal set $\{f_\alpha \otimes g_\beta\}$, which is therefore a basis. $\qquad\square$

In case $\nu = \mu$, this is almost the identification of the Hilbert-Schmidt operators on $L^2(\mu)$ with $L^2(\mu \times \mu)$ mentioned at the end of Appendix 2. To obtain the latter, one merely has to switch $L^2(\mu)$ with $L^2(\mu)^*$ via the antilinear map $h \mapsto \overline{h}$. The Hilbert-Schmidt norm of an operator is then the $L^2$ norm of the corresponding integral kernel.

The proof of the following proposition is similar to the proof of Theorem A.17 but easier. We leave it to the reader.

**A.18 Proposition.** *Let $\mathcal{H}$ be a Hilbert space and $\mathcal{A}$ a set. For each $\alpha \in \mathcal{A}$ let $\mathcal{H}_\alpha$ be a copy of $\mathcal{H}$, and define $\delta_\alpha(\beta) = 1$ for $\beta = \alpha$ and $\delta_\alpha(\beta) = 0$ for $\beta \in \mathcal{A} \setminus \{\alpha\}$. Then the map $(f_\alpha)_{\alpha \in \mathcal{A}} \mapsto \sum f_\alpha \otimes \delta_\alpha$ is a unitary isomorphism from $\bigoplus_{\alpha \in \mathcal{A}} \mathcal{H}_\alpha$ to $\mathcal{H} \otimes l^2(\mathcal{A})$.*

We now consider tensor products of operators. If $S \in \mathcal{L}(\mathcal{H}_1)$, $T \in \mathcal{L}(\mathcal{H}_2)$, and $A \in \mathcal{H}_1 \otimes \mathcal{H}_2$, define the antilinear operator $(S \otimes T)A$ from $\mathcal{H}_2$ to $\mathcal{H}_1$ by

$$(S \otimes T)A = SAT^*.$$

We shall see shortly that $(S \otimes T)A$ belongs to $\mathcal{H}_1 \otimes \mathcal{H}_2$, so that $S \otimes T$ is a linear operator on $\mathcal{H}_1 \otimes \mathcal{H}_2$, called the **tensor product** of $S$ and $T$. (Note that $S \otimes T$ depends linearly on $T$ despite the occurence of $T^*$ in the definition, for $SA(cT)^* = SA(\overline{c}T^*) = cSAT^*$ by the antilinearity of $A$.)

**A.19 Theorem.** *Suppose $S, S' \in \mathcal{L}(\mathcal{H}_1)$ and $T, T' \in \mathcal{L}(\mathcal{H}_2)$. Then:*

a. $S \otimes T \in \mathcal{L}(\mathcal{H}_1 \otimes \mathcal{H}_2)$, and $\|S \otimes T\| = \|S\| \, \|T\|$.
b. $(S \otimes T)(u \otimes v) = Su \otimes Tv$ for all $u \in \mathcal{H}_1$, $v \in \mathcal{H}_2$.
c. $(S \otimes T)(S' \otimes T') = (SS') \otimes (TT')$.
d. $(S \otimes T)^* = S^* \otimes T^*$.
e. If $S$ and $T$ are unitary, so is $S \otimes T$.

*Proof.* To prove (a), first consider the case when $T = I$. If $\{v_\beta\}$ is an orthonormal basis for $\mathcal{H}_2$ we have $\sum \|SAv_\beta\|^2 \leq \|S\|^2 \sum \|Av_\beta\|^2$, so $\|\!|SA|\!\| \leq \|S\| \|\!|A|\!\|$. Now, for $T \neq I$, by (A.12) and what we have just proved we have

$$\|\!|SAT^*|\!\| = \|\!|T(SA)^*|\!\| \leq \|T\| \, \|\!|(SA)^*|\!\| = \|T\| \, \|\!|SA|\!\| \leq \|S\| \, \|T\| \, \|\!|A|\!\|.$$

Thus $(S \otimes T) \in \mathcal{L}(\mathcal{H}_1 \otimes \mathcal{H}_2)$ and $\|S \otimes T\| \leq \|S\| \, \|T\|$. Next, if $w \in \mathcal{H}_2$,

$$(S \otimes T)(u \otimes v)(w) = S(u \otimes v)T^*w = \langle v, T^*w \rangle Su = \langle Tv, w \rangle Su$$
$$= (Su \otimes Tv)w,$$

which proves (b). If we choose $u$ and $v$ to be unit vectors such that $\|Su\| > \|S\| - \epsilon$ and $\|Tv\| > \|T\| - \epsilon$, it follows that

$$\|S \otimes T\| \geq \|(S \otimes T)(u \otimes v)\| = \|Su\| \, \|Tv\| > (\|S\| - \epsilon)(\|T\| - \epsilon),$$

so that $\|S \otimes T\| \geq \|S\| \, \|T\|$, which completes the proof of (a).

(c) is obvious. To prove (d) we use once again the fact that $A \mapsto A^*$ is a unitary map from $\mathcal{H}_1 \otimes \mathcal{H}_2$ to $\mathcal{H}_2 \otimes \mathcal{H}_1$: if $\{u_\alpha\}$ and $\{v_\beta\}$ are orthonormal bases for $\mathcal{H}_1$ and $\mathcal{H}_2$,

$$\langle (S \otimes T)^*A, B \rangle = \langle A, (S \otimes T)B \rangle = \langle A, SBT^* \rangle$$
$$= \sum \langle Av_\beta, SBT^*v_\beta \rangle = \sum \langle S^*Av_\beta, BT^*v_\beta \rangle = \langle S^*A, BT^* \rangle$$
$$= \langle A^*S, TB^* \rangle = \sum \langle A^*Su_\alpha, TB^*u_\alpha \rangle = \sum \langle T^*A^*Su_\alpha, B^*u_\alpha \rangle$$
$$= \langle T^*A^*S, B^* \rangle = \langle S^*AT, B \rangle = \langle (S^* \otimes T^*)A, B \rangle.$$

Hence $(S \otimes T)^* = S^* \otimes T^*$.

Finally, (e) follows from (c) and (d): if $S$ and $T$ are unitary,

$$(S \otimes T)(S \otimes T)^* = SS^* \otimes TT^* = I \otimes I = I,$$

and likewise $(S \otimes T)^*(S \otimes T) = I$. $\qquad\qquad\square$

---

# 4 Vector-Valued Integrals

There are several ways to develop a theory of integrals for functions with values in a topological vector space. We shall adopt the "weak" approach, in which one reduces everything to scalar functions by applying linear functionals.

Let $\mathcal{V}$ be a locally convex topological vector space, and let $\mathcal{V}^*$ be the space of continuous linear functionals on $\mathcal{V}$. Also, let $(X, \mu)$ be a measure space. A function $F : X \to \mathcal{V}$ is called **weakly integrable** if $\phi \circ F \in L^1(\mu)$ for every $\phi \in \mathcal{V}^*$. In this case, if there is a vector $v \in \mathcal{V}$ such that

$$\phi(v) = \int \phi \circ F \, d\mu \text{ for all } \phi \in \mathcal{V}^*$$

(such a $v$ is necessarily unique since the continuous linear functionals separate points on $\mathcal{V}$), $v$ is called the **integral** of $F$, and we write $v = \int F \, d\mu$.

Integrals commute with continuous linear maps, in the following sense. Suppose $F : X \to \mathcal{V}$ is weakly integrable, $\int F \, d\mu$ exists, and $T$ is a continuous linear map from $\mathcal{V}$ to another locally convex space $\mathcal{W}$. Since $\phi \circ T$ belongs to $\mathcal{V}^*$ for every $\phi \in \mathcal{W}^*$, it is clear that $T \circ F$ is weakly integrable and that

$$\phi \circ T \left[ \int F \, d\mu \right] = \int \phi \circ T \circ F \, d\mu \qquad (\phi \in \mathcal{W}^*),$$

which means that $\int T \circ F \, d\mu$ exists and

$$T \int F \, d\mu = \int T \circ F \, d\mu.$$

The following existence theorem is proved in Rudin [123, Theorems 3.27 and 3.29] (although stated there in a slightly different form):

**A.20 Theorem.** *Suppose $\mathcal{V}$ is a Fréchet space and $\mu$ is a Radon measure on the locally compact Hausdorff space $X$. If $F : X \to \mathcal{V}$ is continuous and compactly supported, then $\int F \, d\mu$ exists and belongs to the closed linear span of the range of $F$. Moreover, if $\mathcal{V}$ is a Banach space,*

(A.21)
$$\left\| \int F \, d\mu \right\| \leq \int \|F(x)\| \, d\mu(x).$$

Theorem A.20 is almost sufficient for our purposes, but the condition that $F$ be in $C_c(X, \mathcal{V})$ needs to be relaxed a bit.

**A.22 Theorem.** *Suppose $\mathcal{V}$ is a Banach space and $\mu$ is a Radon measure on the locally compact Hausdorff space $X$. If $g$ is a (scalar-valued) function in $L^1(\mu)$ and $H : X \to \mathcal{V}$ is bounded and continuous, then $\int gH \, d\mu$ exists and belongs to the closed linear span of the range of $H$, and*

(A.23)
$$\left\| \int gH \, d\mu \right\| \leq \sup_{x \in X} \|H(x)\| \int |g(x)| \, d\mu(x).$$

*Proof.* $gH$ is weakly integrable, since $\phi \circ H$ is bounded and continuous and hence $\phi \circ (gH) = g(\phi \circ H) \in L^1(\mu)$ for any $\phi \in \mathcal{V}^*$. Moreover, since $\mu$ is Radon, there is a sequence $\{g_n\}$ in $C_c(X)$ that converges to $g$ in the $L^1$ norm. Then

$$\int \|g_n(x)H(x) - g_m(x)H(x)\| \, d\mu(x) \leq C \int |g_n(x) - g_m(x)| \, d\mu(x) \to 0$$

as $m, n \to \infty$, so by (A.21), the sequence $\{\int g_n H \, d\mu\}$ is Cauchy in $\mathcal{V}$. Denote its limit by $v$; then for any $\phi \in \mathcal{V}^*$,

$$\phi(v) = \lim \phi \left[ \int g_n H \, d\mu \right] = \lim \int \phi \circ (g_n H) \, d\mu = \int \phi \circ (gH) \, d\mu,$$

since

$$\int |\phi \circ (g_n H) - \phi \circ (gH)| \, d\mu \leq C \int |g_n - g| \, d\mu \to 0.$$

In other words, $\int gH \, d\mu$ exists and equals $v$. Moreover, since $\int g_n H \, d\mu$ belongs to the closed linear span of the range of $H$ for each $n$, so does $\int gH \, d\mu$. Finally, since (A3.2) holds with $F = g_n H$, (A.23) holds with $g$ replaced by $g_n$, and it then holds for $g$ by letting $n \to \infty$. $\qquad\square$

When vector-valued integrals arise in the text, $X$ will generally be a locally compact group $G$ and $\mu$ will be Haar measure on $G$. The following situations are the most important ones.

1. *Convolutions.* Let $\mathcal{V} = L^p(G)$, where $1 \leq p < \infty$, and let $L_y$ and $R_y$ be the left and right translation operators defined by (2.5). If $f \in L^p(G)$, the function $y \mapsto L_y f$ is bounded ($\|L_y f\|_p = \|f\|_p$) and continuous on $G$, by Proposition 2.42. Hence, for any $g \in L^1(G)$ we can form the integral $\int g(y) L_y f \, dy \in L^p(G)$. By its definition, this integral satisfies

$$\int \left[ \int g(y) L_y f \, dy \right] (x) h(x) \, dx = \iint g(y) f(y^{-1} x) h(x) \, dx \, dy$$

for every $h \in L^q(G)$, where $q$ is the conjugate exponent to $p$. A simple application of Hölder's inequality and the Fubini-Tonelli theorem shows that the order of integration can be reversed, and since $h \in L^q(G)$ is arbitrary, it follows that

$$\left[ \int g(y) L_y f \, dy \right] (x) = \int g(y) f(y^{-1} x) \, dy = g * f(x).$$

In other words, we have

$$g * f = \int g(y) L_y f \, dy,$$

where the integral can be interpreted either as an $L^p$-valued integral or pointwise (a.e.) as a scalar-valued integral:

$$g * f(x) = \int g(y)L_y f(x)\,dy.$$

If $G$ is unimodular, the same arguments show that

$$f * g = \int g(y^{-1})R_y f\,dy,$$

with the same pair of interpretations. When $p = 1$, this is true even if $G$ is not unimodular. Indeed, the integrand $g(y^{-1})R_y f$ is the product of the functions $y \mapsto \Delta(y)^{-1}g(y^{-1})$ and $y \mapsto \Delta(y)R_y f$; the former is in $L^1(G)$ by (2.32), and the latter is bounded and continuous from $G$ to $L^1(G)$ by (2.25).

2. *Unitary Representations.* Let $\pi$ be a unitary representation of $G$. We wish to form the associated representation of $L^1(G)$,

$$\pi(f) = \int f(x)\pi(x)\,dx \qquad (f \in L^1(G))$$

(see §3.2). Since $\pi$ is only assumed continuous with respect to the strong operator topology, and $\mathcal{L}(\mathcal{H}_\pi)$ is not a Banach space (or even a Fréchet space) with respect to this topology unless $\dim \mathcal{H}_\pi < \infty$, Theorems A.20 and A.22 cannot be applied directly to define $\pi(f)$. However, Theorem A.22 can be used to define $\pi(f)$ pointwise: that is, for each $u \in \mathcal{H}_\pi$, $x \mapsto \pi(x)u$ is a bounded continuous function from $G$ to $\mathcal{H}_\pi$, so the integral $\int f(x)\pi(x)u\,dx$ exists for $f \in L^1(G)$: it is the vector $\pi(f)u \in \mathcal{H}_\pi$ determined by the relations

$$\langle \pi(f)u, v \rangle = \int f(x)\langle \pi(x)u, v \rangle\,dx \qquad (v \in \mathcal{H}_\pi).$$

The map $u \mapsto \pi(f)u$ is easily seen to be linear, and by (A.23),

$$\|\pi(f)u\| \le \int |f(x)|\|\pi(x)u\|\,dx = \|u\| \int |f(x)|\,dx,$$

so $\pi(f) \in \mathcal{L}(\mathcal{H}_\pi)$ and $\|\pi(f)\| \le \|f\|_1$.

# Bibliography

[1] W. Ambrose, Spectral resolution of groups of unitary operators, *Duke Math. J.* **11** (1944), 589–595.

[2] L. Auslander and B. Kostant, Polarization and unitary representations of solvable Lie groups, *Invent. Math.* **14** (1971), 255–354.

[3] L. Baggett, Representations of the Mautner group, *Pac. J. Math.* **77** (1978), 7–22.

[4] L. Baggett, W. E. Mitchell, and A. Ramsay, Representations of the discrete Heisenberg group and cocycles of an irrational rotation, *Michigan Math. J.* **31** (1984), 263–273.

[5] L. Baggett and A. Ramsay, A functional analytic proof of a selection lemma, *Canad. J. Math.* **32** (1980), 441–448.

[6] M. W. Baldoni Silva and D. Barbasch, The unitary spectrum for real rank one groups, *Invent. Math.* **72** (1983), 27–55.

[7] D. Barbasch, The unitary dual for complex classical Lie groups, *Invent. Math.* **96** (1989), 103–176.

[8] V. Bargmann, Irreducible representations of the Lorentz group, *Ann. of Math.* **48** (1947), 568–640.

[9] P. Bernat, Sur les représentations unitaires des groupes de Lie résolubles, *Ann. Sci. Ecole Norm. Sup.* (3) **82** (1965), 37–99.

[10] A. S. Besicovitch, *Almost Periodic Functions,* Cambridge U. Press, Cambridge, U.K., 1932; reprinted by Dover, New York, 1954.

[11] A. Beurling, On a closure problem, *Arkiv för Mat.* **1** (1950), 301–303; also pp. 153–155 in Beurling's *Collected Works*, vol. 2, Birkhäuser, Boston, 1989.

[12] B. Blackadar, *Operator Algebras*, Springer, Berlin, 2006.

[13] R. J. Blattner, On induced representations, *Amer. J. Math.* **83** (1961), 79–98.

[14] R. J. Blattner, Positive definite measures, *Proc. Amer. Math. Soc.* **14** (1963), 423–428.

[15] S. Bochner, Monotone Funktionen, Stieltjessche Integrale, und harmonische Analyse, *Math. Ann.* **108** (1933), 378–410; also pp. 87–119 in Bochner's *Collected Papers*, Part I, Amer. Math. Soc., Providence, R.I., 1992.

[16] H. Boerner, *Representations of Groups* (2nd ed.), North Holland, Amsterdam, 1969.

[17] J. Boidol, H. Leptin, J. Schürman, and D. Vahle, Räume primitiver Ideale von Gruppenalgebren, *Math. Ann.* **236** (1978), 1–13.

[18] N. Bourbaki, *Intégration*, Chapitres 7 et 8 (Eléments de Math., vol. XXIX), Hermann, Paris, 1963.

[19] N. Bourbaki, *Espaces Vectoriels Topologiques*, Chapitres 1 et 2 (Eléments de Math., vol. XV, 2nd ed.) Hermann, Paris, 1966.

[20] T. Bröcker and T. tom Dieck, *Representations of Compact Lie Groups*, Springer-Verlag, Berlin, 1985.

[21] I. D. Brown, Dual topology of a nilpotent Lie group, *Ann. Sci. Ecole Norm. Sup.* (4) **6** (1973), 407–411.

[22] F. Bruhat, Sur les représentations induites des groupes de Lie, *Bull. Soc. Math. France* **81** (1956), 97–205.

[23] H. Cartan, Sur la mesure de Haar, *C. R. Acad. Sci. Paris* **211** (1940), 759–762; also pp. 1020–1022 in Cartan's *Oeuvres*, vol. III, Springer-Verlag, Berlin, 1979.

[24] H. Cartan and R. Godement, Théorie de la dualité et analyse harmonique dans les groupes Abéliens localement compacts, *Ann. Sci. Ecole Norm. Sup.* (3) **64** (1947), 79–99; also pp. 1203–1223 in Cartan's *Oeuvres*, vol. III, Springer-Verlag, Berlin, 1979.

[25] R. R. Coifman and G. Weiss, *Analyse Harmonique Non-commutative sur Certains Espaces Homogènes* (Lecture Notes in Math. 242), Springer-Verlag, Berlin, 1971.

[26] L. W. Corwin and F. P. Greenleaf, *Representations of Nilpotent Lie Groups and Their Applications*, Cambridge U. Press, Cambridge, U.K., 1990.

[27] M. Cowling, The Plancherel formula for a group not of type I, *Boll. Un. Mat. Ital.* (5) **15** (1978), 616–623.

[28] J. Dixmier, Sur les représentations unitaires des groupes de Lie algebriques, *Ann. Inst. Fourier* **7** (1957), 315–328.

[29] J. Dixmier, Sur les représentations unitaires des groupes de Lie nilpotents, V, *Bull. Soc. Math. France* **87** (1959), 65–79.

[30] J. Dixmier, Utilisation des facteurs hyperfinis dans la théorie des C*-algèbres, *C. R. Acad. Sci. Paris* **258** (1964), 4184–4187.

[31] J. Dixmier, *Von Neumann Algebras*, North Holland, Amsterdam, 1981.

[32] J. Dixmier, *C\* Algebras*, North Holland, Amsterdam, 1982.

[33] M. Duflo and C. C. Moore, On the regular representation of a nonunimodular locally compact group, *J. Funct. Anal.* **21** (1976), 209–243.

[34] N. Dunford and J. T. Schwartz, *Linear Operators, Part II: Spectral Theory*, Wiley-Interscience, New York, 1963.

[35] H. Dym and H. P. McKean, *Fourier Series and Integrals*, Academic Press, New York, 1972.

[36] J. Ernest, A decomposition theory for unitary representations of locally compact groups, *Trans. Amer. Math. Soc.* **104** (1962), 252–277.

[37] J. M. G. Fell, The dual spaces of C* algebras, *Trans. Amer. Math. Soc.* **94** (1960), 365–403.

[38] J. M. G. Fell, Weak containment and induced representations of groups I, *Canad. J. Math.* **14** (1962), 237–268.

[39] J. M. G. Fell, Weak containment and induced representations of groups II, *Trans. Amer. Math. Soc.* **110** (1964), 424–447.

[40] J. M. G. Fell and R. S. Doran, *Representations of \*-Algebras, Locally Compact Groups, and Banach \*-Algebraic Bundles*, vol. I, Academic Press, Boston, 1988.

[41] J. M. G. Fell and R. S. Doran, *Representations of \*-Algebras, Locally Compact Groups, and Banach \*-Algebraic Bundles*, vol. II, Academic Press, Boston, 1988.

[42] G. B. Folland, *Harmonic Analysis in Phase Space*, Princeton U. Press, Princeton, N.J., 1989.

[43] G. B. Folland, Harmonic analysis of the de Rham complex on the sphere, *J. Reine Angew. Math.* **398** (1989), 130–143.

[44] G. B. Folland, *Fourier Analysis and its Applications*, Wadsworth and Brooks/Cole, Pacific Grove, Cal., 1992; reprinted by American Mathematical Society, Providence, R.I., 2007.

[45] G. B. Folland, *Real Analysis* (2nd ed.), John Wiley, New York, 1999.

[46] G. B. Folland, Compact Heisenberg manifolds as CR manifolds, *J. Geom. Analysis* **14** (2004), 521–532.

[47] F. G. Frobenius, Über Relationen zwischen den Charakteren einer Gruppe und denen ihrer Untergruppen, *Sitz. Preuss. Akad. Wiss.* (1898), 501–515; also pp. 104–118 in Frobenius's *Gesammelte Abhandlungen*, vol. III, Springer-Verlag, Berlin, 1968.

[48] S. A. Gaal, *Linear Analysis and Representation Theory*, Springer-Verlag, Berlin, 1973.

[49] I. M. Gelfand and M. A. Naimark, Unitary representations of the group of linear transformations of the straight line, *Dokl. Akad. Nauk SSSR* **55** (1947), 567–570; also pp. 18–21 in Gelfand's *Collected Papers*, vol II, Springer-Verlag, Berlin, 1988.

[50] I. M. Gelfand and D. A. Raikov, Irreducible unitary representations of locally bicompact groups, *Mat. Sb.* **13** (55) (1942), 301–316; *Transl. Amer. Math. Soc.* (2) **36** (1964), 1–15; also pp. 3–17 in Gelfand's *Collected Papers*, vol. II, Springer-Verlag, Berlin, 1988.

[51] J. G. Glimm, Type I C* algebras, *Ann. of Math.* **73** (1961), 572–612.

[52] J. G. Glimm, Locally compact transformation groups, *Trans. Amer. Math. Soc.* **101** (1961), 124–138.

[53] R. Godement, Sur une généralisation d'un théorème de Stone, *C. R. Acad. Sci. Paris* **218** (1944), 901–903.

[54] R. Godement, Sur la théorie des représentations unitaires, *Ann. of Math.* **53** (1951), 68–124.

[55] B. Golubov, A. Efimov, and V. Skvartsov, *Walsh Series and Transforms*, Kluwer Academic Publ., Dordrecht, 1991.

[56] F. P. Greenleaf, *Invariant Means on Topological Groups and Their Applications*, Van Nostrand, New York, 1969.

[57] A. Haar, Der Massbegriff in der Theorie der kontinuerlichen Gruppen, *Ann. of Math.* **34** (1933), 147–169; also pp. 600–622 in Haar's *Gesammelte Abhandlungen*, Akadémiai Kiadó, Budapest, 1959.

[58] P. R. Halmos, *Introduction to Hilbert Space and the Theory of Spectral Multiplicity*, Chelsea, New York, 1951.

[59] P. R. Halmos, What does the spectral theorem say?, *Amer. Math. Monthly* **70** (1963), 241–247.

[60] Harish-Chandra, Representations of a semisimple Lie group on a Banach space I, *Trans. Amer. Math. Soc.* **75** (1953), 185–243; also pp. 391–449 in Harish-Chandra's *Collected Papers*, vol. I, Springer-Verlag, New York, 1984.

[61] Harish-Chandra, Representations of semisimple Lie groups III, *Trans. Amer. Math. Soc.* **76** (1954), 234–253; also pp. 490–509 in Harish-Chandra's *Collected Papers*, vol. I, Springer-Verlag, New York, 1984.

[62] Harish-Chandra, The Plancherel formula for complex semisimple Lie groups, *Trans. Amer. Math. Soc.* **76** (1954), 485–528; also pp. 511–554 in Harish-Chandra's *Collected Papers*, vol. I, Springer-Verlag, New York, 1984.

[63] Harish-Chandra, Harmonic analysis on real reductive groups III: the Maass-Selberg relations and the Plancherel formula, *Ann. of Math.* **104** (1976), 117–201; also pp. 259–343 in Harish-Chandra's *Collected Papers*, vol. IV, Springer-Verlag, New York,1984.

[64] S. Helgason, *Differential Geometry and Symmetric Spaces*, Academic Press, New York, 1962.

[65] G. Herglotz, Über Potenzreihen mit positiven Teil im Einheitskreis, *Leipziger Berichte* **63** (1911), 501–511.

[66] E. Hewitt and K. A. Ross, *Abstract Harmonic Analysis*, vol. I, Springer-Verlag, Berlin, 1963.

[67] E. Hewitt and K. A. Ross, *Abstract Harmonic Analysis*,, vol. II, Springer-Verlag, Berlin, 1970.

[68] R. Howe and E. C. Tan, *Non-Abelian Harmonic Analysis*, Springer-Verlag, New York, 1992.

[69] M. Kac, A remark on Wiener's Tauberian theorem, *Proc. Amer. Math. Soc.* **16** (1965), 1155–1157.

[70] J.-P. Kahane, *Séries de Fourier Absolument Convergentes*, Springer, Berlin, 1970.

[71] E. Kaniuth and K. F. Taylor, *Induced Representations of Locally Compact Groups*, Cambridge U. Press, Cambridge, 2013.

[72] I. Kaplansky, The structure of certain operator algebras, *Trans. Amer. Math. Soc.* **70** (1951), 219–255.

[73] Y. Katznelson and B. Weiss, The construction of quasi-invariant measures, *Israel J. Math.* **12** (1972), 1–4.

[74] M. Keane, Sur les mesures quasi-ergodiques des translations irrationelles, *C. R. Acad. Sci. Paris Sér. A-B* **272** (1971), A54–A55.

[75] I. Khalil, Sur l'analyse harmonique du groupe affine de la droite, *Studia Math.* **51** (1974), 139–167.

[76] A. A. Kirillov, Unitary representations of nilpotent Lie groups, *Uspehi Mat. Nauk* **17** (1962), no. 4 (106), 57–110; *Russian Math. Surveys* **17** (1962), no. 4, 53–104.

[77] A. A. Kirillov, *Elements of the Theory of Representations*, Springer-Verlag, Berlin, 1976.

[78] A. Kleppner and R. L. Lipsman, The Plancherel formula for group extensions, I and II, *Ann. Sci. Ecole Norm. Sup.* (4) **5** (1972), 459–516; *ibid.* (4) **6** (1973), 103–132.

[79] A. W. Knapp, *Representation Theory of Semisimple Groups*, Princeton U. Press, Princeton, N.J., 1986.

[80] N. I. Koblitz, *p-adic Numbers, p-adic Analysis, and Zeta Functions*, Springer-Verlag, New York, 1977.

[81] T. W. Körner, *Fourier Analysis*, Cambridge U. Press, Cambridge, U.K., 1988.

[82] B. Kostant, Quantization and unitary representations, pp. 87–208 in *Lectures in Modern Analysis and Applications*, vol. III (C. Taam, ed.), (Lecture Notes in Math. 170), Springer-Verlag, Berlin, 1970.

[83] K. Kuratowski, *Topology*, vol. I, Academic Press, New York, 1966.

[84] L. H. Loomis, *An Introduction to Abstract Harmonic Analysis*, Van Nostrand, New York, 1953.

[85] L. H. Loomis, Positive definite functions and induced representations, *Duke Math. J.* **27** (1960), 569–579.

[86] W. Maak, *Fastperiodische Funktionen*, Springer-Verlag, Berlin, 1950.

[87] G. W. Mackey, A theorem of Stone and von Neumann, *Duke Math. J.* **16** (1949), 313–326.

[88] G. W. Mackey, Imprimitivity for representations of locally compact groups, I, *Proc. Nat. Acad. Sci. U.S.A.* **35** (1949), 537–545.

[89] G. W. Mackey, On induced representations of groups, *Amer. J. Math.* **73** (1951), 576–592.

[90] G. W. Mackey, Induced representations of locally compact groups I, *Ann. of Math.* **55** (1952), 101–139.

[91] G. W. Mackey, Induced representations of locally compact groups II: the Frobenius reciprocity theorem, *Ann. of Math.* **58** (1953), 193–220.

[92] G. W. Mackey, Borel structures in groups and their duals, *Trans. Amer. Math. Soc.* **85** (1957), 134–165.

[93] G. W. Mackey, Unitary representations of group extensions, *Acta Math.* **99** (1958), 265–311.

[94] G. W. Mackey, Infinite-dimensional group representations, *Bull. Amer. Math. Soc.* **68** (1963), 628–686.

[95] G. W. Mackey, Ergodic theory and virtual groups, *Math. Ann.* **166** (1966), 187–207.

[96] G. W. Mackey, *Induced Representations of Groups and Quantum Mechanics*, W. A. Benjamin, New York, 1968.

[97] G. W. Mackey, Induced representations of locally compact groups and applications, pp. 132–166 in F. E. Browder (ed.), *Functional Analysis and Related Fields*, Springer-Verlag, New York, 1970.

[98] G. W. Mackey, *The Theory of Unitary Group Representations* [an updated version of the 1955 "Chicago Notes"], U. of Chicago Press, Chicago, 1976.

[99] G. W. Mackey, *Unitary Group Representations in Physics, Probability, and Number Theory*, Benjamin/Cummings, Reading, Mass., 1978.

[100] G. W. Mackey, Harmonic analysis as the exploitation of symmetry — a historical survey, *Bull. Amer. Math. Soc* (N.S.) **3** (1980), 543–698.

[101] P. Malliavin, Impossibilité de la synthèse spectrale sur les groupes abéliens non compacts, *Publ. Math. Inst. Hautes Etudes Sci.* (1959), 61–68.

[102] F. I. Mautner, Unitary representations of locally compact groups I, *Ann. of Math.* **51** (1950), 1–25.

[103] F. I. Mautner, Unitary representations of locally compact groups II, *Ann. of Math.* **52** (1950), 528–556.

[104] F. I. Mautner, Induced representations, *Amer. J. Math.* **74** (1952), 737–758.

[105] D. Miličić, Topological representation of the group C* algebra of $SL(2, \mathbb{R})$, *Glasnik Mat.* **6** (1971), 231–246.

[106] C. C. Moore, Representations of solvable and nilpotent groups and harmonic analysis on nil and solvmanifolds, *Proc. Symp. Pure Math.* **26** (1973), 3–44.

[107] W. Moran, Ergodic measures for the irrational rotation on the circle, *J. Austral. Math. Soc. (Series A)* **45** (1988), 133–141.

[108] M. A. Naimark, Positive definite operator functions on a commutative group, *Izvestia Akad. Nauk SSSR* **7** (1943), 237–244.

[109] E. Nelson, *Topics in Dynamics*, Princeton U. Press, Princeton, N.J., 1970.

[110] B. Ørsted, Induced representations and a new proof of the imprimitivity theorem, *J. Funct. Anal.* **31** (1979), 355-359.

[111] G. K. Pedersen, *C\*-Algebras and their Automorphism Groups*, Academic Press, London, 1979.

[112] F. Peter and H. Weyl, Die Vollstäntigkeit der primitiven Darstellungen einer geschlossenen kontinuerlichen Gruppe, *Math. Ann.* **97** (1927), 737–755; also pp. 58–75 in Weyl's *Gesammelte Abhandlungen*, vol. III, Springer-Verlag, Berlin, 1968.

[113] L. S. Pontrjagin, The theory of topological commutative groups, *Ann. of Math.* **35** (1934), 361–388.

[114] L. Pukanszky, Representations of solvable Lie groups, *Ann. Sci. Ecole Norm. Sup.* (4) **4** (1971), 464–608.

[115] A. Ramsay, Non-transitive quasi-orbits in Mackey's analysis of group extensions, *Acta Math.* **137** (1976), 17–48.

[116] M. Reed and B. Simon, *Methods of Modern Mathematical Physics, I: Functional Analysis*, Academic Press, New York, 1972.

[117] H. Reiter, *Classical Harmonic Analysis and Locally Compact Groups*, Oxford U. Press, Oxford, 1968; 2nd edition revised by J. Stegeman, Oxford U. Press, Oxford, 2000.

[118] C. E. Rickart, *General Theory of Banach Algebras*, Van Nostrand, New York, 1960.

[119] M. A. Rieffel, Induced representations of C* algebras, *Adv. in Math.* **13** (1974), 176–257.

[120] F. Riesz and B. Sz.-Nagy, *Functional Analysis*, F. Ungar, New York, 1955; reprinted by Dover, New York, 1990.

[121] K. A. Ross, A trip from classical to abstract Fourier analysis, *Notices Amer. Math. Soc.* **61** (2014), 1032–1038.

[122] W. Rudin, *Fourier Analysis on Groups*, Wiley-Interscience, New York, 1962.

[123] W. Rudin, *Functional Analysis* (2nd ed.), McGraw-Hill, New York, 1991.

[124] L. Schwartz, Sur une propriété de synthèse spectrale dans les groupes non compacts, *C. R. Acad. Sci. Paris* **227** (1948), 424–426.

[125] I. E. Segal, The two-sided regular representation of a unimodular locally compact group, *Ann. of Math.* **51** (1950), 293–298.

[126] I. E. Segal, An extension of Plancherel's formula to separable unimodular groups, *Ann. of Math.* **52** (1950), 272–292.

[127] I. E. Segal, Decompositions of operator algebras, I, *Mem. Amer. Math. Soc.* **9** (1951), 1–67.

[128] D. J. Simms, *Lie Groups and Quantum Mechanics* (Lecture Notes in Math. 52), Springer-Verlag, Berlin, 1968.

[129] B. Simon, *Representations of Finite and Compact Groups*, American Mathematical Society, Providence, R.I., 1996.

[130] R. J. Stanton and P. A. Tomas, Polyhedral summability of Fourier series on compact Lie groups, *Amer. J. Math.* **100** (1978), 477–493.

[131] E. M. Stein and G. Weiss, *Introduction to Fourier Analysis on Euclidean Spaces*, Princeton U. Press. Princeton, N.J., 1971.

[132] M. Taibleson, *Fourier Analysis on Local Fields*, Princeton U. Press, Princeton, N.J., 1975.

[133] O. Takenouchi, Sur la facteur-représentation d'un groupe de Lie resoluble de type E, *Math. J. Okayama Univ.* **7** (1957), 151–161.

[134] M. Takesaki, *Theory of Operator Algebras* (3 vols.), Springer, Berlin, 2003.

[135] N. Tatsuuma, Plancherel formula for non-unimodular locally compact groups, *J. Math. Kyoto Univ.* **12** (1972), 176–261.

[136] E. Thoma, Eine Charakterisierung diskreter Gruppen vom Typ I, *Invent. Math.* **6** (1968), 190–196.

[137] E. R. van Kampen, Locally bicompact Abelian groups and their character groups, *Ann. of Math.* **36** (1935), 448–463.

[138] V. S. Varadarajan, *Geometry of Quantum Theory*, vol. II, Van Nostrand, New York, 1970.

[139] D. A. Vogan, The unitary dual of $GL(n)$ over an Archimedean field, *Invent. Math.* **83** (1986), 449–505.

[140] D. A. Vogan, Representations of reductive Lie groups, pp. 245–266 in *Proceedings of the International Congress of Mathematicians 1986*, Amer. Math. Soc., Providence, R.I., 1987.

[141] D. A. Vogan, The character table for $E_8$, *Notices Amer. Math. Soc.* **54** (2007), 1122–1134.

[142] J. von Neumann, Die Eindeutigkeit der Schrödingerschen Operatoren, *Math. Ann.* **104** (1931), 570–578; also pp. 221–229 in von Neumann's *Collected Works*, vol. II, Pergamon, New York, 1961.

[143] J. von Neumann, Almost periodic functions in a group I, *Trans. Amer. Math. Soc.* **36** (1934), 445–492; also pp. 454–501 in von Neumann's *Collected Works* vol. II, Pergamon, New York, 1961.

[144] J. von Neumann, The uniqueness of Haar's measure, *Mat. Sb.* **1**(43) (1936), 721–734; also pp. 91–104 in von Neumann's *Collected Works*, vol. IV, Pergamon, New York, 1961.

[145] J. von Neumann, On rings of operators: reduction theory, *Ann. of Math.* **50** (1949), 401–485; also pp. 400–484 in von Neumann's *Collected Works*, vol. III, Pergamon, New York, 1961.

[146] A. Weil, *L'Intégration dans les Groupes Topologiques et ses Applications*, Hermann, Paris, 1940.

[147] H. Weyl, *Classical Groups*, Princeton U. Press, Princeton, N.J., 1946.

[148] N. Wiener, Tauberian theorems, *Ann. of Math.* **33** (1932), 1–100; also pp. 519–618 in Wiener's *Collected Works*, vol. II, MIT Press, Cambridge, Mass., 1979.

[149] N. Wiener, *The Fourier Integral and Certain of its Applications*, Cambridge U. Press, Cambridge, U.K., 1933; reprinted by Dover, New York, 1958.

[150] E. P. Wigner, On unitary representations of the inhomogeneous Lorentz group, *Ann. of Math.* **40** (1939), 149–204.

[151] H. Yoshizawa, Some remarks on unitary representations of the free group, *Osaka Math. J.* **3** (1951), 55-63.

[152] A. Zygmund, *Trigonometric Series*, Cambridge U. Press, Cambridge, U.K., 1959.

# Index

T - #0013 - 240123 - C0 - 234/156/17 [19] - CB - 9781498727136 - Gloss Lamination